T0179803

GLOBAL NETWORKS

GLOBAL NETWORKS
ENGINEERING, OPERATIONS AND DESIGN

G. Keith Cambron

Former President and CEO, AT&T Labs (Retired), Texas, USA

A John Wiley & Sons, Ltd., Publication

Library of Congress Cataloging-in-Publication Data

Cambron, G. Keith.
  Global networks : engineering, operations and design / G. Keith Cambron.
       pages cm
  Includes bibliographical references and index.
  ISBN 978-1-119-94340-2 (hardback)
 1. Wireless communication systems. 2. Telecommunication. 3. Globalization. I. Title.
  TK5103.C36 2012
  384.5–dc23
                                                              2012022584

A catalogue record for this book is available from the British Library.

ISBN: 9781119943402

Typeset in 10/12pt Times by Laserwords Private Limited, Chennai, India.

Printed and bound in Malaysia by Vivar Printing Sdn Bhd

1   2012

Dedicated to Amos E. Joel, Jr
and the Members of the Technical Staff
at AT&T Labs and SBC Labs

Contents

Part III   TRANSFORMATION

List of Figures

About the Author

Keith Cambron has a broad range of knowledge in telecommunications networks, technology and design and R&D management. His experience ranges from circuit board and software design to the implementation of large public networks.

Keith served as the President and CEO of *AT&T Labs, Inc*. AT&T Labs designs AT&T's global IP, voice, mobile, and video networks. Network technology evaluation, certification, integration, and operational support are part of the Lab's responsibilities. During his tenure AT&T Labs had over 2000 employees, including 1400 engineers and scientists. Technologies produced by Labs ranged from core research to optical transport, IP routing, voice, and video systems.

2003 to 2006 – Cambron served as the President and CEO of *SBC Laboratories, Inc*. The organization, which set the strategic technology objectives of SBC, was structured into four technology areas; Broadband, Wireless, Network Services, and Enterprise IT. SBC Labs led the industry in the introduction of VDSL and IPTV technologies.

1998 to 2003 – Cambron was principal of *Cambron Consulting*, where he provided network and software design consulting services to the telecommunications industry. Working with clients such as SBC, Vodafone Wireless, Coastcom and various enterprise networks, Cambron designed and developed network management systems, a wireless Short Messaging Service (SMS) server, a Service Switching Point (SSP), and an ADSL transmission performance optimization system.

1987 to 1997 – Cambron held leadership positions at *Pacific Bell Broadband*, acting as the chief architect of a network engineering team that developed a 750 MHz hybrid fiber/coax-based network. For this project, Cambron received Telephony's "Fiber in the Loop" design award.

His career started at *Bell Telephone Laboratories* in 1977, where he began as a member of the technical staff. He advanced to Director of Local Switching Systems Engineering and led a team to design automated verification test tools for local digital switch testing. Cambron went on to become Director of Network Systems Verification Testing at *Bell Communications Research*, heading field verification teams in all seven Regional Bell Operating Companies to test "first in nation" technologies, including the first local digital switching systems.

Cambron has been profiled in *Telephony and America's Network*, and has published in *IEEE Communications* and *Proceedings of the IEEE*. He taught Object Oriented Design at Golden Gate University in San Francisco and is a Senior Member of the Institute of Electrical and Electronics Engineers (IEEE).

In 2010, Cambron was named by *CRN Magazine* as one of the Top 25 Technology Thought-Leaders in the world. Keith received IEEE Communications Society's Chairman's Communication Quality and Reliability Award in 2007. He holds ten patents for telecommunications software and systems he designed and deployed.

Cambron received his BS in Electrical Engineering from the University of Missouri, an MS in Systems Management from the University of Southern California, and a Programming Certification from the University of California at Berkeley. He is a retired Commander in the United States Naval Reserve.

Foreword

Networks today are like the air we breathe, so ubiquitous we often take them for granted and in fact don't even realize they're there. Whether we are working, studying, communicating or being entertained, we rely on networks to make whatever we need to happen, happen. This trend is increasing as networks become more and more powerful and reach more deeply into the fabric of our lives.

This reach is not limited to just the wealthy or to developed nations, however, as lower costs and higher capacity extend the power of networks to citizens all around the globe. *That's what makes this book so relevant and so timely.* A clear understanding of these networks is essential for those that would design, construct, operate and maintain them. As Keith points out early in this volume, the growing gap between the academic description of networks and the real world design and operation of these networks, is a key divide that needs bridging. And Keith is in a unique position to do this.

I've known Keith for over 15 years, and have always found him to be a fascinating and indeed remarkable man. His curiosity and intelligence, coupled with a career so deeply involved in network design at AT&T has given him the tremendous insight that he shares in this book. Keith has never been afraid to step outside the accepted norm, if he felt the need, for pursuit of a new area of excellence. This is what makes his knowledge and understanding so valuable and drives the core of this work.

Looking forward, Moore's Law will continue to enable the exponential growth of the value of the underlying technologies, namely processing, memory and optical communications speed, that make these networks tick. The resultant capabilities of the next generations of networks, five years or a decade out, are virtually indescribable today! That in the end is what makes this book so valuable – a thorough understanding of the design principles described herein will allow those that shape our networks in the future to "get it right," enhancing our lives in ways we cannot begin to imagine.

Robert C. McIntyre
Chief Technical Officer, Service Provider Group
Cisco Systems

Preface

When I began my career in telecommunications in 1977 at Bell Telephone Laboratories two texts were required reading, *Engineering and Operations in the Bell System* [1] and *Principles of Engineering Economics* [2]. Members of Technical Staff (MTS) had Masters or PhDs in engineering or science but needed grounding in how large networks were designed and operated, and how to choose designs that were not only technically sound, but economically viable. As the designers of the equipment, systems and networks, engineers at Bell Labs were at the front end of a vertically integrated company that operated the US voice and data networks. Operational and high availability design were well developed disciplines within Bell Labs, and network systems designs were scrutinized and evaluated by engineers practicing in those fields. So ingrained in the culture was the operational perspective, that engineers and scientists were strongly encouraged to rotate through the Operating Company Assignment Program (OCAP) within the first two years of employment. During that eight week program engineers left their Bell Labs jobs and rotated through departments in a Bell Telephone Operating Company, serving as operators, switchmen, installers and equipment engineers. OCAP was not restricted to engineers working on network equipment; members of Bell Labs Research participated in the program. AT&T was not alone in recognizing the value of operational and reliability analysis in a vertically integrated public telephone company, Nippon Telephone and Telegraph, British Telecom, France Telecom and other public telephone companies joined together in technical forums and standards organizations to codify operational and high availability design practices.

After 1982 regulatory, technology and market forces dramatically changed the way networks and systems were designed and deployed. Gone are vertically integrated franchise operators, replaced by interconnected and competing networks of carriers, equipment and systems suppliers, and integrators. Innovation, competition and applications are the engines of change; carriers and system suppliers respond to meet the service and traffic demands of global networks growing at double and even triple digit rates, carrying far more video content than voice traffic. Consumer and enterprise customers are quick to adopt new devices, applications and network services; however, when legacy carriers deliver the service the customers' expectations for quality and reliability are based on their long experience with the voice network. The industry has largely delivered on those expectations because an experienced cadre of engineers from Bell Labs and other carrier laboratories joined startups and their spun off suppliers like Lucent and Nortel. But as time passes, the operational skill reservoir recedes not only because the engineers are retiring, but because of the growing separation between engineers that design and operate networks, and those that design equipment, systems and applications that enter the network. The clearest example of the change is the influx of IT trained software engineers into the fields of network applications and systems design. Experience in the design of stateless web applications or financial systems are insufficient for the non-stop communication systems in the network that continue to operate under a range of hardware faults, software faults and traffic congestion.

My own journey gave me a front row seat through the transformation from a regulated voice network to a competitive global IP network. As a systems engineer in the 1970s I worked on call processing requirements and design for the No. 1 ESS. In the 1980s I led teams of test and verification engineers in the certification of the DMS-10, DMS-100, No. 5ESS, No. 2 STP, DMS STP and Telcordia SCP. I also led design teams building integrated test systems for Signaling System No. 7 and worked for startup companies designing a DS0/1/3 cross connect system and a special purpose tandem switching system. During the last eight years I headed SBC Labs and then AT&T Labs as President and CEO. Working with engineers across network and systems, and spending time with faculty and students at universities I became aware of the growing gap in operational design skills. Universities acknowledge and reward students and faculty for research into theoretical arenas of network optimization and algorithm design. Their research is seldom based on empirical data gathered from networks, and rarer still is the paper that actually changes the way network or systems operate. I chose to write this book to try and fill some of that void. My goal is to help:

- those students and faculty interested in understanding how operational design practices can improve system and network design, and how networks are actually designed, managed and operated;
- hardware and software engineers designing network and support systems;
- systems engineers developing requirements for, or certifying network equipment;
- systems and integration engineers working to build or certify interfaces between network elements and systems;
- operations support systems developers designing software for the management of network systems; and
- managers working to advance the skills of their engineering and operating teams.

The book is organized into three parts; *Networks*, *Teams and Systems*, and *Transformation*. It is descriptive, not prescriptive; the goal is not to tell engineers how to design networks but rather describe how they are designed, engineered and operated; the emphasis is on engineering and design practices that support the work groups that have to install, engineer and run the networks. Areas that are not addressed in the book are network optimization, engineering economics, regulatory compliance and security. Security as a service is described in the chapter on cloud services but there are several texts that better describe the threats to networks and strategies for defense [3, 4].

References

1. *Engineering and Operations in the Bell System*, 1st edn, AT&T Bell Laboratories (1977).
2. Grant, E.L. and Ireson, W.G. (1960) *Principles of Engineering Economy*, Ronald Press Co., New York.
3. Cheswick, W.R. Bellovin, S.M. and Rubin, A.D. (2003) *Firewalls and Internet Security*, 2nd edn, *Repelling the Wily Hacker*, Addison Wesley.
4. Amoroso, E. (2010) *Cyber Attacks: Protecting National Infrastructure*, Butterworth-Heinemann, November.

Acknowledgments

The technical breadth of this text could not have been spanned without the help of engineers I have had the privilege of working with over the years. While I researched and wrote the entire text, these contributors were kind enough to review my material. I am grateful for the contributions of John Erickson, Mike Pepe, Chuck Kalmanek, Anthony Longhitano, Raj Savoor, and Irene Shannon. Each reviewed specific chapters that cover technology within their area of expertise and corrected my technical errors and omissions. They are not responsible for the opinions and projections of future technology trends, and any remaining errors are mine.

I also want to thank the team at John Wiley & Sons, Ltd for guiding me through the writing and publishing process. They made the experience enjoyable and their professional guidance kept me on a sure track.

List of Acronyms

10G	10 Gigabit
100G	100 Gigabit
10GEPON	10 Gigabit Ethernet Passive Optical Network
21CN	21st Century Network
3G	Third Generation Mobile Technology
3GPP	Third Generation Partnership Project
4G	Fourth Generation Mobile Technology
40G	40 Gigabit
400G	400 Gigabit
6rd	IPv6 Rapid Deployment
AAAA	Quad A DNS Record
ABR	Available Bit Rate
ACD	Automatic Call Distributor
ACM	Address Complete Message (ISUP)
ADL	Advanced Development Lab
ADM	Add-Drop Multiplexer
ADPCM	Adaptive Differential Pulse Code Modulation
ADSL	Asymmetric Digital Subscriber Line
ADSL1	Asymmetric Digital Subscriber Line G.992.1 standard
ADSL2+	Asymmetric Digital Subscriber Line G.992.5 standard
AIN	Advanced Intelligent Network
AINS	Automatic In-Service
AIS	Alarm Indication Signal
ALG	Application Level Gateway
ALI	Automatic Line Identification
AMI	Alternate Mark Inversion
AMPS	Advanced Mobile Phone Service
AMR	Adaptive Multi-Rate
API	Application Programming Interface
APN	Access Point Name
APS	Automatic Protection Switching
ARGN	Another Really Good Network
ARP	Address Resolution Protocol
AS	Autonomous System

AS	Application Server
ASON	Automatic Switched Optical Network
ASP	Application Service Provider
AT	Access Tandem
ATA	Analog Terminal Adapter
ATCA	Advanced Telecommunications Computing Architecture
ATM	Asynchronous Transfer Mode
ATSC	Advanced Television Systems Committee
AUMA	Automatic and Manual Service State
AWG	American Wire Gauge
AWS	Advanced Wireless Services
BCP	Business Continuity Plan
BCPL	Basic Combined Programming Language
BGCF	Breakout Gateway Control Function
BGF	Border Gateway Function
BGP	Border Gateway Protocol
BITS	Building Integrated Timing Supply
BLSR	Bi-directional Line Switched Ring
BORSCHT	Battery, Over-voltage, Ringing, Supervision, Codec, Hybrid, Testing
BPON	Broadband Passive Optical Network
BRAS	Broadband Remote Access Server
BRI	Basic Rate Interface (ISDN)
BSC	Base Station Controller
BSD	Berkeley Software Distribution
BSS	Business Support System
BSSMAP	Base Station Subsystem Mobile Application Part
BT	British Telecom
BTL	Bell Telephone Laboratories
BTS	Base Transceiver Station
CALEA	Communications Assistance for Law Enforcement Act
CAMEL	Customized Applications for Mobile network Enhanced Logic
CAS	Channel Associated Signaling
CAT3	Category 3, refers to a grade of twisted pair cable
CATV	Community Antenna Television
CBR	Constant Bit Rate
CCAP	Converged Cable Access Platform
CCIS	Common Channel Interoffice Signaling
CCS	Common Channel Signaling
CDB	Centralized Database
CDF	Charging Data Function
CDMA	Code Division Multiple Access
CDN	Content Delivery Network
CDR	Call Detail Record
CE	Customer Edge
CES	Circuit Emulation Service
CGN	Carrier Grade NAT

CGN64	Carrier Grade NAT IPv6/IPv4
CIC	Carrier Identification Code
CIC	Circuit Identification Code
CLASS	Custom Local Area Signaling Services
CLEC	Competitive Local Exchange Carrier
CLI	Command Line Interface
CLLI	Common Language Location Identifier
CM	Cable Modem
CM	Capacity Management
CMS	Customer Management System
CMTS	Cable Modem Termination System
CNAM	Calling Name Service
CO	Central Office
CONF	Conference Services
CORBA	Common Object Request Broker Architecture
CoS	Class of Service
CPE	Customer Premises Equipment
CPU	Central Processing Unit
CR	Constrained Routing
CRC	Cyclic Redundancy Check
CRS	Carrier Routing System
CSCF	Call Session Control Function
CSFB	Circuit Switched Fallback
CSS3	Cascading Style Sheet 3
CTAG	Command Tag
CURNMR	Current Noise Margin
DA	Distribution Area
DAML	Digitally Added Main Line
DARPA	Defense Advanced Research Projects Agency
DAS	Directed Antenna System
DBMS	Database Management System
DBOR	Database of Record
DCC	Data Communications Channel
DCS	Digital Cross Connection System
DHCP	Dynamic Host Control Protocol
DHCP6	Dynamic Host Control Protocol for IPv6
DLC	Digital Loop Carrier
DLNA	Digital Living Network Alliance
DMS	Digital Multiplex System
DMT	Discrete Multitone
DMTS	Distinguished Member of Technical Staff
DNS	Domain Name System
DNS64	Domain Name System for IPv4 and IPv6
DOCSIS	Data Over Cable Service Interface Specification
DoS	Denial Of Service
DPM	Defects Per Million

DSBLD	Disabled Service State
DSL	Digital Subscriber Line
DSLAM	Digital Subscriber Line Access Multiplexer
DSM	Dynamic Spectrum Management
DSP	Digital Signal Processor
DSTM	Dual Stack IPv6 Transition Mechanism
DSX	Digital Cross Connect
DTAP	Direct Transfer Application Part (SS7)
DTV	Digital Television
DVB	Digital Video Broadcast
DVD	Digital Video Disc
DVR	Digital Video Recorder
DWDM	Dense Wave Division Multiplexing
E911	Enhanced 911
EADAS	Engineering Admin Data Acquisition System
EDFA	Erbium Doped Fiber Amplifier
EDGE	Enhanced Data Rates for Global Evolution
EFM	Ethernet in the First Mile
EGP	External Gateway Protocol
EIGRP	Enhanced Interior Gateway Routing Protocol
EMEA	Europe, the Middle East and Africa
EMS	Element Management System
ENUM	E.164 Number Mapping
EOC	Embedded Operations Channel
EPON	Ethernet Passive Optical Network
ESAC	Electronic Systems Assurance Center
ESME	External Short Messaging Entity
ESS	Electronic Switching System
eTOM	Enhanced Telecom Operations Map
ETS	Electronic Translator System
FCAPS	Fault, Configuration, Accounting, Performance, Security
FCC	Federal Communications Commission
FDD	Frequency Division Duplex
FDMA	Frequency Division Multiple Access
FEC	Forwarding Equivalent Class
FEXT	Far End Crosstalk
FOU	Field of Use
FRR	Fast Reroute
FRU	Field Replaceable Unit
FSAN	Full Service Access Network
FTP	File Transfer Protocol
FTTB	Fiber To The Building
FTTC	Fiber To The Curb
FTTH	Fiber To The Home
FTTN	Fiber To The Node

GEM	GPON Encapsulation Method
GERAN	GSM EDGE Radio Access Network
GGSN	Gateway General Packet Radio Services Support Node
GMPLS	Generalized Multi-protocol Label Switching
GMSC	Gateway Mobile Switching Center
GMSK	Gaussian Minimum Shift Keying
GNOC	Global Network Operations Center
GPON	Gigabit Passive Optical Network
GPS	Global Positioning System
GRE	Generic Routing Encapsulation
GRX	GPRS Routing Exchange
GSM	Global System for Mobile Communications
GTP	GPRS Tunneling Protocol
GTT	Global Title Translation
HD	High Definition
HDSL	High Bitrate Digital Subscriber Line
HDTV	High Definition Television
HFC	Hybrid Fiber Coax
HLR	Home Location Register
HPNA	Home Phone line Networking Alliance
HR	Human Resources
HSDPA	High Speed Downlink Packet Access
HSPA	High Speed Packet Access
HSS	Home Subscriber Server
HSUPA	High Speed Uplink Packet Access
HTML	Hyper Text Markup Language
HTTP	Hyper Text Transfer Protocol
HVAC	Heating, Ventilation and Air Conditioning
IAM	Initial Address Message (SS7)
IAS	Internet Access Service
IBCF	Interconnection Border Control Function
ICMP	Internet Control Message Protocol
IDL	Interface Definition Language
IGMP	Internet Group Management Protocol
IGP	Interior Gateway Protocol
ILEC	Incumbent Local Exchange Carrier
IM	Instant Messaging
IMS	IP Multimedia Subsystem
IMSI	International Mobile Subscriber Identifier
IN	Intelligent Network
IOT	Interoperability Testing
IP	Internet Protocol
IPMI	Intelligent Platform Management Interface
IPTV	Internet Protocol Television
IPX	Internet Protocol Packet Exchange

IRAT	Inter-Radio Access Technology
IRSCP	Intelligent Route Service Control Point
IS	In-Service
ISATAP	Intra-Site Automatic Tunnel Addressing Protocol
ISDN	Integrated Services Digital Network
ISP	Internet Services Provider
ISUP	ISDN User Part
IT	Information Technology
ITP	IP Transfer Point
IVR	Interactive Voice Response
IXC	Interexchange Carrier
IXP	Internet Exchange Point
KPI	Key Performance Indicator
LAN	Local Area Network
LATA	Local Access Transport Area
LCP	Local Convergence Point
LD	Long Distance
LDP	Label Distribution Protocol
LEC	Local Exchange Carrier
LEN	Line Equipment Number
LER	Label Edge Router
LERG	Local Exchange Routing Guide
LFIB	Label Forwarding Information Base
LFO	Line Field Organization
LIDB	Line Information Database
LLDP	Local Loop Demarcation Point
LMTS	Lead Member of Technical Staff
LNP	Local Number Portability
LOF	Loss of Frame
LOL	Loss of Link
LOS	Loss of Signal
LP	Link Processor
LPBK	Loop Back
LPR	Loss of Power
LRF	Location Retrieval Function
LRN	Local Routing Number
LSA	Link State Advertisement
LSDB	Link State Database
LSN	Large Scale NAT
LSP	Label Switched Path
LSR	Label Switch Router
LSSGR	LATA Switching System Generic Requirements
LTE	Long Term Evolution
MA	Manual Service State
MAP	Mobile Application Part

MDF	Main Distribution Frame
MDR	Message Detail Record
MDU	Multiple Dwelling Unit
MED	Multi-Exit Discriminator
MF	Multi-Frequency
MFJ	Modified Final Judgment
MGCF	Media Gateway Control Function
MGW	Media Gateway
MIB	Management Information Base
MIME	Multipurpose Internet Mail Extension
MIMO	Multiple In Multiple Out
MOB	Mobility and Location Services
MME	Mobile Management Entity
MML	Man Machine Language
MMS	Multimedia Message Service
MNO	Mobile Network Operator
MOP	Method of Procedure
MPEG	Motion Pictures Expert Group
MPLS	Multiprotocol Label Switching
MPOE	Minimum Point of Entry
MRFC	Media Resource Function Controller
MRFP	Media Resource Function Processor
MS	Mobile Station
MSC	Mobile Switching Center
MSIN	Mobile Subscriber Identification Number
MSISDN	Mobile Subscriber Integrated Services Digital Subscriber Number
MSO	Multiple System Operator
MSPP	Multiservice Provisioning Platform
MSR	Multi-standard Radio
MSRN	Mobile Station Routing Number
MT	Maintenance Service State
MTS	Member of Technical Staff
MTSO	Mobile Telephone Switching Office
NAP	Network Access Point
NAT	Network Address Translation
NB	Narrowband
NCL	Network Certification Lab
NCP	Network Control Point
NDC	Network Data Center
NE	Network Element
NEBS	Network Equipment Building Standards
NEXT	Near End Crosstalk
NGN	Next Generation Network
NIC	Network Interface Card
NICE	Network-Wide Information Correlation and Exploration

NID	Network Interface Device
NLRI	Network Layer Reachability Information
NMC	Network Management Center
NMP	Network Management Plan
NMS	Network Management System
NNI	Network to Network Interface
NOC	Network Operations Center
NORS	Network Outage Reporting System
NPA	Numbering Plan Area
NPOE	Network Point of Entry
NPRM	Notice of Proposed Rule Making
NR	Normal Service State
NSE	Network Systems Engineering
NSTS	Network Services Test System
NTSC	National Television Systems Committee
NTT	Nippon Telephone and Telegraph
OA&M	Operations, Administration & Maintenance
OEM	Original Equipment Manufacturer
OFDM	Orthogonal Frequency Division Multiplexing
OFDMA	Orthogonal Frequency Division Multiple Access
OID	Object Identifier
OLT	Optical Line Terminal
OMA	Open Mobile Alliance
ONT	Optical Network Terminal
ONU	Optical Network Unit
OOS	Out of Service
ORB	Object Request Broker
ORT	Operational Readiness Test
OS	Operating System
OSA	Open Services Architecture
OSI	Open Systems Interconnection
OSP	Outside Plant
OSPF	Open Shortest Path First
OSS	Operations Support System
OTA	Over the Air
OTDR	Optical Time Domain Reflectometer
OTN	Optical Transport Network
OTT	Over The Top
PAL	Phase Alternating Line video standard
PAT	Port Address Translation
PBX	Private Branch Exchange
PC	Personal Computer
PCEF	Policy Charging Enforcement Function
PCM	Pulse Code Modulation
PCRF	Policy Charging Rules Function

PCS	Personal Communication Service
PCU	Packet Control Unit
PDN	Packet Data Network
PDP	Packet Data Protocol
PDU	Packet Data Unit
PE	Provider Edge
PEG	Public Education and Government
PERF	Policy Enforcement Rules Function
PIC	Polyethylene Insulated Cable
PIC	Primary Inter-LATA Carrier
PIM	Protocol Independent Multicast
PLMN	Public Land Mobile Network
PM	Performance Management
PMTS	Principal Member of Technical Staff
PON	Passive Optical Network
POP	Point of Presence
POTS	Plain Old Telephone Service
PPP	Point to Point Protocol
PRI	Primary Rate Interface (ISDN)
PSAP	Public Service Answering Point
PSL	Production Support Lab
PSTN	Public Switched Telephone Network
PTE	Path Terminating Equipment
PUC	Public Utility Commission
PVC	Private Virtual Circuit
RAB	Radio Access Bearer
RADIUS	Remote Authentication Dial In User Service
RAN	Radio Access Network
RBOC	Regional Bell Operating Company
RCA	Root Cause Analysis
RDC	Regional Data Center
RF	Radio Frequency
RFC	Request for Comment
RFP	Request for Proposal
RIR	Regional Internet Registry
RNC	Radio Network Controller
ROADM	Reconfigurable Optical Add Drop Multiplexer
RP	Route Processor
RRC	Radio Resource Control
RSS	Remote Switching System
RSVP	Resource Reservation Protocol
RTM	Rear Transition Module
RTP	Real-time Transport Protocol
RTT	Round Trip Time
SAI	Serving Area Interface

SAN	Storage Area Network
SBC	Session Border Controller
SCC	Switching Control Center
SCCS	Switching Control Center System
SCE	CAMEL Service Environment
SCP	Service Control Point
SCTE	Society of Cable and Television Engineers
SDH	Synchronous Digital Hierarchy
SDSL	Symmetric Digital Subscriber Line
SDV	Switched Digital Video
SECAM	Sequential Color with Memory (FR)
SEG	Security Gateway
SELT	Single Ended Line Test
SFTP	Secure File Transfer Protocol
SFU	Single Family Unit
SGSN	Serving General Packet Radio Services Node
SGW	Signaling Gateway
SIL	Systems Integration Lab
SIM	Subscriber Identity Module
SIP	Session Initiation Protocol
SLA	Service Level Agreement
SLF	Subscription Locator Function
SMI	SNMP Structure of Management Information
SMIL	Synchronized Multimedia Integration Language
SMPP	Short Messaging Peer to Peer Protocol
SMPTE	Society of Motion Picture and Television Engineers
SMS	Short Message Service
SMSC	Short Message Service Center
SMTP	Simple Mail Transfer Protocol
SNMP	Simple Network Management Protocol
SNR	Signal to Noise Ratio
SOA	Service Oriented Architecture
SOAP	Simple Object Access Protocol
SONET	Synchronous Optical Network
SP	Signaling Point
SPC	Stored Program Control
SPF	Shortest Path First
SPOI	Signalng Point of Interface
SQL	Standard Query Langauge
SS6	Signaling System No. 6
SS7	Signaling System No. 7
SSF	Service Switching Function
SSH	Secure Shell
SSP	Service Switching Point
STB	Settop Box
STE	Section Terminating Equipment

STM	Synchronous Transport Module
STP	Signaling Transfer Point
STS	Synchronous Transport Signal
SUT	System Under Test
TAC	Technical Assistance Center
TAS	Telephony Application Server
TCAP	Transaction Capabilities Part
TCP	Transmission Control Protocol
TDM	Time Division Multiplex
TDMA	Time Division Multiple Access
TE	Traffic Engineering
TID	Target Identifier
TL1	Transaction Language 1
TMF	Telecommunications Management Forum
TMN	Telecommunications Management Network
TNMR	Target Noise Margin
TOD	Time of Day
TRAU	Transcoding and Rate Adaption Unit
TrGW	Transition Gateway
TSD	Technical Service Description
TSI	Time Slot Interchange
TTL	Time to Live
UAS	Unassigned service state
UAT	User Acceptance Test
UBR	Undefined Bit Rate
UDP	User Datagram Protocol
UE	User Equipment
UEQ	Unequipped service state
UHDTV	Ultra-High Definition TV
ULH	Ultra Long Haul
UML	Uniform Modeling Language
UMTS	Universal Mobile Telecommunications System
UNE	Unbundled Network Element
UNI	User to Network Interface
UPSR	Unidirectional Path Switched Rings
URI	Uniform Resource Identifier
URL	Uniform Resource Locator
UTC	Universal Coordinated Time
UTRAN	UMTS Radio Access Network
VBR	Variable Bit Rate
VC	Virtual Circuit
VDSL	Very High Bit Rate Digital Subscriber Line
VHO	Video Home Office
VLAN	Virtual Local Area Network
VLR	Visiting Local Register
VOD	Video On Demand

VP	Virtual Path
VPLS	Virtual Private LAN Service
VPN	Virtual Private Network
VRF	Virtual Routing and Forwarding
WAN	Wide Area Network
WAP	Wireless Application Protocol
WB	Wideband
WLAN	Wireless Local Area Network
WSDL	Web Services Description Language
XML	Extensible Markup Language
YAMS	Yet Another Management System

Part One

Networks

Part One

1

Carrier Networks

We have come a long way in a short time. Instant communication arrived relatively recently in the history of man, with the invention of the telegraph at the beginning of the nineteenth century. It took three quarters of a century before we saw the first major improvement in mass communication, with the arrival of the telephone in 1874, and another half century before a national network and transcontinental communication became common in the US. But what was a middle class convenience years ago is now a common necessity. Today, our worldwide communication Network is a model of egalitarian success, reaching into the enclaves of the wealthy and the street vendors in villages of the developing world with equal ease. Remarkably it remains largely in the hands of the private sector, and is held together and prospers through forces of cooperation and competition that have served society well.

The Network is made up of literally millions of nodes and billions of connections, yet when we choose to make a call across continents or browse a web site in cyberspace we just expect it to work. I use the proper noun Network when referring to the global highway of all interconnection carrier networks, such as Nippon Telephone and Telegraph (NTT), British Telecom (BT), China Telecom, AT&T, Verizon, Deutsche Telekom, Orange, Hurricane Electric, and many others, just as we use the proper noun Internet to refer to the global public Internet Protocol (IP) network. The Internet rides on the Network. If the Network were to fail, even within a city, that city would come to a halt. Instead of purchasing fuel at the pump with a credit card, drivers would line up at the register while the attendant tried to remember how to make change instead of waiting for the Network to verify a credit card. Large discount retail outlets would have their rows of registers stop and for all practical purposes the retailers would close their doors. Alarm systems and 911 emergency services would cease to function. Streets would become congested because traffic lights would no longer be synchronized. So what are the mechanisms that keep the Network functioning 24 hours a day with virtually no failures?

1.1 Operating Global Networks

The global nature of networks is a seismic change in the history of modern man. In the regulated world of the past franchise carriers completely dominated their national networks.

Global Networks: Engineering, Operations and Design, First Edition. G. Keith Cambron.
© 2013 John Wiley & Sons, Ltd. Published 2013 by John Wiley & Sons, Ltd.

Interconnection among networks existed for decades, but carriers did not over build each other in franchise areas. That all changed in the latter decades of the twentieth century as regulation encouraged competition and data services emerged. International commerce and the rise of multinational companies created a demand for global networks operated by individual carriers. Multinational companies wanted a single operator to be held accountable for service worldwide. Many of them simply did not want to be in the global communications business and wanted a global carrier to sort through interconnection and operations issues inherent in far reaching networks.

In parallel with globalization was the move to the IP. The lower layers of the Open System Interconnection (OSI) protocol stack grew because of global scale, and upper layer complexity; the complexity increased with new services such as mobility, video, and the electronic market, largely spurred by Internet services and technology. Operators were forced to reexamine engineering and operating models to meet global growth and expanding service demand. Before deregulation reliability and predictability were achieved through international standards organizations, large operating forces, and highly structured and process centric management regimes. Deregulation, competition, global growth, and service expansion meant that model was no longer economic and could not respond to the rapid introduction of new services and dramatic growths in traffic.

Operating models changed by applying the very advances in technology which drove demand. Reliable networks were realized by reducing the number of failures, by shortening the time for repair, or both. In the old model central offices were staffed with technicians that could respond on short notice to failures, keeping restoral times low. In the new model networks are highly redundant, well instrumented, constantly monitored, and serviced by a mobile work force.

1.1.1   The Power of Redundancy

This section introduces the foundation of global network reliability, redundancy using a simple systems model.

1.1.1.1   Simplex Systems

In the model following, a subscriber at A sends information i_0 to a subscriber at B. The information arrives at B via a communications system S_0 as i_1 after a delay of t (see Figure 1.1).

Subscribers care about two things, the fidelity of the information transfer and transmission time. Fidelity means that the information received, i_1, should be as indistinguishable from the information sent, i_0, as possible. If we assume for simplicity that our communication depends on a single system, S_0, that fails on average once every year, and it takes 4 h to

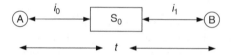

Figure 1.1   Simplex operation.

get a technician on site and restore service, the service will be down for 4 h each year on average, yielding a probability of failure of 4.6×10^{-5}, or an availability of 99.954%. That means we expect to fail about once for every 2000 attempts. For most communications services that is a satisfactory success rate.

But real world connections are composed of a string of systems working in line, possibly in the hundreds, any one of which can fail and impede the transfer. For a linear connection of 100 such systems, our failure probability grows to 4.5×10^{-3} and availability drops to 95.5%. Approximately 1 in 20 attempts to use the system will fail.

1.1.1.2 Redundant Systems

The chances of success can be dramatically improved by using a redundant or duplex system design, shown in Figure 1.2.

In the design two identical systems, S_0 and S_1 are each capable of performing the transfer. One is active and the other is on standby. Since only one system affects the transfer, some communication is needed between the systems and a higher authority is needed to decide which path is taken.

In the duplex system design the probability of failure drops to 2.1×10^{-5} for 100 systems in line, an improvement of more than $100\times$ for an investment of $2\times$. Availability rises to 99.998%. We expect to fail only once in each 50 000 attempts.

Implicit in the model are some key assumptions.

- Failures are random and non-correlated. That is the probability of a failure in S_1 is unrelated to any failure experienced by S_0. Since it's likely the designs of the two systems are identical, that assumption may be suspect.
- The intelligence needed to switch reliably and timely between the two systems is fail-safe.
- When S_0 fails, Operations will recognize it and take action to repair the system within our 4 h timeframe.

1.1.1.3 Redundant Networks

Redundancy works within network systems; their designs have two of everything essential to system health: power supplies, processors, memory, and network fabric. Adopting reliable network systems doesn't necessarily mean networks are reliable. Network systems have to be connected with each other over geographical expanses bridged by physical facilities to build serviceable networks. Physical facilities, optical fiber, telephone cable, and coaxial cable are exposed to the mischiefs of man and of nature. Dual geographically diverse routes

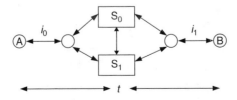

Figure 1.2 Duplex model.

to identical network systems preserve service if the end nodes recognize that one route has failed and the other is viable. Global networks rely on redundant systems within redundant networks. The combination is resilient and robust, providing any failure is recognized early and maintenance is timely and thorough.

The next sections explore this foundational model in more depth in an attempt to understand how it works, and how it can break down in real networks.

1.1.2   The Virtuous Cycle

In the 1956 film *Forbidden Planet*, an advanced civilization called the Krell invents a factory that maintains and repairs itself automatically. In the movie, although the Krell are long extinct, the factory lives on, perpetually restoring and repairing itself. Some academics and equipment suppliers promote this idea today using the moniker "self-healing network." An Internet search with that exact phrase yields 96 000 entries in the result; it is a popular idea indeed. Academic papers stress mathematics, graphs, and simulations in search of elegant proofs of the concept. Yet real networks that perform at the top of their class do so because of the way *people* design, operate, and manage the technology. It is the blend of systems, operations, and engineering that determine success or failure. *Systems and people* make the difference. Figure 1.3 illustrates the Virtuous Cycle of equipment failure, identification, and restoral.

The cycle begins at the top, or 12 o'clock, where the Network is operating in duplex that is full redundancy with primary and alternate paths and processes. Moving in a clockwise direction, a failure occurs signified by the X, and the Network moves from duplex to

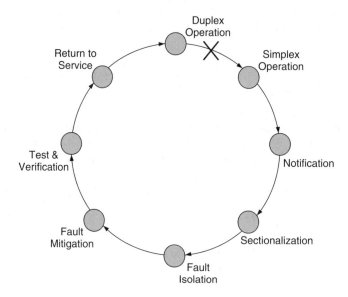

Figure 1.3   Virtuous Cycle.

simplex operation, although no traffic is affected. While the Network is operating in simplex it is vulnerable to a second failure. But before operators can fix the problem they need to recognize it. Notification is the process whereby network elements send alarm notifications to surveillance systems that alert network operators to the situation. Notifications seldom contain sufficient information to resolve the problem, and in many situations multiple notifications are generated from a single fault. Operators must sort out the relevant notifications and sectionalize the fault to a specific network element in a specific location. The failed element can then be put in a test status, enabling operators to run diagnostics and find the root cause of the failure. Hardware faults are mitigated by replacing failed circuit packs. Software faults may require a change of configuration or parameters, or restarting processes. Systems can then be tested and operation verified before the system is restored to service, and the network returns to duplex operation. Later chapters explore these steps in detail.

1.1.3   Measurement and Accountability

The Virtuous Cycle enables highly trained *people* to work with *network systems* to restore complex networks quickly and reliably when they fail. But it does nothing to insure the network has sufficient capacity to handle demands placed upon it. It does not by itself give us any assurance the service the customer is receiving meets a reasonable standard. We can't even be sure network technicians are following the Virtuous Cycle diligently and restoring networks promptly. To meet these goals a broader system of measurements and accountability are needed. Carrier networks are only as good as the measurement systems and the direct line of measurements to accountable individuals. This is not true in smaller networks; when I speak with Information Technology (IT) and network engineers in smaller organizations they view carriers as having unwarranted overhead, rules, and systems. In small networks a few individuals have many roles and are in contact with the network daily. They see relationships and causality among systems quickly; they recognize bottlenecks and errors because of their daily touches on the network systems. Such a model does not scale. Carrier networks have hundreds of types of systems and tens of thousands of network elements. Carrier networks are more akin to Henry Ford's production lines than they are to Orville's and Wilbur's bicycle shop. Quality and reliability are achieved by scaling measurement and accountability in the following ways.

- **End service objectives** – identify measurable properties of each service; commit to service standards, communicate them, and put them into practice.
- **Network systems measurement** – using service objectives analyze each network and network element and set measurable objectives that are consistent with the end to end service standard.
- **Assign work group responsibility** – identify which work group is responsible for meeting each of the objectives and work with them to understand how they are organized, what skills they have and what groups they depend upon and communicate with regularly.
- **Design engineering and management systems** – systems should support people, not the other way round. Find out what systems the teams already use and build on those if at all possible. Don't grow YAMS (yet another management system).

1.2   Engineering Global Networks

Changes in operations as dramatic as they have been are greater yet for the design and engineering of global networks. Carriers in the US prior to 1982 were part of a vertically integrated company, AT&T® or as it was commonly known, the Bell System. AT&T General Departments operated the complete supply and operations chain for the US telecommunications industry. Wholly owned subsidiaries planned, designed, manufactured, installed, and operated the network. AT&T's integrated business included the operations support, billing, and business systems as well. Carriers (the operating companies) had no responsibility for equipment design or selection, and limited responsibility for network design. Today carriers have full responsibility for planning, designing, installing, and operating their networks. They also have a direct hand in determining the functionality and high level design of network systems, and operations and business systems. The sections that follow summarize responsibilities of carrier engineering departments.

1.2.1   Architecture

High level technology choices are the responsibility of Engineering. Engineering architects analyze competing technologies, topologies, and functional delegation to determine the merits and high level cost of each. Standards organizations such as ITU, IETF, and IEEE are forums serving to advance ideas and alternatives. Suppliers naturally promote new ideas and initiatives as well, but from their point of view. Long range plans generally describe the evolution of networks but may not address practical and necessary design and operational transition issues.

1.2.2   Systems Engineering

A wide range of responsibilities rest with systems engineers. They begin with high level architectural plans and translate them into detailed specifications for networks and for the individual network elements. Equipment recommendations, testing, certification, and integration are all performed by these engineers. Operational support, IT integration, and network design are performed by systems engineers as well.

1.2.3   Capacity Management

There are four general ways in which network capacity is expanded. Each is described in the following.

1.2.3.1   Infrastructure Projects

Periodically major network augmentation is undertaken for a variety of reasons.

- Expansion into a new geography is a common trigger. A country adopts competitive rules that enable over building the incumbent.

- Technology obsolescence, such as the shift from Frame Relay to IP networks leads to a phased introduction of new technology. The new networks often must interwork with the legacy technology making the transition more challenging.
- Carrier mergers or acquisitions are followed by network rationalization and integration. The numbers and types of network elements are winnowed out to ease operational demands.
- New lines of business, such as Internet Protocol Television (IPTV) or content distribution, place new demands on the network requiring specialized technology design and deployment.

1.2.3.2 Customer Wins

Major customer contract wins significantly increase demand at large customer locations, rendering the existing capacity inadequate. Sometimes outsourcing of a Fortune 500 company network can be accompanied by an agreement to transfer their entire network, employees, and systems to the winning carrier. If they are of sufficient scope, the accompanying network augmentations are treated as separate projects with dedicated engineering, operations, and finance teams.

1.2.3.3 Capacity Augmentation

By far the most common reason for adding equipment and facilities to a network is the continuous growth in demand of existing services and transport. For decades voice traffic grew at a rate of about 4% each year. Data traffic and IP traffic specifically, have grown at an annual rate of 30–50% for over three decades. With tens of thousands of network systems and millions of facilities, automating demand tracking and capacity management is one of the most resource intensive jobs in engineering.

1.2.3.4 Network Redesign

This is the most neglected, and often the most valuable tool available to network engineers. The demand mechanisms cited above are all triggered by events. Capacity augmentation, the most common engineering activity, is triggered when a facility or network element falls below a performance threshold, such as packet discards or blocked calls. Network engineers generally look at those links exceeding the accepted levels and order augmentation or resizing. If a node nears exhaust, either because of port exhaust or throughput limits, engineers order a new node and rearrange the traffic between that node and adjacent ones. In effect they limit the problem and the solution space to a very narrow area, the particular link or node that exceeded a threshold.

Network redesign broadens the scope to an entire region or community. It is performed by systems engineers, not network engineers. It begins with A-Z (A to Z) traffic demand and uses existing topology, link, and element traffic loads as an informational starting point, not as a constraint. In voice networks Call Detail Records (CDRs) are a starting point since they have the calling party (A) and the called party (Z). In IP networks netflow data, coupled with routing information yield the necessary A-Z matrices. Redesigns are performed far too infrequently and the results often reveal dramatic changes in traffic patterns that no one recognized. Express routes, bypassing overloaded network elements, elimination of elements, and rehoming often result in dramatic savings and performance improvements.

1.3   Network Taxonomy

To better understand network operations and engineering some grounding in networks and systems is needed. Networks are best described as communications pathways that have both horizontal and vertical dimensions. The horizontal dimension encompasses the different types of networks which, when operated in collaboration deliver end to end services. The vertical dimension is two tiered. Network elements, which carry user information and critical signaling information, are loosely organized around the OSI seven-layer model, one of the most successful design models in the last 50 years. As a word of warning, I use the terms *network system* and *network element* interchangeably. *Network system* was in wide use when I joined Bell Telephone Laboratories in the 1970s. *Network element* evolved in the 1990s and is institutionalized in the 1996 Telecommunications Act.

Above the network tier is a set of management systems that look after the health, performance, and engineering of the network tier.

The distinction between network and management systems is almost universally a clear line, as shown in Figure 1.4. Tests used to distinguish between the two systems types are based on how transactions are carried.

1.3.1   Voice Systems

In the first half of the twentieth century transactions meant one thing, a wireline phone call. A wireline call has six distinct stages.

1. The first stage is the service request. For a manual station set, that simply means taking the receiver off the switch hook and listening for dial tone.
2. The second stage is the signaling stage in which the originator dials the called party's number.

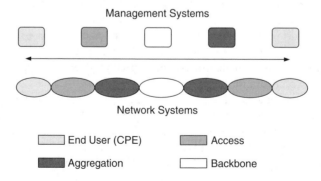

Figure 1.4   Network and management systems.

3. In the third stage, call setup, a signaling connection is established between the originator and the called party and a talking path is reserved.
4. Alerting or ringing the two parties takes place in the fourth stage. Audible ringing, some-times called ringback, is applied to inform the calling party that the call has progressed. Power ringing causes the called station to emit an audible ringing sound alerting the called party.
5. For successful calls, where the called party answers, the fifth stage is the completion of the final leg of a stable two way talking path.
6. In the sixth and last stage, the two parties conclude the call and hang up, after which the end to end path is freed for other users.

These six stages are the same whether a call is originated or terminated by a human or a machine. A wide range of technologies has been used over the years in each stage, but the stages are more or less constant.

For voice services we can then distinguish among systems by applying the following tests:

- Does the system perform a critical function in one of the six stages of call processing?
- If the system is out of service, can existing subscribers continue to place and receive calls?

Network systems when tested yield a yes to the first test and a no to the second. The time frame for applying the tests is important; a reasonable boundary for applying these tests is an hour. A local switching system is the one that gives dial tone and rings wireline phones. If it fails, the effects are immediate. At a minimum no new originations or completions occur because dial tone and ringing are not provided. In severe cases, calls in progress are cut off.

A provisioning system is a counter example. That system is responsible for adding new customers, removing customers, and making changes to existing customers' services. It does not perform a critical function in any of the six stages of call processing. If the provisioning system fails, we simply can't modify a customer's service attributes until the provisioning system returns to service. Existing calls, new originations, and terminations are not affected, so the provisioning system is a management system, not a network system. A second example is billing systems. If a billing system fails on a voice switching system, calls are completed without charge. Unfortunately no one sounds a siren or sends a tweet to let users know the billing system has failed and you can make free calls. The design choice to let calls go free during billing system failure is a calculated economic decision. It is cheaper to let them go free than it is to design billing systems to network system standards. Occasionally losing some revenue is cheaper than building redundant fault tolerant recording everywhere.

But what about the power systems in buildings where communications systems are located? In general network systems operate directly off of DC batteries which are in turn charged by a combination of AC systems and rectifiers. These hybrid power systems are engineered to survive 4–8 hours when commercial AC power is lost. Most central offices have back up diesel generators as well, enabling continuous operation indefinitely, assum-ing the fuel supply is replenished. Cooling systems fall into the same category. These are systems that do not affect the six stages of voice network systems if they remain failed for an hour. So here is a class of systems that if failed, don't affect calls within our hour time frame, but can affect them after a few hours or possibly days, depending on the season. These systems are in a third category, common systems. This is an eclectic group, covering power, cooling, humidity, door alarms, and other systems that if failed, can imperil the network systems within hours under the wrong circumstances.

1.3.2 Data Systems

The original tiered distinction and design for network and management systems came from the wireline voice network, but it applies to data and mobile networks as well. Consider two common data services upon which we can form our definitions, Internet browsing and mobile texting, or Short Messaging Service (SMS). Browsing is generally performed by a subscriber accessing content on the Internet by sending requests to a set of servers at a web site. The subscriber unconsciously judges the service by the response time, measured from the time the return key is stroked until the screen paints with the response. In the case of SMS, the subscriber has no clear way of knowing when the message is delivered, or if it is delivered at all. However, if a dialog between two SMS subscribers is underway, a slow response or dropped message is likely to be noticed.

For mobile subscribers, many of the network systems that carry Internet service and SMS are common. Our criterion for distinguishing between network and management systems is set by the most demanding data service. Before the introduction of 4G mobile services under the banner of LTE, Long Term Evolution, Internet access was the most demanding service. But LTE, unlike prior mobile technologies, uses Voice over Internet Protocol (VoIP) for voice service. With LTE data (VoIP data) delay tolerances become more unforgiving.

For data systems we can use our voice tests to distinguish among systems by applying the same tests, with minor modifications:

- Does the system perform a critical function in the timely delivery of subscriber data?
- If the system is out of service, can existing subscribers continue to send and receive data?

The modifier timely was added to the first test. While it was not included in the comparable test for voice service, it was implied. Recalling the six steps of call processing, untimely delivery of any of the functions is tantamount to failure. If you pick up a wireline receiver and have to wait over 10 seconds for dial tone, it's likely one of two things will occur. If you're listening for a dial tone you may grow impatient and just hang up and try again. If you don't listen and just begin dialing, believing the network is ready, you'll either be routed to a recording telling you the call has failed, or you'll get a wrong number. Consider the case of not listening for dial tone before dialing your friend whose number is 679–1148. You could be in for a surprise. Suppose you fail to listen for dial tone and begin dialing. If dial tone is delivered after the 7, the first three digits the switching system records are 911. Now you will have an unplanned conversation with the Public Service Answering Point (PSAP) dispatcher. When Trimline®[1] phones were first introduced by AT&T they caused a rise in these wrong number events. Trimline was among the first station sets to place the dial in the handset and people did not immediately become accustomed to putting the

---

[1] Trimline is a registered trademark of AT&T.

phone to their ear, listening to dial tone, and then pulling the phone down to dial. Many just picked up the phone and began to dial. Eventually they learned. Users can be trained.

1.3.3 Networks

Our communication infrastructure is actually a network of networks. I mean that in two senses. Networks are different in *kind*, and different by *serving area*.

To say networks differ in kind is an economic and technical distinction. Networks evolve to perform specific roles and the economics, topology, and service demands determine the technologies used and the associated designs. So, local distribution networks that deliver broadband and phone service to homes look far different than backbone networks carrying Internet traffic between major peering points.

Residential distribution networks, whether they are designed by cable providers or telcos tend to be asymmetrical, delivering more bandwidth toward the home, and they are very sensitive economically. If you have to deliver that service to 30 million homes, a $10 saving for each home matters.

Core IP networks carrying petabytes of traffic are at the other end of the technology and economic spectra. They are symmetrical networks that are fully redundant and possess sophisticated mechanisms for rerouting traffic in less than a second in the event of failure. A failure in the core affects all customers and has severe economic impact. Spending more in the core makes sense.

While the first distinction is according to kind, the second is by provider serving area. Each service provider designs, builds, and operates their network, generally under a franchise of the local communications regulator. The goal of universal service was established as U.S. Policy in the Communications Act of 1934 [1]. Two cornerstones of the act were that service should be extended to everyone, and that competing carriers should interconnect, enabling a national network of independent carriers. Prior to regulation in the twentieth century competing carriers often refused to interconnect. After 1934 interconnection and cooperation became common practice in the industry. It naturally extended to the Internet, although the U.S. Communications Act of 1934 does not directly apply to that network.

International cooperation and carrier interconnection are remarkable and beneficial practices that emerged from our twentieth century industrial society. Railroads in that era by comparison are a different story. Different gauges continued well into the twentieth century in Europe [2], inhibiting travel and commerce. When travelers reached an international border they often disembarked from one train and loaded aboard a different train because of the differences in railroad gauges. We take for granted our ability to place a call anywhere in the world, access any Internet site, and send e-mail and text messages to anyone anywhere. Interconnection only becomes news when it is taken away, as some Middle Eastern countries experienced in the Arab Spring uprisings. We'll explore network interconnection in depth in a later chapter.

1.3.4 Network Systems

Network systems support the Virtuous Cycle in the following ways.

- **Duplex operation** – achieving non-stop processing in the face of internal and external failures is founded upon redundant operation, and it requires the following functionality.

- **Field replaceable units (FRUs)** – failures occur on components with active electronics and components with software. Active electronic components are housed in circuit packs on assemblies that can be replaced in the field by technicians.
- **Hot swappable cards** – this takes FRUs one step further. Critical cards, such as system controllers and network fabric cards must be able to be removed and inserted while the system is in service, without affecting system performance. This is far more difficult to design than one might think. We will explore the necessary design steps.
- **Fault detection and spare switching** – systems must detect faults and re-route traffic to functioning cards, once the card is ready to process the load. Craft should also be able to return traffic to the primary card once it has been verified as restored.

- **Notification and sectionalization** – effective implementation is contingent upon trouble identification, correlation, and communication. Identification is achieved through hardware and software monitoring of internal hardware and software systems, and external circuits. Off-normal conditions must be resolved in a hierarchical structure that reports the most likely cause to the upper layers. Sectionalization resolves off-normal conditions as local or far-end related.
- **Fault isolation and mitigation** – fault detection and re-routing must happen automatically in milliseconds under the control of the system. However isolation and mitigation are performed by craft and rely on the software management of boards and ports. Object management is closely tied to the notification process and the implementation of administrative and operational state management.
- **Test, verification, and return to service** – object management again plays a key role, coupled with on board diagnostics.

The following chapters describe how network hardware, software, management systems, and work group standards and practices make this cycle successful.

1.4 Summary

Networks expanded across the entire globe from humble beginnings 100 years ago. Today's networks interconnect billions of people with highly reliable voice, Internet, and video services around the clock. Redundancy, fault tolerant systems, and operators and management systems working together in the Virtuous Cycle detect and resolve failures before they affect customer service. Network engineers monitor network elements and links, adding capacity, and augmenting the networks on a daily basis. Systems engineers extend networks with new technology and certify technology for seamless introduction into live networks. Network systems can be upgraded and repaired in the field without service interruptions. Hardware and software designs are defensive, meant to continue to operate in the face of failures and traffic overload. As we'll see in the next chapter, network systems are designed to very different standards than support systems.

References

1. United States Federal Law Enacted as Public Law Number 416, Act of June 19 (1993).
2. Siddall, W.R. (1969) Railroad gauges and spatial interaction. *Geographical Review*, **59** (1), 29–57.

2

Network Systems Hardware

The history of communications systems is rich and storied [1] but not broadly known in today's engineering community. When I meet with university students I find they know the origins of the Internet [2] but know little of the history of fiber optic communications. Certainly taking nothing away from Vint Cerf or Robert Kahn, I can safely say that without the laser, optic cable, and the Erbium Doped Fiber Amplifier (EDFA) the Internet would not exist in anything approaching its current form. The laser and fiber optic cable were invented at Bell Telephone Laboratories (BTL) and the EDFA was invented by Dr David N. Payne and Dr Emmanuel Desurvire. While the EDFA was not invented at BTL, they were quick to adopt it [3].

The global communications industry continues to spur invention and innovation with suppliers, universities, and carriers around the globe. Just as the global communications network provided the foundation for the Internet, we are now seeing computing and communication technology merging to the point that the hardware systems are indistinguishable.

2.1 Models

There are three common models for network systems hardware.

1. *The telco systems model* is the oldest and most common one found in carrier networks. It evolved from designs by the major carriers and equipment suppliers of the mid-twentieth century. It is purpose built for each specific application. Physical characteristics such as rack width, height, and operating temperature range were standardized early on. Later specifications were expanded to include earthquake survivability, electromagnetic susceptibility, and heat generation [4].
2. *Open modular computing* efforts began in the early 1990s. The goal was to develop a standardized chassis that conforms to telco requirements but is flexible enough to support a wide range of network systems and applications. The first widely adopted standard was CompactPCI®; today the AdvancedTCA® or ATCA chassis is more widely used. These systems promote innovation by solving the difficult hardware engineering problems, laying a ready built foundation for software developers to focus their talents on new services and systems.

Global Networks: Engineering, Operations and Design, First Edition. G. Keith Cambron.
© 2013 John Wiley & Sons, Ltd. Published 2013 by John Wiley & Sons, Ltd.

3. *The blade center model* is used for all-IP applications where general purpose servers, or hosts, implement all facets of a service. Network systems built around this model can use servers that meet the Network Equipment Building System (NEBS) standards [5], or they can be built using data center standard servers. The latter are less expensive, but require a more expensive common systems design to maintain narrower data center environmental standards.

Table 2.1 is a summary of the attributes and applications of the three models. I've chosen attributes that I believe represent the mainstream development and deployment of these systems. It's quite possible that someone has developed an Integrated Services Digital Network (ISDN) card for a modular computing system or even a general purpose host machine, but that is an outlier and I haven't represented those here (see Table 2.1).

These three models are discussed in the following sections.

2.2 Telco Systems Model

Network systems came of age in the second half of the twentieth century as voice switching systems. Electromechanical systems, such as step-by-step and panel switches, adopted and standardized in 1919 [6], were largely designed with distributed control. In the period 1950–1980 stored program control, or computer driven systems became the reference design for network systems and we have never looked back [7]. Figure 2.1 is a reference diagram of a stored program control network system. The diagram can be used as a reference for the Western Electric No. 1 ESS [8], a voice switching system introduced in 1963, or for Cisco's carrier grade router, the CRS-3 [9], introduced in 2010.

2.2.1 Form and Function

Hardware form follows function in these systems. In the following sections distinctions are drawn between telco systems and more general packet systems. While much of the functionality is common for carrier grade systems, differences emerge for these two general types of systems. Other differences are driven by the scope of deployment or the operations and engineering model of the carrier.

2.2.1.1 Concentrators and Aggregators

Switches can be concentrators or aggregators; in both cases the fabrics are designed to gather lower speed traffic and combine it onto higher speed links. A fine distinction can be made between concentrators and aggregators. Aggregators are non-blocking; there are an equal number of incoming channels and outgoing channels. An example is an M13 multiplexer. It accepts 28 incoming DS1s and aggregates them onto a single DS3 carrying the 28 DS1s in a different format. Concentrators have fewer outgoing channels than incoming channels, or at least the potential to have fewer. Fabrics for both are often M x N switches, where M lower speed ports are connected to N higher speed ports. The ratio of M to N varies from 2 : 1 to 500 : 1. A typical voice switch uses three stages of concentration. The first stage connects M line input output (I/O) ports via concentration to the second stage. Some concentration is achieved between the first and second stage, and then the rest of the concentration is achieved between the second and third stages.

Table 2.1   Network systems models

Attribute	Telco system	Modular computing	Blade center
Connectivity	TDM, ATM, Ethernet, Analog, SONET/SDH, Frame Relay, ISDN, FTTH, xDSL, DOCSIS, and SS7	TDM, ATM, Ethernet, Analog, SONET/SDH, SS7, Fiber Channel, and InfiniBand	Ethernet, Fiber Channel, and InfiniBand
Applications	Multiplexers, TDM voice systems, DWDM systems, ROADMs, radio systems, Network terminating equipment (NTE), edge IP routers, core P routers, media gateways, and mobility systems	Multiplexers, TDM voice systems, DWDM systems, ROADMs, media gateways, radio systems, mobility systems, voice response systems, and conferencing systems	All management system: EMS, NMS, OSS, BSS, and billing systems. All IP applications: VoIP soft switches, feature servers, and media servers. Network databases: SCP/NCPs, HLRs, HSS, DNS, LDAP, and ENUM. Content delivery and web services applications. Data center optimized routers.
Hardware	Generally single sourced from the supplier since it is purpose built. There are some exceptions, such as D4 channel bank cards which have multiple sources. Often ASICs are developed specifically for these systems, and so the hardware is not easily replicated.	Chassis and circuit cards are generally available from a range of suppliers who comply with the ATCA/PCI standards. Specialized cards may be developed for a small specialized market. There is greater use of FPGA technology and less use of ASICs to build on common hardware.	Generally hardware is specific to each supplier, but because these are general purpose host machines, supplier interchangeable hardware matters less. Some effort has been made to standardize and support multi-vendor designs[a] Economics are driven by the server market, not the telecommunications market.
Software	The admin or management processor may use a small version of Unix such as BSD, or Linux. All the other cards are likely to be embedded systems without an operating system, or with a real-time OS such as QNX or VxWorks to achieve a more deterministic work schedule.	A wide range of software is available. Operating systems, middleware, protocol stacks, I/O boards with embedded system APIs, and applications suite for management are all available. Integration is the challenge.	Software ranges from complete system solutions for OSS/BSS systems to extensive middleware for web services, databases, messaging busses, adaptors, and protocol stacks.

(continued overleaf)

Table 2.1   (*continued*)

Attribute	Telco system	Modular computing	Blade center
Key advantages	Ideal for high volume single purpose systems, such as subscriber multiplexers (DSLAMS and CMTS). Also well suited to very high capacity systems such as 100G/Terabit routers where proprietary ASICs are critical to scale. Full supplier support for these systems is an advantage.	The flexible backplane design with both TDM and Ethernet pathways make it well suited for low and mid volume time to market solutions and for multi-purpose platforms. Systems are more likely to ride Moore's law than telco systems.	Best suited for all IP stateless applications, such as web services and databases. Economics and availability are superior if the solution can be completely software based. These systems ride Moore's curve, and so you can count on performance and economic improvement over time.
Disadvantages	Time to market can be long if you are the first customer. The economics for anything other than high volume markets are difficult. There is no re-use of these systems and performance improvements come slowly, if at all.	Software integration can take longer than planned and achieving the desired performance may take time. Lifecycle support can also be challenging if multiple parties are providing different parts of the solution and the integrator is not committed and capable.	Stateless systems work well in general, but stateful systems can be problematic, particularly at scale. This approach can lead to poor functional distribution and that in turn causes availability and performance shortcomings that are difficult to resolve.

[a] See www.blade.org
EMS: Element Management System; NMS: Network Management System; OSS: Operations Support System; BSS: Business Support System; SCP: Service Control Point; NCP: Network Control Point; HLR: Home Location Register; HSS: Home Subscriber Server; DNS: Domain Name System; DWDM: Dense Wave Division Multiplexing; ENUM: E.164 Number Mapping; OS: Operating System; API: Application Programming Interface; CMTS: Cable Modem Termination System; FTTH: Fiber To The Home; BSD: Berkeley Software Distribution.

Examples of Time Division Multiplex (TDM) concentrators are class 5 switching systems, digital cross connect (DSX) systems such as the Tellabs® Titan® series and TDM multiplexers, such as the Alcatel DMX-3003.

In packet networks concentrators and aggregators generally serve as a point of traffic or service management as well as aggregation. Load balancers can be included in this general category, as well as layer 3/2 switches, session border controllers (SBCs) and broadband remote access servers (BRASs).

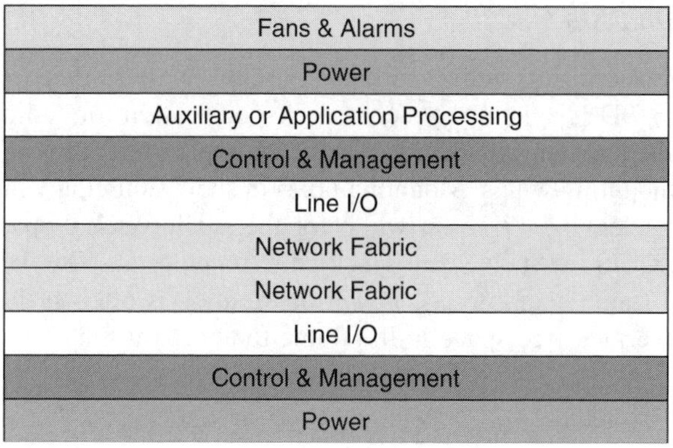

Fans & Alarms
Power
Auxiliary or Application Processing
Control & Management
Line I/O
Network Fabric
Network Fabric
Line I/O
Control & Management
Power

Figure 2.1 Telco system model.

2.2.1.2 Multi-Service Platforms

Multipurpose Systems are valuable in smaller carrier networks or in smaller offices of larger carriers because they offer a variety of interfaces and services from a common platform. These systems achieve flexibility by either a midplane design or a separate shelf for lower speed I/O. They tend to be somewhat future proof since they have high speed passive data and control plane designs coupled with flexible processor and I/O. New cards can be developed and take advantage of legacy cards and software. They can be repurposed easily if they are no longer needed in a particular application.

2.2.1.3 Add Drop Multiplexers

SONET/SDH (Synchronous Optical Network/Synchronous Digital Hierarchy) and Reconfigurable Optical Add Drop Multiplexers (ROADMs) have a primary transport high bandwidth east–west path, and a flexible shelf arrangement for lower speed circuits being delivered or joining the east–west path. ROADMs add and drop optical wavelengths while SONET/SDH systems add and drop electrical payloads, typically in Synchronous Transport Signal (STS) or Synchronous Transport Module (STM) formats.

2.2.1.4 Distribution Multiplexers

These systems include DSLAMS (Digital Subscriber Line Access Multiplex System) and DOCSIS [10] (Data Over Cable Service Interface Specification) multiplexers. They are deployed in large numbers and are purpose built with specific configurations and few card types, quite the opposite of multipurpose systems. They are designed with a view toward optimization, mass production, and simplicity of installation and operation. Distribution multiplexers are concentrators with high concentration ratios, but also have line interfaces with transmission modes suited to the particular distribution network, such as xDSL for telco networks and Quadrature Amplitude Modulation (QAM) for cable networks.

2.2.2   Frames and Shelves

While many North American systems continue to use the Western Electric standard of 23 in. wide shelves, EIA 310D is a common mechanical specification for communications frames today. EIA 310D shelf spacing is specified as 19 in., which refers to the distance from the outer edges of the mounting wings. Mounting holes in shelf mounting wings are horizontally spaced at 18 5/8 in., leaving 17.72 in. width for the shelf. Vertical spacing is a repeating sequence of 1/2, 5/8, 5/8 in. Shelves are specified in units, or Us. One U is 1.719 in. Seven feet is the common frame height today. Power distribution is often at the top of the frame, reducing available vertical free space in the frame to about 6ft 8in.

2.2.3   Chassis

Chassis incorporate one of more shelves and provide power, grounding, air plenums, and cabling routes. Separate frame and power grounds are generally provided and power is distributed on a shelf basis, with individual fuses for each shelf and power bus. Shelf and frame alarms communicate the status of the highest outstanding alarm in those units. Visible red alarm lights on shelves and frames are a great aid for line field technicians that have been dispatched to an unfamiliar office. They can find the failed pack much more quickly and reliably if they get visual confirmation from frame and shelf alarm lights.

In telco systems metallic I/O cables are typically terminated on fixed connectors on the back of the shelf and fiber connections are mounted on the front of the optical circuit cards that occupy the shelves. In TDM systems where copper pairs are used, D type Amphenol connectors are often fixed on the backplane and exposed at the shelf rear. Compatible cables are connected to the shelf and the other ends of the cables are connected to cross connect fields on DSXs or the horizontal of intermediate or main frames. Backplanes can span an entire chassis or an individual shelf. Two common designs are full backplanes and midplanes.

2.2.3.1   Full Backplanes

Full backplanes accommodate a single card in each slot and have fixed I/O connectors on the back, which means slots are dedicated for designated front facing circuit packs, with some flexibility within a family. In some cases a double sized card occupies two slots. Although backplanes can extend across two or more shelves, it's also common to dedicate a backplane to a shelf and design shelf extension connectors and cables to enable inter-shelf communication. Bus extender cables extend not only data planes, but synchronization and control as well (see Figure 2.2).

2.2.3.2   Midplanes

Midplane designs are more flexible, but more expensive and more complex. A connection plane near the center of the chassis terminates front facing cards and rear facing cards. Midplanes can extend to the full height of the shelf, or can occupy a portion of the shelf, as shown in the drawing in next page. Midplanes separate front facing processing from rear facing physical I/O. A typical use of a midplane design is to enable multiple low speed or

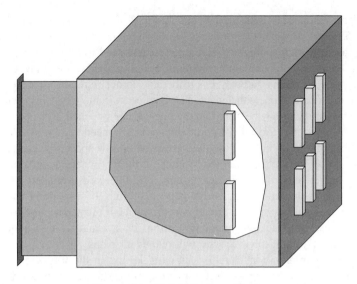

Figure 2.2   Full backplane chassis.

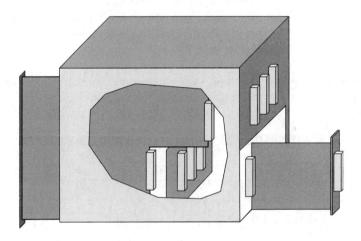

Figure 2.3   Midplane chassis.

a high speed termination on a given line card. A deep packet inspection card might support two different I/O cards, a GigE or multiple 100 Base-T terminations. In the latter case an Ethernet aggregator would be built into the rear facing I/O card so the pinouts into the midplane are identical for either configuration (see Figure 2.3).

2.2.4   Line I/O

Front-facing line cards support ports with physical interfaces by providing the necessary signaling, impedance, decoding, and encoding. Other functions may include test points, loopbacks, or maintenance capabilities. Line cards must convert data on the terminating port to the form compatible with the network fabric. For an Asynchronous Transfer Mode (ATM) port, 53 octet cells must be processed through an adaption layer specific to the ATM option being used. Data mapping and signaling control need to be routed to the bearer and control paths, with the path or channel routing information used to steer the cells through the switch fabric. A VoIP media gateway has to be able to accommodate a wide range of line I/O cards for interconnection with legacy TDM voice switching systems. Multi-frequency receivers and transmitters may be built into the line card or switched in as needed. Continuity tones, echo cancelers, and test access points are generally needed as well.

Automatic sparing under failure is seldom provided on low speed line cards, but has been designed into some connection oriented systems, typically where a port terminates a T3 (45 Mbps) or higher data rate. In those systems, one line card in a shelf is designated as a hot spare and is switched into the active path if a failure of a working card is detected. The physical ports do not terminate directly on the card circuitry, but rather on a protection switch with a bypass rail enabling fault protection. IP systems are connectionless and so they generally achieve fault tolerance through redundant paths.

There are some common considerations in card design. In the following narrative a T1 line card is used as an example to discuss these considerations. Figure 2.4 is a high level design of a T1 line card.

This design is similar to a D4 channel bank line interface unit. In the T1 line card example connectors are segregated for the various functions for illustration, although they would be combined into a single connector in most designs. The goal of a hot swappable circuit card means the following requirements must be met.

- **Hitless pack removal and insertion** – when circuit packs are inserted or removed, they cannot inadvertently affect power, address, data, control, or clock busses.

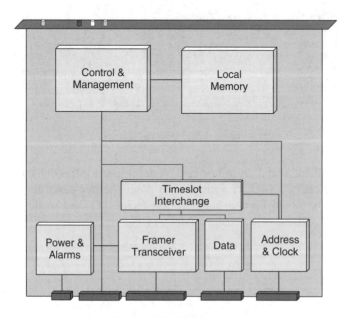

Figure 2.4   T1 line card.

- **Auto configuration** – I/O terminates on the backplane of these systems and so configuration information is associated with the slot, not with the circuit pack. When a faulty pack is removed and a new pack of the same type is inserted, the new pack should automatically be configured to the data set of the previous pack.
- **Automatic protection switching** – faults can be isolated and switching can automatically bypass a faulty link or circuit pack.

2.2.4.1 Mechanics

Connectors and cards are normally keyed, meaning that mechanical stops are designed to prevent the insertion of a card in any slot other than a compatible one. Keyed connectors prevent craft from mating symmetrical connectors with the wrong polarity. That doesn't stop people from trying, however.

Cards have grown in size over time. Western Electric D4 channel banks have been in service for decades. There are dozens of card types and many have Light Emitting Diode (LED) indicators on the faceplate to aid in diagnostics. Millions of D4 cards have been produced with dimensions of 6.1 in. deep, 1.4 in. wide, and 5.6 in. high. Most of these cards have a locking lever at the bottom of the card which is engaged when the card is fully inserted. In contrast, Cisco's latest core router, the CRS-3, is a midplane design with front chassis cards that are 20.6 in. in height and 18.6 in. deep. About 20 times larger than D4 cards, CRS-3 cards have substantial physical guides and dual locking levers to support insertion and removal.

Attention to detail and human factor analyses make a significant difference in the operational effectiveness of these systems. Placing the pack bar code where it can be easily scanned and updated is an example. Plainly marking the face of the pack with the family code and series prevents many craft errors in the event of a circuit pack recall. In the case of duplex packs in a chassis, LEDs indicating which is active and which is in standby or out of service prevents a common error; instead of replacing the failed pack, craft pulls the last working pack. Sticking with red, yellow, and green LEDs wherever possible is preferable to LCDs with alpha numeric displays. Circuit packs are generally replaced by line field technicians who cannot possibly be trained on the thousands of pack types they may be called upon to service. Field craft are dispatched by operations center technicians and only given a location and pack identifier. Imagine coming into a central office and finding two identical circuit packs in a shelf, one with an LCD displaying E4 and the mate pack displaying F2. One has failed and the other is in service. It's not likely you'll be able to reach anyone at the operations center that can definitively tell you what the two codes mean. If one has a green LED lit and the other has a red LED lit, chances are you should replace the unit with the red LED.

Cable and fiber fields also need care in design. Fiber bend radius limits must be carefully observed so clearances to equipment covers have to be checked. Coaxial cable bend radii are also constrained; exceeding design guidelines puts strain on the connector and sometimes on the backplane. Metallic connectors with screw terminations need to leave room for a craft's hands to use a screw driver. Wire wrap pins are often used to bring out telemetry alarm points or shelf options to the backplane. In cramped spaces it's possible those pins will be broken when wraps are applied and if there's no provision for replacement, such as a daughter board, the backplane may need to be replaced. Some packs still have reset buttons to restart hung processes or return the pack to the original factory configuration. These reset buttons need to be recessed, preferably with a design that only allows them to be activated with a small screw driver. I have seen pack designs with the reset button flush on the face plate, and have seen craft lean on the shelf when talking with a friend and reset the pack. That's not only the fault of the technician; it's the fault of the designer.

2.2.4.2   Power and Initialization Sequence

The sequence of power and initialization for circuit packs is controlled by the physical design of the connectors and the power and alarm module on the circuit card. Varying the designed pin length is used to control the power up and initialization sequence. Shelf guide rails engage with the circuit pack guide and faceplate first, and usually act as part of the power sequence by grounding electronic static discharge (ESD) as cards are inserted into the chassis. Long pins engage next. At first contact with the backplane, longer pins activate ground, and power control logic first. Power control circuitry will not supply power to the full board until the short pins are engaged, which are the last to make physical contact. Before the short pins are engaged, pull up, and pull down resistors are biased through ground and supply to insure a predictable start up. Critical busses are buffered with tri-state devices that are in high impedance states at start up.

The power and alarm function also guards against surges on the backplane power bus, and insure local supply levels are within limits before they are activated. When two power busses are available, load may be shared across both.

2.2.4.3   Local Control and Management

In the T1 example, on power up local control and management initialize registers in the framer and associated integrated circuits or ASICs. Interrupt masks and registers are set early in the startup. Counters and registers are cleared and the central controller is notified, generally at the next poll. Configuration information and connection instructions are downloaded from the central controller and used to set the framer and local control registers. The administrative and operational states are also copied from the controller. If indicated by the controller, an attempted link alignment is undertaken.

2.2.4.4   Local Memory

Locally memory is a combination of non-persistent and persistent memory. Configuration information may be stored in persistent memory. On start up local control and management can check the slot address and Common Language Location Identifier (CLLI) code which is unique to a shelf or system. If they match the stored code, the configuration can be used. Otherwise the information is invalid and cannot be used.

Local persistent memory can also be used to store "last gasp" information in the event of a local crash. If an out of bounds interrupt is generated by an invalid pointer, or a local watchdog timer expires or dynamic memory exhausts, the servicing interrupt routine may be able to write critical register information or even a stack trace into a protected persistent memory location. This is extremely valuable in post mortem analysis and can be the only real clue in finding the source of relatively rare but catastrophic events.

2.2.5 Power Supply Cards

AC and DC supplies are available with most equipment. Traditional telco TDM systems are likely to have −48 VDC as the standard supply with AC as an option. Systems that cross into enterprise and telco markets, like mid-range Ethernet switches and routers, are likely to have both options. Some systems accommodate both supplies at the same time, using AC when active and switching without interruption to DC when AC is not available. Power systems are full duplex, with dual busses, fusing, and alarming. System power supplies generally rectify AC into a common DC supply voltage, such as 24 V or lower, which is distributed throughout the shelves. DC to DC converters are often provided locally on each card to suit the needs of the chips and solid state devices on the card. Redundant system power supplies have sufficient filtering and capacitance to provide a safe window for switching if a disruption in the primary source is detected.

Power and fans are often combined into single units, and sometimes a remote telemetry alarm function is added. The design for remote telemetry is derived from a legacy system named E2A, which preceded SNMP, TL1 and modern monitoring systems. Remote telemetry monitoring survives because it uses out of band communication, typically voice band modems, for notification. That means it is likely to survive any data or bearer path related outage. Combined power and fan units are often located at the top or bottom of a shelf, augmenting natural convection to improve system air flow.

Power systems should be sophisticated enough to share load in normal circumstances, but be able to carry the entire load on one supply during times of stress. They should also block transients, and protect against foreign power crosses and out of bounds voltage and current limits. DC power feeds can be particularly dangerous if automatic low voltage cutoffs are not part of the design. DC to DC converters tend to overdrive when the supply voltage drops, so it is better to shut down the system, as unpleasant as that may be, and save the converters from high current failure. Intermediate thresholds should invoke alarms when supplies begin to move outside of norms. Design of the pin insertion and detection circuitry on power supply cards is particularly critical since the supplies must be hot-swappable, or at a minimum field replaceable; it may be acceptable to change out a power supply and turn it on only when it is fully inserted. Low DC voltages typically distributed on the power plane mean that high currents are required, so the supplies must be fully engaged before they begin to assume current load. Control circuitry must be engaged first and confirm the power pack is fully inserted before load can be transferred.

2.2.6 Network Fabric Cards

Network fabric cards are the pathways through the system for traffic arriving and leaving via the line I/O cards. Fabric cards are often redundant with in-service non-stop routing on failure. In systems with full redundancy, or N + N protection, if one card fails, the mate network provides full pathways with no loss of capability. In some systems an N + 1 sparing strategy is used. A single spare network switches in when any of the primary N networks fail. Switch capacity is often expanded by adding network fabric cards or modules, increasing the line I/O capacity of the system. The technology used in the switching fabric depends on the system and line cards. In TDM systems the fabrics are synchronous time division or a combination of time division and space division interchanges. Smaller TDM systems, DS3 and below, can achieve full connectivity without separate network cards; they use high speed backplanes with time slot interchanges. Packet data systems network fabrics are generally either hubbed Ethernet switches or Banyan switches. Separate Fiber Channel interfaces may be used to access storage area networks (SANs). Electrical or optical components can be used in the fabric. Packet data fabrics are often a combination of time and space division elements with buffering, traffic policing, and fault management.

2.2.6.1   System Control and Management

Control and management communications are isolated electrically and logically from the bearer plane fabric. If the bearer plane is used for these critical functions, system behavior during overload can be unpredictable. System upgrades and maintenance work on the fabric are also problematic since interruptions of the fabric can remove critical maintenance and control functions. Figure 2.5 is a functional diagram of a control and management board.

2.2.6.2   Control Bus

A low speed control and management bus is usually designed into the backplane, with addressable access to all cards. The bus is generally passive, with no backplane electronics and tri-state devices on each board. Cards are polled by the active controller and their identity and state are verified. A simple bus can be implemented with discrete logic, or via a micro controller or on a card processor. A two-wire industry standard invented by Phillips is often used [11]. A hard reset or card interrupt line may also be provided, so that any card that fails to respond to a poll, or experiences a watchdog timeout can be reset by the controller.

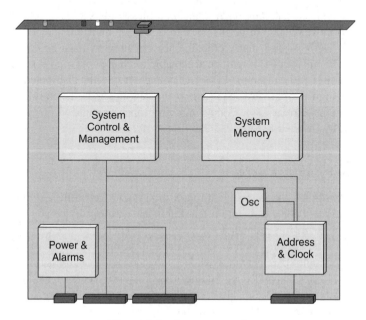

Figure 2.5   System controller.

2.2.6.3 Address and Clock Bus

In TDM systems and some packet systems a timing bus travels the backplane. Generally multiple clocks are transmitted, based on a single master clock, or set of clocks. Synchronous systems follow a hierarchy and stratum protocol. In these systems, multiple local oscillators are needed to meet drift, jitter, and wander requirements for the designated strata for plesiochronous operation. TDM systems of DS3 and higher speeds have the following priorities for deriving clocking.

1. Building Integrated Timing Supply (BITS) – all Central Offices have a BITS system homed on a stratum 1 clock through a timing chain.
2. Loop timed, generally on a stratum 3 or higher source. The timing source may be one of the switched T1 lines or may be an external T1 line. Usually two loop timed circuits can be designated, a primary and a secondary.
3. Local oscillator.

The system time of day (TOD) clock is usually located on the controller board, sometimes with local battery backup.

2.2.6.4 Management Functions

The acronym FCAPS (Fault, Configuration, Accounting, Performance, Security) categorizes the management functions of the system, as described below. ITU formally defined FCAPS within the framework of the Telecommunications Management Network (TMN) as Management Functional Areas. These definitions have gained wide acceptance in the standards community and apply to network and management systems [12].

Fault Management
The system controller monitors the operational state of all of the ports, circuit cards, and other resources in the system, real, and virtual. Circuit cards usually have a watch dog timer, either in hardware on the card or a software timer on the controller. The timer must be reset before it expires or an exception is raised to the controller, where it is assessed. As soon as the watch dog is reset, it restarts a countdown. If the watch dog expires and raises an exception, the controller may reset the errant card, remove it from service, or execute another recovery procedure. The controller may also generate a notification or alarm, notifying craft of the fault. This periodic check of each card by the controller is sometimes called an "all seems well" or a "heartbeat" function.

Configuration
Setting and maintaining the operational parameters of the system including routing tables, and line card settings is a control function. Routing and fault management are often closely bound; when a port fails or is restored to service, it is automatically changed to reflect the new operational state. Routing changes should also be verified before they are put into service. If craft tries to change routing toward a port that is not provisioned or in service, the configuration management subsystem should deny the change.

Accounting

Industry definitions for this function vary, but generally it is used to cover non-performance related measurement functions such as message detail recording and engineering. Message detail records (MDRs) are generated and collected as subscriber usage information, often resulting in a customer charge. Engineering metrics are collected to monitor the use of critical resources, such as port utilization, file space exhaust, and processor usage. Equipment engineers are responsible for monitoring the metrics and adding capacity before the system or resource exhausts, creating service failures.

Performance

Each network system plays a part in the delivery of an end to end service. Network systems have to perform within well-defined thresholds to insure acceptable service margins. Typical performance objectives for network systems include lost packets, cutoff calls, failed attempts, packet jitter, and misroutes. When the relevant performance metrics can be measured and delivered directly they are, but often surrogate measures or indicators of performance are delivered instead. For example, during receive buffer overflow it may not be possible to know how many packets were lost, but the overflow event and duration can be reported. Performance thresholds are often set to cause the performance subsystem to invoke a notification, delivered by the fault management subsystem.

Security

This increasingly important topic does not always receive the attention it needs. In addition to obvious design requirements, such as using Secure Shell (SSH) for secure telnet, using secure File Transfer Protocol (FTP), and using secure versions of Simple Network Management Protocol (SNMP) (version 3 is recommended), much more should be done. Setting user privileges to the minimum level and protecting critical databases, both network and subscriber, are foundational.

2.2.7 Application Processing

Application or auxiliary processing has become more common as the network has taken on additional responsibilities and features and as Moore's law has increased the amount of processing engineers can design into one system. Examples are firewalls, session border control, deep packet inspection, content delivery, spam filtering, and lawful intercept. In most cases it is not practical to burden line cards with these functions. The functions aren't needed everywhere and are processing intensive, making it much more difficult to manage the performance of a line card, and the management overhead is significant. By placing the function on a separate card in the network system the function can be applied where needed; management is centralized and the system can use the existing backplane and network fabric to route to and from the application card.

A design choice has to be made by the systems engineer when placing these functions. That is, should the function be embedded in the network system or dedicated as a separate system. Firewalls illustrate the choices. Firewalls can be integrated into gateway or edge routers, or they can be a separate appliance sitting between the gateway and the Local Area Network (LAN). Companies like Barracuda specialize in building network appliances that offer a wide range of security services; router suppliers offer integrated solutions but also have network appliances as an option. For carriers operation and administration is often the deciding factor when choosing. The solution that scales, offers nonstop in-service upgrades, and lowers network complexity is likely to dominate over price.

2.3 Modular Computing – Advanced Telecommunications Computing Architecture (AdvancedTCA™)

The Advanced Telecommunications Computing Architecture (ATCA) specification was written and approved under the sponsorship of the PCI Industrial Computer Manufacturers Group (PICMG®), a consortium of over 250 companies. The specification and conforming systems incorporate telco standards into a platform that is produced by a number of hardware manufacturers, relieving system developers of the difficult and expensive task of designing a dedicated telco conforming hardware system for each application. System developers focus on application specific development and cut development time significantly by starting from a solid base. Carriers can also reduce integration time by testing ATCA hardware platforms once for NEBS and other requirements. After that only the specific applications need to be verified and integrated. Common OA&M software frameworks are also available that build upon the standardized hardware.

2.3.1 Chassis

I simplified the architecture of ATCA in Figure 2.6 in order to draw the parallels between ATCA and the telco system model. A faithful overview of the base ATCA architecture can be found in the PICMG 3.0 specification [13]. Sub specifications include PICMG 3.1 for Ethernet, PICMG 3.2 for InfiniBand, PICMG 3.3 for StarFabric, PICMG 3.4 for PCI Express, and PICMG 3.5 for Rapid I/O.

The standard chassis is 12 U high while the cards are 8 U high. Fourteen or 16 card slots are typically available for 19 and 23 in. chassis respectively. A European standard 600 mm chassis is also part of the ATCA standard. Other shelf sizes are permitted by the standard, including "pizza box" designs, provided they maintain the front card and Rear Transition Module (RTM) specifications.

2.3.1.1 Back and Midplane

Flexibility is a cornerstone of the design, incorporating three distinct zones of connectivity. Zones 1 and 2 are designated as the backplane and have a common passive design for all ATCA systems, although networking, which uses Zone 2, is quite flexible, supporting a wide range of fabric options. Zone 3 is application specific. A midplane can be used in the application or front boards can connect directly with RTMs for rear facing I/O.

- **Zone 1**
  Telco heritage is fully reflected in this section of the backplane. Dual −48 VDC power, metallic test and ringing, and slot address and management functions are incorporated into the bus via the P10 connector. Like the telco model, dual power feeds originate

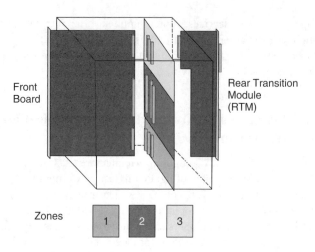

Figure 2.6 ATCA chassis.

at the top of the frame, optionally with a signal conditioning panel for additional noise suppression and filtering. A and B −48 VDC busses are extended across the slot positions with fusing on each shelf to prevent power crosses extending beyond the shelf. System management is supported with slot identification and an Intelligent Platform Management Interface (IPMI) bus. IPMI capable cards communicate via this bus with shelf controllers.

- **Zone 2**
  Data transport is assigned to Zone 2. Up to 5 40 pair connectors designated P20–P24 populate the backplane, with P20 being the uppermost connector and P24 the lowest. Four separate interfaces are provided via these connectors, a Base Interface, a Fabric Interface, an Update Channel Interface, and a Synchronization Interface.
  - **Base interface** – Intended for shelf management control, this interface uses slots 1 and 2 as Ethernet hubs and the remaining slots as nodes in a dual star configuration.
  - **Fabric interface** – a wide range of communication topologies, including full mesh, dual star, and dual-dual star fabric card interconnections can be supported by this interface. The interface can be partitioned to support multiple interconnection arrangements.
  - **Update channel interface** – typically used to support direct communication between redundant boards; four pairs can be used for high speed communication between boards.
  - **Synchronization interface** – used for TDM or other synchronous communications.

2.4   Blade Center Model

The blade center model has grown dramatically in two areas of carrier networks. Data centers supporting cloud services have grown over the last five years and the trend is growing. The growth is in the network services arena, not systems. Content delivery, compute and storage as a service, and virtual computing are all examples of data center enterprise services that are offered by carriers. Carriers have an advantage where they use the diversity and flexibility of the network, and integrate it with the service. Virtual Private Networks (VPNs) are a good platform for these services because they begin with a secure managed foundation. Carriers have global presence enabling them to distribute and manage low latency services that are resilient.

The second area where the blade center model is emerging is in the replacement of legacy telco technology with general purpose hosts. The IP Multimedia Subsystem (IMS) architecture is an example of this trend. Mobility voice in 2G and 3G networks is TDM based, with Mobile Switching Centers (MSCs) that were designed around their wireline TDM counterparts. With the introduction of Long Term Evolution (LTE) technology into mobile networks, voice moves from TDM to VoIP based on the IMS architecture. As the move is made to all IP networks the elements themselves move to general purpose hosts.

Both trends change operations models and economic models. Functions that transition readily to the blade center model are often stateless and transaction oriented. Web-based services are an example, as are routing functions which generally involve using a source set of data to derive a route from a single look up. Stateful functions require more care in design because of the distributed nature of the blade center model, and because guaranteeing routing in a connectionless network is more problematic.

Figure 2.7 is typical of the current generation of high end blade center systems. The diagram is based on the HP Blade System c7000, but IBM's Blade Center HT, Oracle's Sun Blade 6000 and Dell's PowerEdge M1000e have many of the same design features.

Here is a synopsis of some of the features that make these systems well suited for carrier operations. The functional diagram in Figure 2.8 is an aid to the descriptions.

2.4.1 Midplane Design

The chassis support a variety of full height, half height, and mezzanine cards that mate with a highly available midplane signaling and data field. Either a passive design is used on the midplane or active components are duplex hot swappable, virtually eliminating the possibility of midplane failure. These midplanes are fundamentally different than the designs

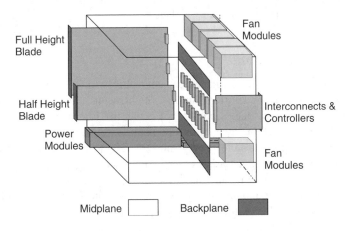

Figure 2.7 Blade center chassis.

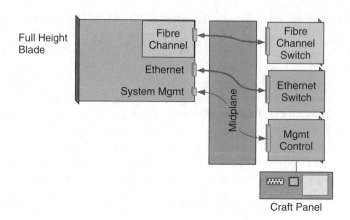

Figure 2.8 Functional blade center diagram.

in the telco and ATCA systems, which have protocol specific backplanes. The blade center system backplanes have a common Ethernet pathway and in addition they have protocol agnostic pathways or lanes between the server blades and the I/O or interconnect modules, where the switching fabrics reside.

2.4.2 Flexible High Speed Interconnection

The I/O or interconnect modules fall into two general categories, pass through and switched. Pass through modules are just that; the midplane provides a physical pathway from the front side server blade to the rear I/O connector, possibly with some electrical protection and minor buffering in the interconnect module. Switched interconnect modules include Ethernet switches, Fiber Channel switches and InfiniBand Switches. The interconnect switches connect via midplane lanes to native interfaces on the front side server blades, or to compatible mezzanine cards mated to the server blades. Ethernet is the common I/O on all blades for control and management. Express PCI or other formats may be used as the interface between the mezzanine card and the server blade.

2.4.3 Management Controller

Most systems have a mandatory management controller with an optional redundant controller. Some systems support a virtual fabric, which means there is an abstraction layer between the front side server blades and the physical and logical assignments of the rear side I/O ports. The abstraction layer allows logical switching in the event of a failure. That enables hot swapping and auto provisioning of failed units. Management controllers aggregate all of the blade administration onto a single card, significantly reducing cabling if nothing else. All of the systems also have local serial and RJ-45 Ethernet access to the blades via the controller. Optionally the systems have a mini-LCD panel with diagnostics and status information on the health of the system.

2.4.4 Power and Fans

Three phase redundant AC power is the standard, with −48 VDC available as an option. Unlike telco and ATCA systems, power management is critical in this model. Power consumption and the management of air flow need careful consideration because of the heat generated by a fully populated system. Special attention is also needed to place filler plates when a card is removed from the front or rear-connect slots. A missing filler plate, even in the midplane, can dramatically change airflow and quickly lead to hot spots in the chassis.

2.5 Summary

We can expect the blade center model to dominate in IP services and networks in the future. The move to VoIP and web services means that legacy telco systems will be replaced by the blade center model over the next decade or so, particularly in the core of the network. The operational model for these systems is still evolving. They have adopted important attributes of telco systems, such as field replaceable units (FRUs), redundant power, and hot swapping. But the management models are still based largely on the IT systems environment, not the network systems environment. The direction these systems are taking is encouraging, with innovations such as virtual fabrics which lay the groundwork for automatic fault protection and reconfiguration, but the alarming and managed object aspects of the systems are immature.

Traditional telco designs still have an advantage in connection oriented services and distribution services where loop transmission relies on specific transmission technologies, such as xDSL and QAM. They also have an advantage in the outside plant environment. It's hard to imagine placing a blade server in a remote terminal in Loredo Texas in mid-July. Apart from the extreme environmental conditions, remote terminals have to serve legacy TDM services and it is not economic to try and build these services into a blade center system.

ATCA systems have potential in the middle of these substantial competitors. Unlike blade center systems they meet telco standards for the operational model as well as the physical characteristics. They readily support TDM services as well as IP services and are a solid foundation for developing any new network system. They will be considered in telco applications destined for central office environments or remote terminals, and should compete well where time to market is a priority and the number of systems is fewer than a thousand systems. Above that, a purpose-built telco system is likely to be a better fit.

References

1. Joel, A.E. (1982) *A History of Engineering and Science in the Bell System, Switching Technology (1925–1975)*, Bell Telephone Laboratories.
2. Cerf, V. and Kahn, R. (1974) A protocol for packet network intercommunication. *IEEE Transactions on Communications*, **22** (5), 637–648.
3. Desurvire, E., Simpson, J.R., and Becker, P.C. (1987) High-gain erbium-doped traveling-wave fiber amplifier. *Optics Letters*, **12** (11), 888–890.
4. Telcordia Technologies (2012) GR-63and GR-1089.

5. Bellcore (2012) NEBS Level 3 is Defined in Bellcore Special Report SR-3580.
6. Joel, A.E. (1982) *A History of Engineering and Science in the Bell System, Switching Technology (1925–1975)*, Bell Telephone Laboratories, p. 22.
7. Joel, A.E. (1982) *A History of Engineering and Science in the Bell System, Switching Technology (1925–1975)*, Bell Telephone Laboratories, p. 199.
8. Joel, A.E. (1982) *A History of Engineering and Science in the Bell System, Switching Technology (1925–1975)*, Bell Telephone Laboratories, p. 275.
9. Cisco Systems (2010) Cisco CRS-3 4-Slot Single-Shelf System, http://www.cisco.com/en/US/prod/collateral/routers/ps5763/CRS-3_4-Slot_DS.html (accessed May 31, 2012).
10. DOCSIS 3.0 (2006) is the current ITU-T standard J.222. *Data Over Cable Service Interface Specification (DOCSIS) developed by CableLabs (Cable Television Laboratories, Inc.) and standardized by ITU*.
11. UM120204 (2007) I2C Bus Specification User Manual, Rev. 03–19 June 2007.
12. ISO 10040. *TMN Forum Recommendation on Management Functions (M.3400)*.
13. PCI Industrial Computers Manufacturing Group (2003) PICMG 3.0 Short Form Specification, January 2003.

3

Network Systems Software

This section covers network systems software design and operation. It is a black box examination of how network systems should perform. At different points the chapter looks inside the box to explain common software design issues, but this is not a treatise on software design. There are good texts dedicated to that topic [1, 2]. Performance or system behavior, is dominated by software, how it's designed, how it is written, and how it is managed. Traditional telco systems were developed by companies with deep talent in designing and delivering real-time systems that scaled and had critical features such as in-service software upgrades. Lessons learned from hard experience were passed on from developer to developer within those companies, but many of those companies and developers are gone. Modern software design techniques, such as object oriented design, help in many ways, but they also create traps for developers designing their first carrier grade system. The texts referenced earlier and the information in this section may help guide those developers.

3.1 Carrier Grade Software

What makes carrier system software different than IT systems or Internet applications?

3.1.1 Real-Time

Most Internet applications and many IT systems have a human being engaged in a session with the software application. Human beings are forgiving when it comes to timing. They will wait until the system responds and they have no concrete expectation of when that will happen. Network systems are I/O driven and have hundreds or thousands of sessions in progress and must service each protocol associated with those sessions on a tight schedule. They have to manage critical service schedules while performing other administrative or operational tasks at the same time. Real-time means there is a work scheduler that manages critical tasks and guarantees they are met, in priority order, within concrete and unforgiving service windows. That is very different from a threading model that offers priority treatment, but with no guarantees or ways to even measure what priority means.

Global Networks: Engineering, Operations and Design, First Edition. G. Keith Cambron.
© 2013 John Wiley & Sons, Ltd. Published 2013 by John Wiley & Sons, Ltd.

3.1.2 Reliable

Carrier system software must operate non-stop, all the time, no excuses. Peoples' lives depend upon it. Yet at the end of the development process when a system build is complete and we test the full system, scores of bugs are found that have to be fixed. Testing is a statistical process; if a hundred bugs are found, there must be others that were not found and testing can't go on forever. The product must be shipped at some point. Until a system can be designed and built from scratch that goes through full verification testing without a single bug, we need to acknowledge the inherent risks in new systems. New systems are shipped with bugs. Reliable systems are achieved through defensive, in depth, software strategies, not through perfect processes and infallible developers.

3.1.3 Scalable

This is the most complex attribute to describe because it touches so many aspects of the complete system. As a starting point, network systems can never be software bound. They can be processing bound or I/O bound but not software bound. By software bound I mean a system with an indeterminate transaction capacity that will only be discovered in deployment when the software fails. The capacity of such a system can't be predicted nor is it consistent, varying from incident to incident because the designers did not structure the software in a way that makes performance predictable under stress.

3.1.3.1 I/O Bound

These systems can processes as many transactions as can be carried with all of the terminated links running at full capacity. You can't increase the number of transactions because you can't add more links. They are predictable.

3.1.3.2 Process Bound

The system has an upper limit on the number of transactions it can process in a given period of time. Call that the system capacity. As the number of offered transactions reaches the system capacity, the system invokes measures to preserve throughput and maintain stability. Those measures include:

- Invoking flow control, notifying senders to back off on their requests.
- Suspending non-critical work functions. That can include provisioning, measurements, and administration, but not operational management.
- Discarding requests. This usually involves completing transactions in process at the expense of new requests.

A well-designed process bound system continues to process at system capacity even if the offered load is several times the capacity. These systems are predictable.

3.1.3.3 Software Bound

Over the years I have worked with many developers and learned much. In one conversation I had with a developer moving from an IT environment to a network system I asked about the capacity of the system he was developing. He gave me a number. I could see the system was not I/O bound so I asked how the system would perform in our network if demand exceeded that limit. He replied "you (the network operator) shouldn't let that happen." He wasn't sure what the system would do beyond the limit he cited. He thought his job was to design a system that could process a certain number of transactions, and it was the carrier's job to make sure customers didn't demand more than that.

Software bound systems exhaust on logical or physical resources and either crash or worse, limp along creating havoc in the network. Their transaction processing rate drops dramatically as they exceed design capacity. Those systems are described as poorly behaved, as shown in Figure 3.1.

Here are examples of how these systems exhaust:

- **Stack overflow** – either recursive calls or serving processes are not scheduled in a timely way.
- **Buffer overflow or underflow** – a lack of synchronization or a lack of predictable scheduling under overload.
- **Heap exhaust** – often caused by a poor object creation regime. Object orientated design is great, but flyweight patterns or simple structures handle real time processing better than heavy weight objects. Object pooling is another useful technique.
- **Database corruption** – records are locked for update, but transactions fail and they are not backed out and cleaned up.
- **Message corruption** – marshaling, segmentation, and reassembly processes break down. Invalid messages can be sent downstream, propagating the problem.
- **Out of (pick one: file descriptors, sockets, security keys, virtual identifiers, table space)** – the system exhausts on an invisible resource.
- **Synchronization** – semaphore and signal interlocks break down because threads become stalled and don't release critical objects.
- **Exception log or syslog overflow** – yes, it happens. The logs fill and the system halts on a blocking write.

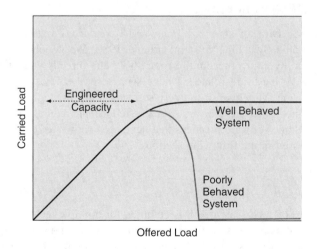

Figure 3.1 Poorly behaved systems.

These systems are unpredictable. Keep in mind we're not talking about the system encountering a bug. *They are designed and shipped with no idea of what will happen under stress.*

3.1.4 Upgradable and Manageable

Real people manage these systems, and systems and networks change. A carrier grade network system has to keep processing, even into the future with software and hardware changes the original designer cannot envision. IT developers often have weekly, or even daily maintenance windows during which systems are taken completely down and significant changes are made and the systems restarted. Not so with carrier systems. They must continue to operate during and after the change.

- **System configuration** – configuration files in IT systems are often read at system startup. If you want to change the configuration, you restart the system. Network systems must be able to accept new configurations on an in-service basis.
- **Firmware upgrades** – network systems must have support for upgrading firmware on line cards, or system firmware on an in-service basis. Line cards may go out of service during the firmware upgrade, but not the system.
- **Patches** – they must be loaded on a dynamic basis. Dynamic libraries are used with great care taken not to break the Application Programming Interfaces (APIs) exposed by the old library.
- **Protocol upgrades** – when changing a protocol between two systems it is unreasonable to expect craft to be able to convert all of those systems at the same time. Network systems should be bi-lingual on any interface, supporting the current protocol and at least one earlier version. Version information should be exchanged during session establishment to enable adaptation and prevent two systems on different protocol versions from aligning and passing garbled data.

3.2 Defensive Programming

Dennis Reese, now a Director at AT&T Labs, and I conducted a study of outages on the No. 1 ESS (Electronic Switching System) around 1980. We concluded that 40% of all outages were caused by craft procedural errors, 40% were directly attributable to software errors, and 20% were hardware related. I have seen several studies since and the results are still about the same. Moreover, it is likely software related outages are under estimated because well designed software can dramatically reduce procedural, and to some extent hardware outages. Some software practices that have been shown to improve overall system reliability are discussed in the following sections.

3.2.1 Are You Really Sure?

Carrier networks are of such scale that it is common for multiple technicians to be working on the same or interconnected systems at the same time and not know it. They are in different work centers or central offices. One technician may be accessing the system directly via a console in the central office. Others may be logged into the system remotely. I have seen significant outages occur because one craft took down a redundant link to a critical system to perform maintenance, and a second craft took down the remaining the last working link, isolating the system or the office. Those outages are recorded as procedural errors, but in many cases there is sufficient knowledge in the system to know that craft is attempting to take down the last working facility. Any command that has the potential to cause great harm should force the craft to issue the command a second time, after a warning is posted with as much relevant information as possible.

3.2.2 Default Parameters

Modern communications systems can have thousands of parameters that can be set. Most developers do a good job of setting the parameters to a default that at least does no harm, but there are exceptions. Some systems have a parameter with a range of $0-N$, and the developer sets the default to 0 because they are lazy when that value can do harm. Bounds checking parameters in the context of the current configuration is a defensive practice that requires extra effort. Timers, link speed, and buffer sizes are often related. The ranges of timers and buffers should be checked against the configured line speed.

3.2.3 Heap Management

Heap exhaust and memory leaks are still a problem in modern systems. In Java garbage collection timing is not guaranteed since it is system specific, but it's particularly hard to understand why heap exhaust occurs in systems written in C or C++. Certainly there are many tools on the market that claim to detect the leaks. If the tools are not completely effective, designs can and should pre-allocate heap memory to isolate any processes or code that manages it poorly or has a memory leak. With dedicated heap space the violator will soon be discovered.

3.2.4 Exception Handling and Phased Recovery

My premise is that all systems are shipped with bugs, so exception handling should be designed with the idea that it will be executed and integration testing should create thrown exceptions to verify the design. The design should have successive levels of exception handling invoking procedures that are ideally localized to the source of the trouble. In the example that follows, I use a card with four Ethernet ports. An escalating exception handling strategy might isolate a single port and reset it as a first step in recovery and then escalate on failure and finally reset the card, dropping all four ports.

The design of each recovery phase begins by defining an assumption set stating which system resources are assumed to be stable and uncorrupted and which are suspect. For example, system hardware may be assumed to be stable and system memory may be assumed to be corrupted in a lower level recovery design. Under that set of assumptions system configuration, connections, and state information can be rebuilt from hardware registers and hardware memory. If the phased recovery is successful, the card may be restored without affecting service. Service often depends on the hardware state, not necessarily the state of memory or processes. Data records on the system controller can play a role in the card recovery as well. The process of recovery design may even uncover flaws in the processes and structures of the base design.

3.2.5   Last Gasp Forensics

I mentioned this earlier but it is worth repeating. Many systems have exception handing and interrupt handling routines that when invoked should write as much forensic information as possible to flash or other persistent memory before reset. Register information, stack traces, recent logs, anything that is deemed relevant can help discover root cause later.

3.2.6   Buffer Discards and Dumps

Shared buffers can create a good deal of harm if there is no service protection designed into the buffer management. Picture a busy roundabout you've driven in on your automotive travels. If one of the egress routes becomes congested the shared pathway begins to block all traffic throughout the system. In a network system a single blocked egress can take out the entire system if no strategy is in place. Ideally flow control messages back to the sources reduce the flow until it can be served. But system integrity may not wait for that to happen. A strategy is needed to discard undeliverable traffic and quite possibly all traffic on the shared facility to preserve system integrity.

3.3   Managed Objects

A managed object is a network system, circuit card, port, link, or other physical telecom resource that is managed by telco operations.[1] The concept when extended across all network elements is a common pattern for managing communications networks and systems. Telco craft can perform basic operations on a system they haven't worked on in a long time, or have never even seen. In implementation it is hierarchical, lending structure to administration and operational management. The concept is implemented through software integrated with hardware components and it underpins the entire system design. At the center of the concept is a set of state machines. Objects move between states based on events. Events can be generated by craft, by the system itself, or by external real world events such as link failures.

This section uses another example to make the discussion of managed objects concrete. Figure 3.2 illustrates a four port Ethernet Network Interface Card (NIC) designed for a hypothetical system.

The four-port NIC implementation has six managed objects, four Ethernet ports, the NIC card, and the network element, or node. The hierarchy of managed objects is shown in Figure 3.3.

The node has a specific name GTWNTXWPMN1, called a common language location identifier (CLLI). It is composed of four fields:

- A four letter city identifier – GTWN for Georgetown
- A two letter state identifier – TX for Texas

---

[1] http://www.atis.org/glossary/definition.aspx?id=3928.

- A two letter building identifier – WP for Woodland Park
- A three character identifier for the specific element in the building – MN1 for managed node 1.

Managed objects use aggregation, not composition. The NIC card is not composed of four Ethernet ports, it has four Ethernet ports. Every managed object has two distinct states, an administrative state and a service state. Service states are sometimes called operational states. There are two names for the same type of state because Telco specifications, such as those

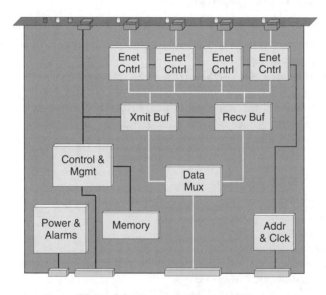

Figure 3.2   Four port Ethernet NIC.

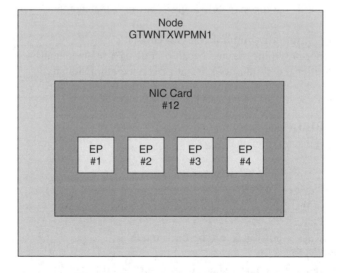

Figure 3.3   Managed object hierarchy.

written by Telcordia use the term "service state." IETF uses the term "operational state." The administrative state reflects what we would like the object to be; the service state is what it actually is. States have a primary and secondary identifier. Identifiers have been standardized by Telcordia and ITU [3], but practical implementations use only a small subset of the standards; state transition models are left up to the equipment designer, as it should be.

3.3.1   Administrative States

Our example system uses two administrative states.

- **In service (IS)**
  The object is fully provisioned and should be placed into service. IS is the primary state. There are no secondary states.
- **Out of service disabled (OOS-DSBLD)**
  The object is out of service and not ready for service. It may not be provisioned or the physical circuit may not be connected to the object. OOS is the primary state. DSBLD is the secondary state. Remember, the administrative state is set by operations, not by the system.

3.3.2   Service States

Our system uses four primary service states. In this case the dash – separates the primary service state from a qualifier.

- **In service normal (IS-NR)**
  The object is in-service and operating normally, within service limits.
- **Out of service automatic (OOS-AU)**
  The object is not providing service because the system took it out of service automatically.
- **Out of service manual (OOS-MA)**
  The object is not providing service because craft or a management system took it out of service.
- **Out of service automatic-manual (OOS-AUMA)**
  The object is not providing service because craft or a management system marked it as out of service after the system took it out of service automatically.

Our system uses seven secondary service states.

- **Disabled (DSBLD)**
  The object was disabled manually.
- **Fault (FLT)**
  The object is disabled and in an alarm condition.
- **Automatic in service (AINS)**
  The object is in the process of performing checks and attempting to return to service.
- **Loopback (LPBK)**
  The object (always a port) is in a loop back mode.
- **Maintenance (MT)**
  The object is in a maintenance state. It still carries traffic but the traffic is maintenance traffic and alarms may not be generated.

- **Unassigned (UAS)**
  The object is not fully provisioned.
- **Unequipped (UEQ)**
  The object, typically a card, is not physically present.

3.4 Operational Tests and Fault Conditions

So how and why do managed objects change service states? The general answer is through success or failure of operational tests. Operational tests are sets of measurable conditions with procedures that determine the readiness of a link, card, or system for service. They are usually performed in a consistent linear fashion to verify the physical object is ready for service. As each test is completed, if the tests pass the next test is invoked and when all tests are complete the physical object is ready for service; the managed object can then be placed into a proving state or directly into service. If the physical object fails the test the managed object remains in an OOS-AU state. Operational tests are invoked under four conditions as described below.

3.4.1 Service Turn Up

When the administrative state of a managed object is changed from OOS-DSBLD to IS by craft the managed object verifies that its parent's service state is In-service Normal (IS-NR) and if so, it begins a set of operational tests to return to service. As an example, an Ethernet port cannot be returned to service if the card it resides on is not IS-NR. During a prove-in period the state of the managed object is OOS-AU with a secondary state of AINS.

3.4.2 Interrupt or Fault Induced

When a physical object detects an off normal or fault condition it notifies fault management of the condition. A common occurrence of this sequence is when an interface chip detects an error and invokes a hardware interrupt. The interrupt handler diagnoses the chip via registers and determines what set of operational tests to run to restore the chip to service. The response may be as simple as clearing a buffer or it may require a complete chip reset and complete sequence of operational tests to prove in the circuit. If a fault is detected the managed object moves to an OOS-AU state with a secondary state of FLT. When a managed object begins an operational test sequence to return to service, the secondary state moves to AINS.

3.4.3 Out of Service Retries

Most link failures are caused by outside or inside plant facility failures. Someone digs up a fiber or craft moves the wrong circuit on a frame or cross connect bay. When either happens interface circuits detect the failure and go from IS-NR to OOS-AU with a secondary cause that reflects loss of signal. Links are usually configured to begin operational tests to re-establish link integrity. Periodically they start a retry sequence, testing for a received link signal and framing. The retry sequence may include periodic self-tests to insure the trouble is remote and not local. Assuming the managed object detects no local faults it remains in a state of OOS-AU with a secondary state of AINS while it is in a prove-in mode.

3.4.4   On Demand

Craft can generally invoke operational tests on demand. Some tests can be performed when the managed object is in an IS-NR condition; others require the object be OOS-MA or OOS-AUMA. Two secondary states can be used by craft in the OOS-MA mode. If the circuit is looped to aid the far end in a remote test sequence, the near end secondary state is set to LPBK. Often craft places a circuit in a maintenance mode to verify continuity but does not place live traffic on the circuit because the full path is not provisioned. Under those circumstances the circuit is placed in an OOS-MA state with a secondary state of MT.

3.5   Alarms

Network systems are designed to constantly check and verify their operation. From a design point of view, alarms are *objects* that are created when managed objects transition from an in-service state to a maintenance state. Notifications are *events* generated by alarms but alarms themselves persist as long as the out of service or maintenance condition persists; active alarms must be consistent with the managed object service state. Engineers sometimes use the terms notifications and alarms synonymously but they are different. An alarm can exist without a notification and will continue to exist until it is retired. A single alarm may create many notifications during its lifetime; some systems require notifications to repeat until the alarm is acknowledged by an Element Management System (EMS) or Network Management System (NMS). Figure 3.4 illustrates the processes that make up the alarm function in a network element.

At the center of the alarm architecture is an alarm manager. Managed objects create alarms and forward notifications to the alarm manager. The alarm manager is aware of the managed object hierarchy and filters notifications to management agents, such as Simple Network Management Protocol (SNMP), for further forwarding to EMSs and NMSs. The alarm manager also arbitrates polling requests to retrieve alarms that arrive via the same management agents. The following sections describe this process in more detail.

3.5.1   Notifications

Autonomous messages are generated by the network system and forwarded to an EMS, an NMS, or both. In this text I use the term notification to include all autonomous messages generated by network elements. There are two types of notifications, alarm notifications and informational messages. Notifications are distinct from log messages or polled information.

3.5.2   Severity

Alarms are further divided into three categories [4] based on the severity or seriousness of the event:

- **Critical** – indicates a severe service affecting event has occurred. Immediate corrective action is required.

- **Major** – used for hardware or software conditions that indicate a serious disruption of service or the malfunctioning or failure of important circuits. The urgency is less than critical because of a lesser or immediate effect on service or system performance.
- **Minor** – used for troubles that do not have a serious effect on service to customers or for troubles in circuits that are not essential to network element performance.

These descriptions are repeated in other Telcordia specifications, such as the LATA Switching System Generic Requirements (LSSGR) [5] but are not definitive. In practice a critical alarm is an out of service condition, usually because of complete (duplex) system failure or node isolation. Major alarms are generated when a network event forces a network system into simplex operation, often because of a protected node or protected link failure. An example is loss of a single route from an edge router to a core router. The edge router still provides service via the mate core router, so it is not service affecting, but needs immediate attention because of simplex operation.

3.5.3 Scope

The size of the circuit or system failure also matters. Critical and major alarms should not be generated on smaller circuits. Telcordia defines a switched circuit size of 128 DS0s as the threshold for alarm generation. A practical definition is an unprotected DS3 or larger circuit for a Time Division Multiplex (TDM) system. For a packet system that translates roughly to a 100 Mbps link or higher. Critical alarms are reserved for service affecting conditions. If a link is protected, a major or minor alarm should be generated, not a critical alarm.

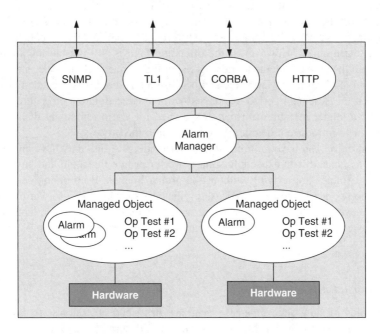

Figure 3.4   Network element alarm processes.

To put this topic in perspective we need to understand where the network element sits in the carrier network and what responsibilities have been relegated to the EMS and NMS. Some network elements, like Signaling Transfer Points (STPs) have a designated mate and keep track of the status of that mate, and of the mate's connectivity to the rest of the network. When an STP detects a Service Signaling Point (SSP) can only be reached via its mate, it issues a transfer restricted notification, which generates a major alarm. If the STP determines neither it nor the mate can reach the SSP, a critical alarm notification is generated. So the network element, an STP in this case, is capable of recognizing network isolation of another network element and issuing the appropriate notification. In many cases the responsibility for recognizing network isolation falls on the NMS, or possibly EMS.

3.5.4 Creation and Persistence

From an object design viewpoint, alarms are objects created by a managed object when it transitions from one service state (IS-NR) to another (OOS-AU). They are usually created in response to an operational test and do not necessarily reflect the exact failure that caused their creation. Creation of an alarm causes a notification to be issued to relevant network element agents via the alarm manager as shown earlier; those agents then transmit notifications to subscribed managers. Alarms can also be created when operational thresholds associated with a managed object are exceeded. For example, a link may have an error threshold of 10^{-4}. If in a set measurement period such as one minute, the error rate exceeds that value, a minor alarm is created and a notification issued; the managed object may also transition to an OOS-AU state if the error threshold was an out of service threshold rather than a warning threshold. Once created, an alarm object will persist until it is retired or cleared. Alarms are retired when the fault condition no longer exists. Since most alarms are created when an operational test fails, passing that test will retire any outstanding alarms associated with the test. Alarms can be cleared by authorized craft through a management agent. In effect that destroys the alarm object. However, if the offending fault is still present the alarm will be recreated when the network element attempts to put the device back into the service because the operational test will fail. Alarms created because a threshold was exceeded will only be retired when a measurement period has passed without exceeding the error threshold. Most systems have a longer measurement period specified for alarm retirement than they do for error detection to introduce a degree of hysteresis into the process.

A managed object can have multiple alarms outstanding. A link may have an outstanding minor alarm associated with exceeding some lower error rate threshold and have a framing alarm because frames are arriving without the proper bounds. Both operational tests have to pass for the two alarms to be retired. Alarm managers usually only forward notifications for the most severe of the alarms associated with a given managed object. They also filter alarms for lower tier managed objects if a parent has an outstanding alarm of the same or higher severity.

3.5.5 Ethernet NIC Example

Three examples of how state changes occur are described as follows. In the examples administrative states are indicated by gray circles and service states are indicated with clear circles.

3.5.5.1 Link Failure

In this example an Ethernet port is in an administrative state of in service, denoted by the gray circle in the center of Figure 3.5. Beginning at the top of the diagram, the service state at 12 o'clock is IS-NR, meaning the port is carrying traffic.

A fault occurs on the physical link, denoted by the X, and the system takes the link out of service, moving it to an OOS-AU, FLT service state. A major alarm is generated with an accompanying notification. Craft sees the service state and marks the circuit as manual out of service, which is indicated by the new service state, OOS-AUMA, DSBLD. The reason a circuit that has already been taken out of service automatically by the system is marked by craft is that other craft may be examining the link as well. By marking it, everyone can see the trouble is being worked and they need not interfere. A second reason for moving the link to a manually disabled state is that links have a tendency to bounce in and out if craft is working on a physical circuit. That disrupts the customer more than temporarily removing it from service by marking it and returning it when the work is completed and verified.

Once craft receives an indication the physical link is repaired, they can remove the manual maintenance state, moving the link to OOS-AU. That should cause the system to try to put the link back into service automatically, moving it to OOS-AU, AINS. That service state notifies everyone the system is trying to restore the port to service. So field forces can see the link is attempting to restore. When the link is completely restored it moves back to an IS-NR state, and issues a clearing notification referencing the original alarm.

The state machines in the examples are not comprehensive. They illustrate one set of transitions. A complete set of state machines would occupy several pages. Let's look at another example.

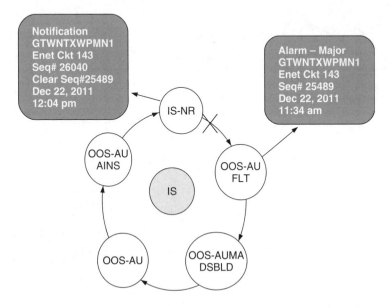

Figure 3.5 Link out of service example.

3.5.5.2   Provisioning a New Port

A newly provisioned port is being placed in service. The provisioning system sets the parameters of the port and then issues a change to move the administrative state from OOS-DSBLD to IS, shown by the gray circles in Figure 3.6. When the port is in the administrative state OOS-DSBLD, the service state is OOS-MA, DSBLD.

When the provisioning system changes the administrative state to IS, the system begins an initialization process to try and put the port into service, but in our example craft has not arrived at the office and equipped the system with the necessary circuit pack. So the service state moves from OOS-MA, DSBLD to OOS-AU, UEQ and an alarm is created and a notification forwarded. When craft arrives at the office and puts the correct pack into the right system and slot, shown as a diamond on Figure 3.6, the system recognizes the card insertion, checks the administrative state, which is IS, and starts a prove-in process to bring the link into service. It simultaneously moves the service state to OOS-AU, AINS. When the link senses alignment with the far end, it moves the service state to IS-NR; the alarm is retired and a notification forwarded clearing the original alarm.

3.5.5.3   Card Failure

If there is a hard failure on the circuit card and it fails to pass the system heartbeat or otherwise respond to central controller commands, the system marks the card's service state as out of service, automatic OOS-AU, FLT. All four ports are also marked as out of service by the central controller. Their service state is now OOS-AU, FLT. It is possible the ports are still working, but only the communication to the controller has failed. But the controller will mark the lines as out of service because the parent managed object, the card, is out of service.

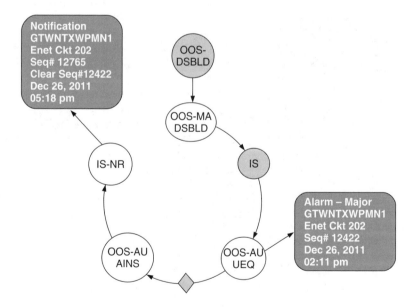

Figure 3.6 Provisioning a new port.

3.6 Network System Data Management

Managed objects describe the state of a system or port at a specific point in time, but give no history or trend information. An object is either IS, or it is not. Data are collected on the objects and as discussed later, used by different work groups for different purposes. Network element data management designs normally have the following structure.

- Low level embedded software manages the registers and interrupts on the communication chips on the line cards. Information from those low level routines is filtered and aggregated on line cards by a local control function that maintains state information for the ports and card.
- State information, critical events, performance data, and engineering data are forwarded to the system management controller for filtering and storage. System controllers maintain data for at least 48 h and more typically 96 h or more.
- System controllers implement northbound protocols such as SNMP, File Transfer Protocol (FTP), Hypertext Transfer Protocol (HTTP), Extensible Markup Language/Simple Object Access Protocol (XML/SOAP), Common Object Request Broker Architecture (CORBA), and Transaction Language 1 (TL1) to enable data collection.
- System controllers also implement northbound protocols for autonomous messages, such as alarm notifications, using SNMP and TL1. They also have filters and threshold setting mechanisms that allow performance data to generate alarms or other notifications.

In designing the example to carrier standards we'll use the hierarchical data collection and management approach illustrated in Figure 3.7.

Network data are organized in three distinct categories: operations data, engineering data, and services data. Three organizations and three sets of systems manage the data. This section

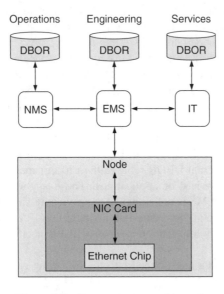

Figure 3.7   Data management hierarchy.

Table 3.1 Data responsibilities

Ownership	Data description	Transaction model
Operations	Alarm notifications, logs, system configurations, network topology, inventories, trouble tickets, traffic, and performance data	Operations has read-write privilege across the network and access to engineering and IT DBORs
Engineering	System configurations, network topology, traffic, and performance data	Engineering has read privileges for most of the data in the network
Information technology	Subscriber information, billing information, service definitions, and inventories	IT has read privileges for most of the data in the network and write privileges for service provisioning

describes the types of data managed by each group, how access is managed and how network elements are designed to support the structure. Later chapters describe how work groups use the data. The characteristics of the three databases of record (DBORs) are shown above. Ownership of a dataset does not imply development responsibility. Many Operations systems are built by IT or Engineering (see Table 3.1).

Collection begins at the chip or card level. Data collected there are in the process of accumulation, meaning the data are not complete. Periodically the card moves data from accumulation registers to hold registers and either notifies the system controller or awaits polling for the data to be moved to the system controller. Data often persists less than 15 minutes on the card before it moves to the system controller. The system controller tabularizes the data and retains it in a rolling data store for 48–96 h. In that period an EMS or other systems poll the node and retrieve the data. The EMS feeds the data to a DBOR, usually an Structured Query Language (SQL) compliant system that serves operations and engineering systems which are described in later sections.

Having a hierarchy serves several purposes. An EMS is designed by the node supplier and so data retrieval can be optimized for minimal disruption of the network element. If applications query network elements directly a standard protocol must be used and that can tax the network element Central Processing Unit (CPU). Where there is no suitable EMS, Secure File Transfer Protocol (SFTP) is the preferred method for tabular data retrieval. Transferring data to an external DBOR relieves the EMS of managing archives and reduces the burden of serving a variety of applications. Those applications may need particular indexing or views that the EMS is not equipped to provide, or can only provide by burdening the CPU. Finally, by collecting like data into a single DBOR applications can create joins and views that pull topology information together with other information from disparate but related systems.

Collected or managed data are grouped into the following categories.

3.6.1   Management Information Bases (MIBs)

Management Information Base (MIB) refers to data declared under a formal structure defined by IETF as management objects [6], not to be confused with managed objects described earlier for managing ports and systems. All SNMP data are defined in MIBs, but not all MIBs are collected and managed with SNMP. Here are some MIBs appropriate for our example system.

3.6.1.1   Usage and Traffic Information

In our Ethernet example the primary relevant MIBs are MIB-2 [7] and IF-MIB [8]. These two MIBs have object sets for collecting data on Internet Protocol (IP) interfaces, like Ethernet. Some of the data we can collect using these two MIBs are shown in Table 3.2.

IETF MIBs are helpful in defining traffic datasets for collection and serve enterprise applications well. But the scale of carrier networks imposes limitations on what can be achieved with SNMP. Some counters defined by IETF MIBs are rolling counters and do not give a measure of traffic over a fixed time period which is what is required for traffic

Table 3.2   MIB-2 excerpt

Data object	Access	Privilege	Description
ifInOctetsCounter	Read-only	All	A count of the number of received octets (bytes).
ifInDiscards Counter	Read-only	All	A count of the number of received discarded packets that were not erred. These packets are generally discarded for system reasons, such as buffer overflows.
ifInErrorsCounter	Read-only	All	A count of the number of received discarded packets because they failed checksum or other sanity tests and could not be passed to upper layers.
ifInUnknownProtos Counter	Read-only	All	A count of the number of received discarded packets because the protocol is not supported by our system.
ifOutOctets Counter	Read-only	All	A count of the number of outbound octets (bytes)
ifOutDiscardsCounter	Read-only	All	A count of the number of outbound discarded packets that were discarded for system reasons.
ifOutErrorsCounter	Read-only	All	A count of the number of outbound discarded packets because they failed a sanity tests and could not be passed to upper layers.
ifOutUnknownProtos Counter	Read-only	All	A count of the number of outbound discarded packets because the protocol is not supported by our system.

engineering. That means frequent polling is necessary to get meaningful granular periodic data. SNMP is also CPU intensive, particularly for tabular data, so SNMP polling is not the best option for bulk data collection across hundreds or thousands of nodes on a daily basis.

3.6.1.2 Accounting and Configuration Information

Industry standard MIBs, such as IF-MIB, generally need to be augmented with data sets definitions unique to the system. These MIBs are referred to as enterprise MIBs and may follow SNMP data definition rules spelled out in the SNMP Structure of Management Information (SMI) rule set. An enterprise MIB is likely to contain configuration and inventory information beyond that found in IF-MIB or MIB-2. An enterprise MIB should include the identity, version, and firmware information on every card in the system. It may also contain state information and environmental information.

Accounting and configuration information is controlled by provisioning systems or by Operations. System configuration is controlled by Operations and port information tends to be more service related, so it tends to be controlled by provisioning systems, with Operations able to override. Examples of system and port configuration information are shown in Tables 3.3 and 3.4 that follow.

Note that card information is populated by the network element and is read only.

3.6.2 Syslog

The term *syslog* has a generic and a specific meaning, much like the term internet. IETF standardized a syslog format and protocol [9] but many systems use the Berkeley Software Distribution (BSD) syslog format or have a format of their own. Cisco® has its own format and defines severities for syslog messages with the following scheme,

- Emergency: **0**
- Alert: **1**
- Critical: **2**
- Error: **3**
- Warning: **4**
- Notice: **5**
- Informational: **6**
- Debug: **7**.

The scheme is fine for enterprise operation, but does not map directly into telco alarm definitions and carriers make important distinctions between syslog messages, alarms, and notifications. Syslogs are written to a file and are not regularly reviewed. Scripts can be written to generate alarm notifications from syslog messages, but that is a work around and not a preferred solution. Mixing informational and debug messages with critical alarm messages is not practical for surveillance in a carrier network. System logs are used for forensic work after service has been restored. Logs are also used for analyzing chronic troubles to see if a pattern can be discerned. Recording alarms in system log files is helpful because time of day clocks often differ. If an NMS receives a critical alarm and the same alarm is recorded in a system log it is easy to see what informational messages were

Table 3.3   Example of system configuration data

Data object	Access	Privilege	Description
hostName	Read-write	Ops Tier3	The host name of the system
tl1Tid	Read-write	Ops Tier3	The TL1 target identifier
externalAlarmPt1active	Read-write	Ops	External alarm #1 active
externalAlarmPt1severity	Read-write	Ops	External alarm #1 severity
externalAlarmPt1description	Read-write	Ops	Description of alarm #1
slot10cardID	Read-only	Network element	Card family code
slot10cardVersion	Read-only	Network element	Card hardware version
slot10cardFirmware	Read-only	Network element	Card firmware version
slot10cardEquipped	Read-only	Network element	Slot is equipped

Table 3.4   Example of port configuration data

Data object	Access	Privilege	Description
fullDuplex	Read-write	Ops	Determines half or full duplex operation
macAddress	Read-only	Network element	Mac address is set by the network element
autoNegotiate	Read-write	Ops	Is line speed negotiated (10/100/1000T)
shortFramePadding	Read-write	Ops	Ensures a minimum frame size of 64 bytes
flowControl	Read-write	Ops	IEEE 802.3x compliant flow control enabled
crcEnabled	Read-write	Ops	A 32 bit CRC is appended to each frame
ipAddress	Read-write	Provisioning	IP address assigned to this interface
subNetMask	Read-write	Provisioning	Subnet mask
bridgingEnabled	Read-write	Provisioning	Enabled bridging of IP packets
spanningTreeProtocol	Read-write	Provisioning	The protocol used for this bridge
bridgeGroup	Read-write	Provisioning	Specifies which interfaces belong to a specific group
multiCastEnabled	Read-write	Provisioning	Enabled multicast on this interface

CRC = Cyclic Redundancy Check.

recorded that led up to the incident by aligning the time frames of the alarm notification and the syslog messages.

3.6.3   Audits

Network element audits are processes that run on a schedule or on demand that verify the data integrity of the system. Audits check for data corruption, data semantics, and data consistency across a system. It is an important defensive programming tool that is often overlooked or simply not considered by some developers. Network elements often have custom designed data structures because the Operating System (OS) they are using is either custom or does not support a commercial Relational Database Management System (RDBMS). That is particularly true at the card level. Embedded systems used on cards may use file structures that lack complete transactional integrity and data checking. Audits run under the supervision of the network system can compare the controller's copy of data with card data files. Comparisons can also be made between redundant controllers. Card audits can verify the hardware configuration of an associated integrated circuit (ASIC) or port against the configuration stored in persistent memory.

A variety of fault conditions can be avoided by running audits in background on a regular basis. Transients generated when circuit packs are inserted into a system can corrupt data or cause the on-board hardware to fail to initialize properly. Power surges or power system transients can cause resets of system controllers. Interrupted processes may have been writing data structures to file and so the data store becomes corrupted. Audits should be flexible, capable of being invoked automatically, or at least scheduled automatically on system restart. That practice catches corruption caused by unplanned system failures and shutdowns.

A final benefit of audits is that they catch mistakes made during software upgrades. When a system controller or card receives a patch or point release there is generally a lower level of system testing performed by the supplier. System audits serve as a backstop by checking data structures after the new load is implemented. IT system errors are also caught by audits. If a provisioning system is upgraded and a mistake is made that creates an out of bounds or semantically invalid order, an audit can catch the mistake even if the network element allowed the bogus record to be written to the local database.

3.7 Summary

Reliable networks begin with reliable network systems. The previous chapter described how hardware developers have evolved their designs over decades and reached a high level of art in the production of nonstop always-on systems. Network system software has evolved as well, but the application of defensive software design across the industry is more problematic and uneven. Proven practices such as the implementation of managed objects, phased recovery, audits, and last gasp forensics are sometimes missing in newer systems, particularly those produced by companies without a history of designing network systems. Approximately 40% of all outages are caused by software bugs or a lack of defensive design. It is a mistake to believe quality and system integrity can be infused into a system through rigorous testing and patching. Design matters.

References

1. Greg, U. (2005) *Robust Communications Software, Extreme Availability, Reliability and Scalability for Carrier-Grade Systems*, John Wiley & Sons, Inc., New York.
2. Rising, L. (ed.) (2001) *Design Patterns in Communications Software, (SIGS Library)*, Cambridge University Press.
3. Telcordia (2010) GR-1093-CORE Issue 2 and ITU-T X.731.
4. Telcordia Technologies (2011) OTGR-000474.
5. Telcordia Technologies (1984) LSSGR-00064.
6. McCloghrie, K., Perkins, D., and Schoenwaelder, J. (1999) IETF RFC 2578, *SMIv2*, April 1999.
7. McCloghrie, K. and Rose, M. (eds) (1991) MIB-2, IETF RFC 1213, March 1991.
8. McCloghrie, K. and Kastenholz, F. (1997) IF-MIB, IETF RFC 2233, November 1997.
9. Gerhards, R. (2009) IETF RFC 5424. *The Syslog Protocol*, March 2009.

4

Service and Network Objectives

I was once having a conversation with a senior engineering executive who said his application took several seconds to download data on his mobile device. I was in charge of the Labs and so he then looked at me expecting an explanation. I asked, "what is the objective, how is it measured, and by whom?" In other words, if there isn't a service objective and someone responsible to meet it, it's not a problem. System performance did not meet his expectation, but there was no objective or measurement, so no one cared. Before beginning a discussion on network operation, engineering, and management, the stage needs to be set by defining what service objectives are, and how to translate them into meaningful metrics. Service examples are used throughout the discussion, first for a consumer wireline voice service and a second for enterprise Voice over Internet Protocol (VoIP) service. Only network service objectives are discussed, but it's important to remember there are other aspects of a service that are important as well. If voice quality is excellent but the bill is wrong each month, customers are not going to remember the great voice quality when they think of the service.

4.1 Consumer Wireline Voice

There is a rich history of technical work for this century-old service so there is much to draw upon. Recall the six stages of a phone call.

4.1.1 Service Request

For a manual station set, that simply means taking the receiver off the switch hook and listening for dial tone. Dial tone is applied to a line when the switching system is able to collect digits. The objective is that 95% of the attempts receive dial tone with 3 s during the busiest hour of the day. It is measured by the originating switch. In older switching systems failure to meet the objective meant Engineering had not provided enough digit receivers.

Global Networks: Engineering, Operations and Design, First Edition. G. Keith Cambron.
© 2013 John Wiley & Sons, Ltd. Published 2013 by John Wiley & Sons, Ltd.

4.1.2 Address Signaling

This is the stage in which the originator dials the called party's number. There are a number of objectives for this stage. Partial dial timeout is a count of the number of times an incomplete address is registered because the subscribers did not dial enough digits. Partial dial abandon is a count of the number of times subscribers hang up in the middle of dialing. Both measures can be symptomatic of problems in the switching system. If these measures are exceeded, it may mean the routing tables are incorrect.

4.1.3 Call Setup

During call setup a signaling connection is established between the originator and the called party, and a voice path is reserved. Failed attempts are measured in two ways. Failures in call setup caused by call processing or system errors are recorded as equipment irregularities. Failures because there aren't enough circuits are recorded as blocked calls. If the former is excessive, Operations is notified. If the latter exceeds the allowed margin, 1% of calls in the busiest hour of the day, Engineering is notified to add more circuits.

4.1.4 Alerting

Audible ringing, sometimes called ringback, is applied to inform the calling party that the call has progressed and the called party is being alerted. Power ringing causes the called station to emit a ringing sound alerting the called party. There aren't any counters associated with this stage of the call, although the called line is checked for foreign voltages before the line is rung and an error message is generated if it fails.

4.1.5 Call Completion

For successful calls where the called party answers, a stable two-way talking path is established all the way to the called party. Each switch along the way periodically tests talking path continuity during call setup to insure the path is valid. If these tests fail, error messages are generated and if enough failures are recorded, the circuit is taken out of service. Attempts and completed calls are measured as "pegs" and the time the circuit is in use is measured as "usage." Once the voice path is established and the two parties begin conversation, voice quality is in play. Two measures which have an influence on voice quality are loss and voice path delay. Volumes have been written on end-to-end voice objectives, how to measure performance, and how to allocate loss and delay across the network [1]. Echo and noise also affect voice quality, but in digital networks these impairments occur at specific points in the network, and so objectives are not allocated across the network.

4.1.6 Disconnect

The two parties conclude the call and hang up, after which the end to end path is freed for other users. Short holding time calls and long holding time calls, calls that are statistically aberrant, are also measured since they may be an indication of chronic failures.

In addition to these measures there are measures for inter-system loss (attenuation), echo and noise, and the matching impedance of individual lines is periodically measured. There are scores of measurements made for what seems to be the most basic of services. Each measurement is compared against a standard and if it is out of bounds, an engineering, operations, or care work group receives a trouble ticket or other notification to look into the situation. But where do these measurements occur and which work group is responsible? The answer to that starts with end to end service objectives, and impairment allocations to each network element.

4.1.7   Network Service Objectives

As an example I've set a short list of end-to-end objectives for wireline voice service in Table 4.1.

4.1.8   Consumer Wireline Voice Network Model

Figure 4.1 illustrates how the consumer voice impairments can be allocated across the example network.

Loss, blocking, and voice path delays are caused by the circuits between switching systems. Call setup delays are incurred by the switching systems themselves. In our example performance against the objectives can be guaranteed in the following ways.

Table 4.1   End-to-end voice service objectives

Attribute	Objective	Comments
Loss	12–24 dB	Loss is measured at 1 kHz. Station sets have a measure of automatic gain control so this dynamic range is acceptable
Blocking	2%	Measured in the busiest hour of the day. Outside of the busy hour the probability of blocking drops to near 0%
Voice path delay	50 ms	Maximum one way delay is generally specified between 100 and 150 ms. We'll use 50 ms for this simplified example because the number of switch hops is small. An international call or mobile call would use the full 150 ms
Call setup delay[a]	1.2 s	Measured from the end of dialing until a call progress tone is heard
Total system downtime	3 min/year/ switching system	A long term average over many systems

[a]Also called post dial delay.

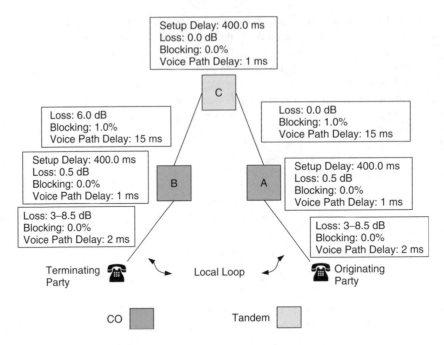

Figure 4.1   Network impairment allocations.

4.1.9   Local Loops

Local copper loops were designed using $1300\,\Omega$ design rules, meaning the size of the cable was set to be the highest gauge (smallest diameter) that could reach the customer with $<1300\,\Omega$ loop resistance. Maintaining supervision and ringing were the principle considerations, not insertion loss or voice path delay. However, copper loops transmit voice at $0.7 \times$ speed of light and have insertion loss in the range of $1-8.5\,dB$. Very short loops are automatically padded by the serving central office, so we can meet our voice path delay, loss, and blocking objectives based solely on the design of existing copper loops. No additional measurements need to be put in place.

4.1.10   Originating Office A

The originating central office A receives dialed digits from the subscriber and starts the call toward office C using a toll connect group, A–C.

4.1.10.1   Setup Delay

Timing allocated for a call setup delay of $400\,ms$ begins after the final dialed digit is recognized. Four hundred milliseconds is the specified call setup delay for class 5 switching systems [2], but a counter is needed for calls exceeding our specification since it is an indicator of more serious problems. If the counter exceeds a predetermined threshold, for example, 1 call in 1000, during a specified measurement interval, usually 15 minutes, Operations should be notified.

4.1.10.2 Loss

Virtually all North American switching systems are time division G.711 Pulse Code Modulation (PCM) systems, so a loss of 0.5 dB is the default. That loss is incurred in the line, or BORSCHT circuit. No measurements are needed.

4.1.10.3 Blocking

For our example we'll assume our switching systems are non-blocking. Class 5 systems can have concentration ratios of up to 8 : 1, lines to trunks, but choices are made based on local traffic characteristics; they are designed to be non-blocking for that community. A measure of calls that are blocked because of the switch fabric should be made. Operations should be notified if it exceeds a preset threshold.

4.1.10.4 Voice Path Delay

Voice path delay in a time division switching system is determined by the number of timeslot interchanges (TSIs) that are traversed. The interchanges operate at T1 line speeds, so it takes up to 0.125 ms for a voice channel to traverse a single TSI. The allocation of 1 ms can be met by any system with eight or fewer TSIs, which is well within reasonable design constraints. No measurements are needed.

4.1.11 Toll Connect Group A–C

4.1.11.1 Loss

Toll connect groups have a loss of 0 dB as the default. No measurements are needed since we are assuming digital transport.

4.1.11.2 Blocking

Busy hour blocking is set for 1%. That can be best managed by using one way trunk groups and making the measurements at the originating end, office A. Peg, usage, and overflow measurements are needed at Office A. Measurements should be collected daily and forwarded to Engineering.

4.1.11.3 Voice Path Delay

Voice path delay in a time division transport system is governed by the number of transport system (Digital Access Carrier Systems: DACs and Multiplexers: MUXs) TSIs traversed and the propagation properties of fiber and copper. Even if our circuit traverses 24 TSIs the delay is less than 3 ms. An allocation of the remaining 12 ms to the physical circuit means it could be up to 2000 km long and still be within specification. So we are well within the design constraint and no measurement is needed.

4.1.12 Tandem Office C

The requirements and analysis for office C are identical to office A, except for loss. Tandem offices have no BORSCHT circuit so there is no need for an allocation of 0.5 dB loss.

4.1.13 Toll Completing Group C–B

4.1.13.1 Loss

The Public Switched Telephone Network (PSTN) fixed loss plan prescribes a 6 dB loss on toll completing groups. Loss is inserted electronically at the tandem office so no variance is experienced. No measurements are needed.

4.1.13.2 Blocking

Busy hour blocking is set for 1%. Again, that is best managed by using one way trunk groups. Peg, usage, and overflow measurements are needed at Office C. Measurements should be collected daily and forwarded to Engineering.

4.1.14 Terminating Office B

The requirements and analysis for office B are identical to office A.

4.1.15 Long Term Downtime

The industry standard for Time Division Multiplex (TDM) switching system is that the long term average must be less than 3 minutes per year. That average is being met today because the systems underwent continuous improvement over the years, but more directly because the software hasn't been significantly changed in over 10 years. Historically, software and craft procedures cause 80% of outages. Since no one is taking these systems through yearly upgrades, they virtually never have a complete outage.

The Federal Communications Commission (FCC) also has regulations on reporting complete outages on these systems [3], so a measurement regime is in place, though seldom exercised these days.

4.1.16 Measurement Summary

Table 4.2 is a summary of the measurements that are needed in our example, and the work groups responsible for meeting them. Measurements are only needed where usage may affect fixed resources, such as trunk groups, or equipment irregularities or congestion that can create delays or failures.

Table 4.2 Network systems measurements

Office	Measurement	Frequency	Work group
All offices	Setup delay	15 min interval	Operations
	Peg (calls carried)	15 min interval	Engineering
	Usage (Erlangs)	15 min interval	Engineering
	Overflow (blocked)	15 min interval	Engineering

4.2 Enterprise Voice over IP Service

Now we'll explore how service objectives are set in enterprise networks, but first a bit more history is in order.

4.2.1 Five 9's

One misplaced idea that gets quoted often but has little relevance in a discussion of carrier networks is "five 9's reliability." The reference is usually to a service, or system, that is 99.999% reliable. The term is usually quoted as a definition of carrier grade networks, yet it has no real relevance and can't easily be translated into a meaningful service or network objective.

So, where did "five 9's" come from? I believe I know. When the voice network began moving from electromechanical systems to stored program control systems in the late 1960s and early 1970s it soon became painfully clear that a system with a central processor could have a failure that took the entire switching system out of service. That simply did not happen with electromechanical systems. Bell Labs engineers had to set an objective against which stored program control switching systems could be designed and measured. After all, the culture was to measure *everything* associated with service. I believe Hans Oehring and Al Hood[1] at Bell Telephone Laboratories (BTL) came up with the idea that the No. 1 ESS (Electronic Switching System) shouldn't be out of service for more than 2 h in 40 years over a cup of coffee, or perhaps something stronger. At that time switching systems like panel and step by step lasted 40 years. Two hours in 40 years equates to 3 min a year, or an up time reliability of 99.9994%. Voila, an industry standard was created that survives long after the No. 1 ESS, but without any analytical or empirical foundation.

4.2.2 Meaningful and Measurable Objectives

In the wireline voice example end to end objectives were translated into network element and transport objectives that can be monitored and managed. To achieve that for enterprise VoIP the starting point is to understand the makeup of the network and then develop objectives based on what the technology can achieve. The process facilitates the introduction of a metric that is far more meaningful in an Internet Protocol (IP) network, defects per million (DPMs).

---

[1] I met Hans briefly but I knew Al well as my department head and one of the brightest and best mentors anyone could have.

4.2.2.1   Defects per Million (DPM)

Converting network objectives to DPMs and allocating them establishes a lingua franca for networks, network elements, work groups, and services. It enables goals to be set for network elements that can then be summarized as network goals. Likewise service DPM objectives can be computed from the aggregate of the different kind and number of network elements needed to deliver the service. Lastly, work groups can be measured as DPMs, giving direct accountability that translates into network and service quality.

$$\text{DPMs} = 1 \times 10^6 \times \text{number of defects/number of chances}$$

The application of DPMs varies.

For a packet switch a count of the number of valid packets that are delayed, dropped, or otherwise fail routing is divided by the total number of valid packets received for routing, normalized to a total number of a million packets.

For a network of such switches the number of defects is the total count of defects across the population and the number of chances is also the total valid packets received across the population.

For a service the computation requires some granularity in measurement when common equipment, such as a packet switch is encountered. It is the number of packets for that service that are to be counted, not the total of all packets.

4.2.2.2   Total vs Partial Outages

The calculus of DPM can also have beneficial effects in placing special attention on total outages. FCC-reportable outages are those affecting at least 30 000 users for at least 30 minutes [4] so large scale outages are judged to be serious. When a critical system like a Domain Name System (DNS) fails, all system defects resulting from that failure are ascribed to the failed system, DNS. So DNS isn't charged just with the number of queries it estimates it might have missed; it is charged with the best available estimate of all lost traffic in the IP/MPLS (Multiprotocol Label Switching) network. If the outage time exceeds caching timers, that is a much larger number. There is also fairness in this approach. The IP/MPLS work group should not be charged with failures caused by DNS.

4.2.2.3   Enterprise VoIP Network Model

Figure 4.2 contains the different networks and subsystems that are necessary to place a call from a VoIP PBX (Private Branch Exchange), on the lower right, to a wireline phone on the lower left.

This diagram is a simplification. Redundant systems, such as the mate P routers, Signaling Transfer Points (STPs), and databases are not shown and the call is one of the most basic. Think of someone calling from their office to home. As services become all IP, including voice and video, there is a need to change the way network objectives are set based on end to end service objectives.

A remark often heard is that carrier VoIP services should be as reliable as the core optical and IP networks. An examination of the diagram makes it clear enterprise VoIP depends directly on the optical transport and IP/MPLS network, but it also depends on the carrier

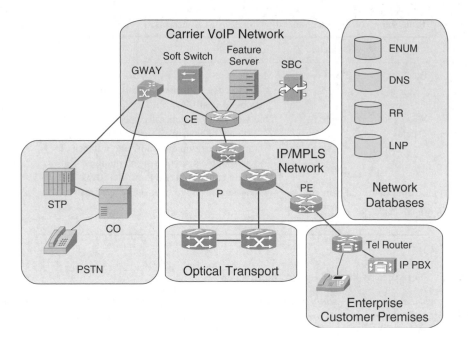

Figure 4.2   Enterprise VoIP network.

VoIP network systems and the E.164 Number Mapping (ENUM) database.[2] Unless those additional systems are designed and operated to perfection, never failing, mathematically VoIP service cannot be as reliable as transport and basic IP services. These networks are described in more detail in later chapters; for now the goal is to set network objectives for this complex service.

How, then, are meaningful network objectives set for VoIP service? Recalling the essential elements of a measurement plan from Chapter 1, each of the following sections examines their applicability to the IP/MPLS network.

4.2.2.4   Develop End to End Service Objectives

Identify measurable properties of each service, commit to service standards, communicate them, and put them into practice.

Many objectives used for consumer wireline voice are also valid for enterprise voice. The objectives that apply for a carrier VoIP service carried by an IP/MPLS network are repeated in Table 4.3.

Voice band loss isn't relevant for an IP network if a VoIP phone is used at call origination or termination. Call setup delay is also different in a VoIP network than that of a TDM wireline network because the IP network treats call control messages the same way it treats bearer path voice. The bearer path voice delay requirement is more stringent than the set up delay requirement, so it tends to bound the signaling delay.

---

[2] ENUM is not in universal use, but should be within a few years.

Table 4.3   VoIP service objectives for IP/MPLS networks

Attribute	Objective	Comments
Blocking	2%	Measured in the busiest hour of the day. Outside of the busy hour the probability of blocking drops to near 0%
Voice path delay	50 ms	One way delay is generally specified between 100 and 150 ms. We'll use 50 ms for this simplified example because the number of switch hops is small. An international call or mobile call would use the full 150 ms
Total system downtime	3 min/yr/ switching system	A long term average over many systems

4.2.2.5   Network Systems Measurement

Using service objectives, analyze each network element and set measurable objectives for that network or element that are consistent with the end-to-end service standard.

Call Path Delay

In the enterprise network pictured in Figure 4.2, the delay or latency in the IP/MPLS network begins at the customer edge router (Tel router) and ends at the media gateway (MGW). Assuming the VoIP network equipment is collocated with the IP network and allowing 8 ms delay for the final PSTN leg and a 2 ms delay for the VoIP Session Border Controller (SBC) we have a 40 ms objective for the IP/MPLS network. While our allocation is an example, it is consistent with major carriers like AT&T [5].

Blocking

Following the consumer wireline model we allocate 1% blocking to the terminating end office, leaving 1% for the VoIP network. But the 1% is a busy hour measure. A rule of thumb, long used, is that there is the equivalent of 10 busy hours of traffic in a business day. So the blocked call rate for the entire day is on the order of 0.1%, assuming no blocking occurs outside the busy hour. That translates into a total objective for VoIP of 1000 DPM. Allocation of the budget against the individual systems is beyond the scope of the example, but the exercise begins by using existing DPM objectives for networks like the optical and IP/MPLS networks. Then the state of the art and empirical data are brought in to estimate what is possible with the other systems. Such analyses are beneficial, possibly leading to changes in architecture or network design as the sensitivity of each system is examined. For example, rather than designing a single large ENUM complex a more distributed approach may increase the availability of such a critical system.

Interestingly, Packet Cable has an ineffective attempt objective of 500 DPM [6] for VoIP, close to our example. They also compute a cut off call objective, which I have left out of this analysis, but warrants a full study. Their objective is 125 DPM. That objective would not apply to all systems in our example, only the optical network, the IP/MPLS network, the media gateways, and possibly the SBCs, since they are the ones that are likely to affect calls in progress by disrupting the bearer path.

Down Time

Given the DPM approach, downtime used as a separate metric is of marginal value.

4.2.2.6   Assign Work Group Responsibility

Identify which work group is responsible for meeting each of the objectives and work with them to understand how they are organized, what skills they have and what groups they depend upon and communicate with regularly.

The divisions illustrated in Figure 4.2 align directly with Operations work groups in a large carrier. Later those work groups are discussed in more depth. Engineering also has a role to play in managing to the DPM targets. If any of the systems or transport facilities is under engineered, lost transactions begin, affecting the DPM results. DPM system objectives must then be divided between Operations and Engineering and a measurement regime put in place to fairly allocate between the two.

4.2.2.7   Design Engineering and Management Systems

Systems should support people, not the other way round. Find out what systems the teams already use and build on those if at all possible. Don't try to sell YAMS (yet another management system).

In the example of adding VoIP to a carrier network it is likely that the IP/MPLS and optical transport networks already have suitable systems in place to report DPM measurements. Latency, or bearer path delay, can be measured either by probes or by some routers directly. Cisco, for example, in their IOS-XR operating system enables one way or two way latency measurements (OWMaxSD) [7]. A system design supporting VoIP aggregates these measurements and VoIP-layer performance measurements, and presents them visually to the respective work groups they are focused on meeting their particular goals and measures. It also summarizes the results into Key Performance Indicators (KPIs), meant for higher management to present a general state of the network and services.

4.3   Technology Transitions

The examples illustrate how objectives have been established in carrier networks and how they are likely to continue to evolve as new services emerge. We like to think of technology as moving quickly and having the inherent ability to move aside older designs to introduce new services. But new network technologies are like the new kid at school. They have to fit in before they can make a mark. To illustrate the point consider adding VoIP in the subscriber loop in the consumer voice example. To keep our end to end service requirements intact, the new VoIP system must meet the requirements of the system it is replacing, the local copper loop. Loss and blocking objectives can easily be met, but that's not the case for call path delay.

If the 2 ms loop call path delay objective is devoted entirely to gathering VoIP PCM samples based on G.711, 16 samples can be collected at an 8 kHz sampling rate. With PCM that means 16 bytes of data are forwarded in a single protocol data unit (PDU). If Adaptive Differential Pulse Code Modulation (ADPCM) is used instead of PCM, 8 bytes of data are forwarded in a PDU. The data has to be transmitted to the central office in an IP packet. VoIP uses Real-time Transport Protocol (RTP) over User Datagram Protocol (UDP) over IP. So the overhead is quite high, about 40 bytes per packet; for PCM sampling 56 byte packets are sent with 16 bytes of data, with an overhead of 71%, and it must be sent 500 times a second. If we have a broadband service with 500 kbps upstream, it will take about 1 ms for the data to be transmitted after it is sampled, meaning the delay objective will be exceeded by 1 ms. Moreover, in the example a single voice call takes about 200 kbps, 40% of the upstream bandwidth. Certainly if the 2 ms call path delay objective is relaxed the efficiency of VoIP on the broadband link is improved.[3]

This simple example illustrates how new technologies have to conform with legacy network objectives if end to end service quality is to be maintained. In the example only call path delay is examined. Quality of service, security, and other requirements must be met by new VoIP technology as well [8]. New network technology does not often find a "green field" where it can be deployed without interworking with the existing network. The new technology must also fit within the operational and administrative models of the carrier work groups, as described in later chapters.

4.4   Summary

Every service needs objectives that reflect a satisfactory experience. Those end to end objectives are translated and allocated across network systems into a measurement plan using the existing metrics of the systems, or adding probes or additional system metrics to collect relevant measures. Each work group affecting network service is charged with fulfilling their responsibilities by meeting the objectives relevant to that group. Network and management system designs are put in place to support the work groups and deliver relevant reports to focus their attention on out of bounds or marginal performance.

Legacy services, voice in particular, have detailed network objectives that define the service and the expectation of subscribers. New technologies, like VoIP, have to meet established network standards. While new technologies have economic and functional advantages, they may find it difficult to meet existing standards for delay, reliability, and availability. More distributed designs and greater investment in common infrastructure can move new technologies closer to historical standards.

References

1. Bellcore (1990) *Telecommunications Transmission Engineering*, vols. 1–3, Bellcore Technical Publications.
2. Bell Communications Research (1987) LSSGR, TR-TSY-000511 (Issue 2), Section 11, July 1987.
3. FCC (2009) Reportable Outage User Manual, Version 6, April 2009.

---

[3] In practice samples are forwarded at intervals of 20–30 ms to improve efficiency, at the cost of voice path delay.

4. FCC (1994) Second Report and Order 94–189, CC Docket No. 91–273 (9 FCC Record 3911), adopted July 14, 1994.
5. AT&T (2011) Business Service Guide, SLA 3.2, AT&T Network Latency Performance Objectives.
6. Cisco IP (2003) Solutions for Cable High Availability Networks, Cisco Whitepaper dated 2003.
7. Cisco IOS XR (2009) System Monitoring Command Reference for the CRS-1, Software Release 3.9, p. MNR-234.
8. Pramode, K.V. and Wang, L. (2011) *Voice over IP Networks, Quality of Service, Pricing and Security*, Lecture Notes in Electrical Engineering, Springer Verlag.

5

Access and Aggregation Networks

Carrier networks started as purpose built networks, one network for voice, and another for data. In today's "everything over" Internet Protocol (IP) network, three distinct layers have emerged. Access and aggregation networks, shown as the lower tier below, provide connectivity, and gather and concentrate traffic for delivery to high capacity high speed backbone networks. Services are instantiated in data center complexes or central offices (COs) connected to the access networks via the backbone. Peers and partners connect through the backbone in a controlled and carefully managed way (see Figure 5.1).

Digital Subscriber Line (DSL), Passive Optical Network (PON), and Hybrid Fiber Coax (HFC) access networks are the networks that serve consumers and small businesses. They terminate on the side of homes in a Network Interface Device (NID). Enterprise access networks terminate in the network point of entry (NPOE), typically a wiring closet. Access networks are "last mile" networks that reach out into neighborhoods and city centers connecting and extending the network into those homes and work places. Aggregation networks sit above access networks and below the core. They collect and distribute traffic between the core and the access networks. Core and aggregation networks tend to have similar designs but access networks vary significantly. Access network technology reflects the history of the service provider and new communication modalities, mobility, and broadband. This chapter describes a range of technologies used in access, xDSL, HFC, fiber to the curb (FTTC), fiber to the home (FTTH), and mobile wireless. Each has different service capabilities and has to be managed with an understanding of the limitations and impairments inherent in the design. A discussion of the relative economies of the technologies is beyond the scope of this text, but some general observations are included in the descriptions.

There are common characteristics across all of the access technologies. They all carry voice services and broadband service. Cable began as a broadband network and added voice. Wireline phone service started out with voice and added broadband, as did mobile wireless. Carriers have had a common objective of pushing fiber deeper into neighborhoods for more than two decades, but the relative advantages of different designs are still debated. Table 5.1 is an aggregate of statistics taken from US financial reports and industry sources in 2011, such as National Cable & Telecommunications Association (NCTA). The data are

Global Networks: Engineering, Operations and Design, First Edition. G. Keith Cambron.
© 2013 John Wiley & Sons, Ltd. Published 2013 by John Wiley & Sons, Ltd.

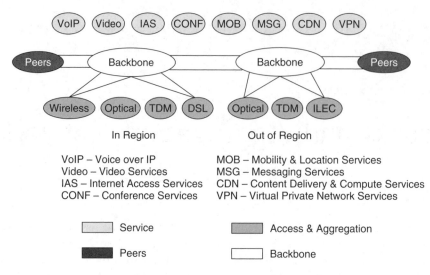

Figure 5.1   Carrier functional model.

Table 5.1   US Communication subscribers

Subs (xM)	Total	Telcos	Cable MSOs	Satellite TV	Others
Wireline voice	111	82	24	–	5
Broadband	77	31	46	–	–
Video	98	7	60	31	–
Wireless	329	257	–	–	72

not precise because of differences in how companies account for subscribers and because some carriers are private and don't publish subscriber data. Each device or connection counts as a subscriber, so the total of all subscribers is greater than the US population.

5.1   Wireline Networks

Phone networks were designed to carry voice services in the late nineteenth and early twentieth century. Initially open wire, bare copper, was used. Today, twisted pair copper cable is the dominant distribution medium, but fiber is moving deeper into the network, even to the home. Wireline access and aggregation networks were designed for voice services, but have evolved to carry data and video services.

5.1.1   Voice Services

Voice networks have clear demarcation between a distribution network, the network connecting the CO to the home, and a Local Access Transport Area (LATA) network, connecting a cluster of COs to Points of Presence (POPs). At the POP interexchange carriers (IXCs) interconnect with local exchange providers.

5.1.1.1 Local Access Transport Area (LATA)

The term "central office" is overloaded, sometimes used to mean a local switching system and sometimes used to refer to a wire center (see Figure 5.2). A CO in the broadest sense is a carrier facility center, a building where communications systems and facilities terminate. Serving Area Interfaces (SAIs) placed near the center of a Distribution Area (DA) are connected via feeder cable to a CO that is the wire center home. Wire centers vary in size from a 100 subscribers up to hundreds of thousands. Connectivity for voice services is through the local switching system, or class 5 switching system as described in Chapter 4. In the US Bell System network prior to 1982 there were five levels of switching in a hierarchical national network. There were 12 regional centers with class 1 switching systems at the top of the hierarchy and thousands of local switching systems, class 5 switching systems, at the bottom of the hierarchy. Tandem switching systems, interconnecting other switches were designated class 2, class 3 (primary), and class 4 (toll center) systems. Only class 5 switching systems served consumers directly, with dial tone and other services.

Mapping wire centers to switching systems is not direct. The size of a wire center reflects the topology and size of the community it serves. Switching systems are placed according to traffic demand, the number of call attempts and the usage. Large wire centers, such as those found in Manhattan, have enough demand to warrant multiple local switching systems. Very small wire centers, like my home town in Morehouse, Missouri have fewer than 1000 subscribers and don't warrant even a single local switching system. They can be served by a remote terminal with an umbilical to a local switching system in a neighboring city.

Prior to 1982 the mapping of exchanges to switching system was a many-to-one unique format. Phone numbers in the US are of the format NPA–NXX-XXXX where NPA is the numbering plan area; N is any decimal digit other than 1 or 0, and X, P, and A are any digit from 0 to 9. Exchanges (NXXs) are mapped distinctly to one local switching system. A switching system can serve more than one NPA and of course, NPAs map to many local switching systems. With that convention across the US, the serving class 5 office could be directly determined from the called NPA-NXX, without further processing.

The Modified Final Judgment (MFJ) of August 1982 which settled a long running antitrust complaint the Justice Department filed against AT&T, changed the rules of interconnection

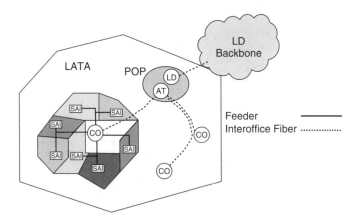

Figure 5.2   Local access transport area.

of the US voice network in a far reaching way. The MFJ forced the breakup of AT&T, separating design (Bell Telephone Laboratories), manufacturing (Western Electric), and long distance (LD) (AT&T Long Lines) from the 23 local telephone companies. LATAs did not exist, at least in a formal or regulatory sense before divestiture. Changes brought about by divestiture introduced a regulatory mandated network to network interconnection between class 4 and class 3 switches at a POP. Consequently the architecture of local networks within LATAs is now driven by regulation and does not vary to any significant degree across local carriers. Backbone LD networks have fewer regulatory requirements and their designs are more variable.

Within LATAs interoffice facilities ride synchronous optical network (SONET) redundant rings so no single outside plant failure should diminish connectivity or capacity. Because it is a connection oriented network, voice switching systems use load sharing interconnection across trunk groups and alternate routing via other switching systems. Unlike IP and DSL networks, Time Division Multiplex (TDM) voice networks use full duplex 64 kbps symmetrical connections. A conversion from four wire (separate transmit and receive paths) to two wire (a single pair carrying transmit and receive audio) is performed at the class 5 switching system via a hybrid circuit in the line card. Hybrid circuits have balance networks designed to approximate the impedance of the loop, but some echo from the receive path is often present on the transmit path. Echo cancellation is the responsibility of the LD carrier and it is often deployed at the first LD switching system beyond the local access network.

5.1.1.2 Local Signaling Network

Soon after local voice networks moved to TDM switching, sometimes called digital switching, they moved to common channel signaling using the Signaling System No. 7 (SS7) standard [1]. Signaling networks overlay the voice switching network and act as the control plane, setting up and tearing down all calls on the network. Switching systems, called Service Signaling Points (SSPs) in SS7 standards language, connect to two Signaling Transfer Points (STPs) via SS7 data links. STPs also provide connectivity to common databases, called Service Control Points (SCPs) (see Figure 5.3).

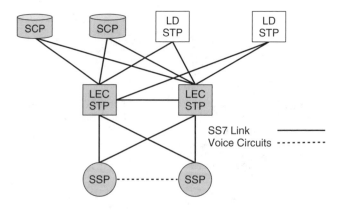

Figure 5.3 Local exchange SS7 network.

SS7 interconnection to LD STPs occurs at the same POPs where voice trunk groups are interconnected. SS7 is a packet protocol that is quasi associated with the voice bearer path. In a non-associated design control signals travel completely independent paths from the bearer. In fully associated designs there is a one to one relationship between control and bearer path. In SS7 signaling paths can be mapped to a set of bearer paths. In Figure 5.3, calls originating or terminating in this LATA must travel via the four links interconnecting LD STPs with Incumbent Local Exchange Carriers (ILECs) STPs, but any link can be used for a particular call. Moreover, in the absence of failure, once a call is initiated on a particular link, it will continue to use that link, reducing the probability of a race condition where the message sequence may be violated.

As services became more complex, it became apparent that a good deal of information needed for routing calls was common, and independent of where the call originated. Toll free, or 800 Service, is a prime example. Toll free numbers are associated with, and complete to a Plain Old Telephone Service (POTS) routable number. Rather than replicate a toll free database in every switching system, SSPs can send a query to a designated toll free SCP, retrieve the POTS routable number and route the call directly. An SCP is a real time transaction oriented database accessible via SS7. Today SCPs, sometimes called Network Control Points (NCPs) are used for a range of services including third party billing (LIDB, Line Information Database)[1] calling name display (CNAM), local number portability (LNP), and virtual private networks (VPNs). LNP is particularly critical; since number portability was mandated in the US by the Federal Communications Commission (FCC), calls terminating within a LATA are routed there based on the NPA-NXX of the dialed number. But with number portability the NPA-NXX no longer reliably maps to the serving office of the subscriber. Before sending the call to the local carrier for completion, IXCs must query an LNP database to find the correct serving class 5 switching system. The LNP SCP returns a routing NPA-NXX which may be different than the dialed NPA-NXX. An SS7 initial address message (IAM) containing both the routing NPA-NXX and the dialed number are forwarded to the serving class 5 switch for completion.

5.1.1.3 Distribution Area (DA)

Unlike the interoffice network, essential wire center concepts have not changed in over a century, but design rules and technology have. Economics dictate that the network be managed with three separate designs, the feeder network, the distribution network, and the drop. Figure 5.4 illustrates the designs. The drop, not shown on the diagram, connects the terminal to the NID on the side of the home.

The polygon in the illustration is a DA. DAs serve between 200 and 600 residential units; they are designed based on geographic constraints and projected population growth [2]. A feeder cable, represented by the dotted line brings service from the CO to the SAI, which is sometimes called a B Box. SAIs are cabinets that can be mounted on poles, at the base of poles, in a median or underground in some cases. Cross connects between feeder and distribution cables are made at the SAI. Outside plant engineers try to place the SAI as close to the center of the DA as possible to eliminate extreme loop lengths. Distribution cable begins at the SAI and ends at a terminal, either aerial or a pedestal in someone's yard.

---

[1] Line Information Database.

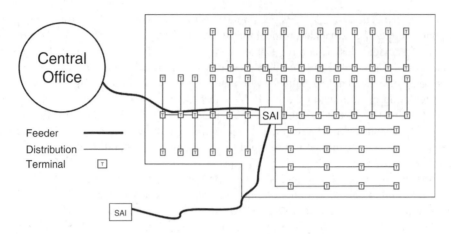

Figure 5.4   Distribution area concept.

Terminals serve from 6 to 20 lines. Drops connect from the serving terminal to the NID on the side of the home.

Feeder cable is usually 26 gauge and is Polyethylene Insulated Cable (PIC) or pulp (paper wrapped copper pairs). Distribution cable is almost always PIC. Broadly speaking, about half of all distribution cable is 26 gauge and half is 24 gauge, with a few designs in rural areas using 22 gauge or even 19 gauge. Loading coils are also used in some rural areas to flatten frequency response, but these loops are disappearing as fiber is pushed deeper into the network and those areas begin to be served by remote terminals. All elements of this design were put in place to support 3 kHz voice and the associated signaling and supervision of the CO.

5.1.2   Broadband Services

Public data services first appeared in the voice network in 1958 with the introduction of the 110 baud Bell 101 modem. Private line (Teletypewriter Exchange) (TWX) data service was introduced much earlier, in 1938. Voice band modems continued to improve through the early 1990s, claiming data rates of up to 56 kbps.

National Integrated Services Digital Network − 1 (ISDN−1) basic rate ISDN (BRI) service was introduced in the fourth quarter of 1992 [3]. BRI service supports two symmetrical bearer (B) channels and a symmetrical data (D) or signaling channel. The combination yields a 128 kbps bearer data and a 16 kbps signaling channel, sometimes referred to as 2B + D. Primary Rate Interface (PRI) was introduced around the same time as an enterprise service, supporting symmetrical data service (23B + D) over a T1 transport. BRI service is virtually out of use today, but PRI is still used in Private Branch Exchange (PBX) and call center services for network connectivity.

5.1.3   DSL

The first true broadband consumer service in the US emerged in the 1990s with the ratification of the Asymmetrical Digital Subscriber Line (ADSL1) standard by ANSI as

T1.413-1995. A family of DSL services have been introduced beginning with ADSL1, which is still the most widely deployed of the DSL technologies. ADSL1 takes advantage of the unused spectrum in copper loops by going well beyond the original 3 kHz voice band-pass design, using an extended bandpass of 25 kHz to 1.1 MHz. With Discrete Multitone (DMT) modulation data rates up to 8.192 Mbps are possible in the downstream (toward the home) direction; 6 Mbps is a practical limit for relatively short loop lengths. Data rates of up to 640 kbps are achieved in the upstream (toward the network) direction [4]. Since loss increases with the approximate square of frequency, practical upstream rates are close to the theoretical maximum.

Deployment of ADSL1 in carrier networks is very different from ISDN deployment in an important way. ISDN uses a baseband encoding scheme, 2B1Q and ADSL uses an adaptive multitone scheme, DMT. When installing ISDN craft was either successful, or they weren't. Customers achieved 144 kbps or they did not. Data rates for ADSL vary according to the transmission characteristics of the individual loop. For customers and for carrier sales agents that presented an entirely new way of defining and managing a consumer service. ADSL1 data rates cannot be predicted precisely in advance, so a conservative estimate is made of the available data rate based on outside plant (customer) loop records. The qualification process relies on empirical results from similar loops in the network. Tiered services are sold as banded rates, with a minimum and maximum downstream rate. The highest tier service for ADSL1 generally has a maximum data rate of 6 Mbps and a minimum of 3 Mbps. The experience of deploying ADSL1 and developing analytical models for loop qualification based on empirical measurements of the ADSL1 population led to the introduction of two new terms to ADSL service design, structural variability and conditional variability.

5.1.3.1   Structural Variability

Structural variability is a probability distribution capturing differences in performance of subscriber lines whose designs appear identical. For example, subscriber loops with 26 gauge underground feeder and distribution cable with a total loop length of 8 kft may experience downstream data rate differences of 20% above or below the mean rate for those loops at a 90% confidence level. Differences can be caused by differences in splices, binder placement, induced AC or differences in cross connects at the SAI or CO. Figure 5.5 shows pictures of a cross connect field in an SAI and a serving terminal. They illustrate how pairs are managed in the field and how differences occur in what appear to be identical pairs.

5.1.3.2   Conditional Variability

Conditional variability captures changes in loop performance over time. A single loop may operate at one data rate during the day and another slightly different rate at night because it is in an aerial cable and as the cable warms in the day the impedance of the cable increases. Radio Frequency (RF) noise sources, such as vacuum cleaners or light dimmers can also cause changes in line speed.

5.1.3.3   ADSL1

ADSL1, legacy ADSL, is a good choice for Internet access and home office access to the office via VPN. It is a marginal choice for Voice over Internet Protocol (VoIP) because of

Figure 5.5   SAI cross connect and serving terminal.

the additional latency experienced at the lower ADSL speeds, and because ADSL1 does not support quality of service (QoS).

5.1.3.4   ADSL2+

ADSL2+ extends the bandpass of ADSL to 2.2 MHz and improves performance through improved framing and mandatory Trellis coding [5]. Compatibility with ADSL1 is achieved by a common spectral band plan, 998. Data rates up to 24 Mbps downstream and up to 1.3 Mbps upstream are possible on very short loops. With higher data rates and QoS, ADSL2+ supports VoIP and Internet access. It can support Internet Protocol Television (IPTV), but only on relatively short loops or through bonding. The option for Ethernet framing also improves throughput by eliminating the Asynchronous Transfer Mode (ATM) cell tax incurred by ADSL1.

5.1.3.5   VDSL2

VDSL2 (Very-High-Bit-Rate Digital Subscriber Line 2) [6] extends the usable bandpass out to 8 or 12 MHz, depending on the implementation. Data rates of 25 Mbps can be achieved in most loop plant at distances of about 3 kft of 26 gauge cable (American Wire Guage: AWG). Upstream data rates at that distance are between 1 and 2 Mbps. When bonded VDSL2 is used, the range extends to 4 kft. Superior upstream bandwidth, 2 Mbps versus <1 Mbps for ADSL1 and ADSL2+, enable VoIP, IPTV, and improve Transmission Control Protocol (TCP) throughput by virtue of better acknowledgment speeds. Like ADSL2+, VDSL2 has QoS and Ethernet framing.

5.1.4   DSL Design and Engineering

DSL deployment requires changes in how loop plant is managed, with special care needed to avoid degrading existing services. Spectrum management, loop conditioning and attention to bonding and grounding are needed at deployment, and for the life of the service.

5.1.4.1 Spectral Management

The physical characteristics of copper loops make it well suited for DMT transmission, but it requires careful design and management. Telephone cable is made up of unshielded twisted pairs. Pairs are twisted to reduce noise and induced AC from nearby power lines. Pairs are organized into binders with typically 25 pairs to a binder in PIC cable and 25–100 pairs in a pulp cable binder. Binders are then organized into cable, with up to 3000 pairs in a cable, although counts above 1800 pair are rare. The numbers of twists per meter within a binder vary by design to improve crosstalk immunity, but since the pairs are not shielded crosstalk has to be managed using several techniques. Pairs within binders are more likely to experience interference from each other because of their physical proximity. Increasing the complexity of binder management is the inventory of digital services that ride in the cable. Figure 5.6 illustrates the potential for crosstalk or interference from different services.

Services occupying the same spectrum have the potential to interfere. Interference is more severe if the service's relative power spectral density (PSD) are significantly different and coupling increases with the square of frequency, so VDSL is more susceptible than ADSL. Power levels are controlled by two factors, the technology used, as shown in Figure 5.6, and the launch point of the transmitter, illustrated by the example in Figure 5.7.

In this example VDSL service is introduced into a serving area where ADSL is already in service. VDSL and ADSL coexist in the same cable from the SAI to the homes on the right. By the time the ADSL signal reaches the SAI it will have been attenuated by 7 kft of cable. The ADSL signal will be reduced by 40–50 dB by the time it reaches the SAI but

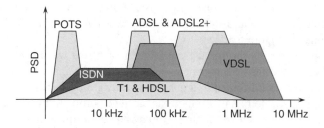

Figure 5.6 Power spectral density for digital services.

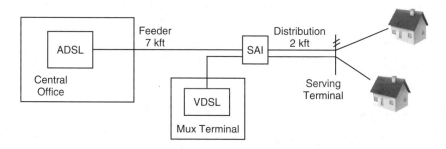

Figure 5.7 ADSL – VDSL cable sharing.

the VDSL signal will be launched at full power creating the possibility of the VDSL signal interfering into the ADSL signal. Coming from the home the opposite problem occurs. The ADSL modem in the home will launch at full power to overcome the 9 kft of loop loss. The VDSL modem in the neighbor's home will transmit at lower power because it has a short distance to reach the VDSL terminal. It is possible then that the ADSL signal will interfere into the VDSL upstream signal. The potential for crosstalk can be mitigated by using separate binders for ADSL and VDSL or by swinging ADSL circuits onto the VDSL terminal. VDSL line cards are generally compatible with ADSL.

Incompatible services, such as T1 and High Bitrate Digital Subscriber Line (HDSL) may need to be moved into a dedicated binder or at least separated from DSL service. T1 Alternate Mark Inversion (AMI) is the dominant interferer into ADSL downstream if binder management is not enforced. ADSL is less likely to interfere with T1 because T1 is repeatered. HDSL is a source of near end crosstalk (NEXT) and so also requires management [7]. HDSL management was made much more complex when in 1996 the FCC ordered loop unbundling, allowing Competitive Local Exchange Carriers (CLECs) to lease loops in shared facilities. With multiple companies provisioning services in the same cable, spectrum management became more difficult. AT&T and other ILECs work with the CLECs using rules designed to enable new services while protecting existing services.[2]

5.1.4.2 Bridge Taps

Distribution cable is not dedicated to households. They are reused and when customers disconnect that distribution cable may end up serving a neighbor or someone down the street or a few blocks away. In some cases distribution cable is intentionally spliced by adding a tap cable so the same distribution cable serves two neighborhoods. In both cases the result is similar to Figure 5.8.

The shaded area under the bridge tap is a quarter of a sine wave illustrating that if the bridge tap is open at the far end, it appears as a short on the main line for the frequency where the bridge tap length is equal to a quarter wavelength of the affected signal. In effect the signal from the VDSL terminal is split at the tap. The portion of the signal that travels on the tap is reflected by the open circuit and returns $180°$ out of phase, canceling the original signal, or at least a significant portion of it. A reflected out of phase signal is experienced for every odd multiple of a quarter wavelength, but as the tap length increases the effect is reduced because of the greater attenuation caused by the longer tap. The impedance of the tap at all frequencies begins to approach the characteristic impedance of the cable as the tap lengthens. The effect is more pronounced in VDSL than ADSL because the broader spectrum of VDSL is affected by a broader range of taps and the services delivered under VDSL are more demanding. So, tap removal is part of the loop conditioning recommended for VDSL deployments.

5.1.4.3 Other Interferers

Serving terminals, aerial and in ground, are not shielded for RF interference. VDSL, ADSL2+, and to a lesser extent ADSL occupy spectrum used by AM radio, 550 kHz

---

[2] For an explanation of the history of ILECs and CLECs in the US, see Chapter 8.

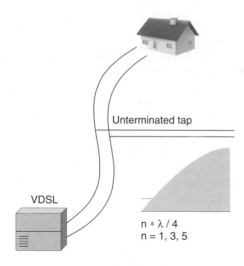

Unterminated tap

VDSL

$n * \lambda / 4$
$n = 1, 3, 5$

Figure 5.8   Bridge tap.

to 1.8 MHz. Stations with sufficient signal strength interfere directly with DSL. DMT is designed to skip those frequencies or assign more robust margins to tones overlapping with the interferers.

Impulse noise is less predictable. It comes and goes and can be generated from a wide range of sources, from air conditioners to light switches. It can be broader in spectrum and DMT is not a direct defense. Interleaving and forward error correction can mitigate impulse noise impairments, depending on the severity of the noise and the duration. Interleaving in effect spreads the noise, giving forward error correction a better chance of working without breaking down. Interleaving, particularly at ADSL speeds, comes at a price. Delays of up to 20 ms can be added to the normal latency. VoIP and gamers may experience degradation.

5.1.5   DSL Operations

DSL deployment, particularly ADSL2+ and VDSL are craft intensive. There are two craft titles that work on consumer DSL installation and maintenance, premises technicians and outside plant technicians, although some carriers combine the two jobs. Premises technicians install equipment and local networks in homes after outside plant technicians have turned up and verified service to the NID on the side of the home. The two jobs are often distinct because outside plant technicians seldom need entry to the home and so can schedule their work strictly based on demand. Having service completed to the side of the home also means the customer will have a shorter wait on the day of install; only the premises technician needs to complete their work before full service is available. To prepare for and bring up service loops often need conditioning, removing bridge taps, bonding and grounding, and binder management.

Special support systems were developed for DSL outside plant technicians.

- **Loop qualifications** – loop records are notoriously poor. New systems were built to determine if a customer was within the service area when they called to ask about DSL service, and an estimate of the expected data rates was needed so the widest range of services could be offered.

- **Single ended line tests (SELTs)** – carriers worked with DSL equipment providers to develop SELTs that are performed from the DSL multiplexer before service is attempted. These tests are able to predict service availability, quality, and speed. The tests can be done remotely with no need for craft dispatch.
- **Transmission optimization** – DSL technologies have a range of parameters that enable customization of the protocol on an individual loop basis. Systems were developed to gather transmission performance information over several weeks. The systems implement algorithms to optimize individual line parameters to improve stability and increase line speeds within the boundaries of the service. Optimization routines are run periodically to compensate for conditional variability.
- **On demand transmission analysis** – craft dispatched for maintenance or installation can spend a good deal of time disconnecting the customer modem and terminating a test set, waiting for the DSL line to resynch on the test set, and then gathering information to diagnose the line. Moreover, the test set performs differently than the customer modem so time can be wasted when the test set works but the customer modem does not. Web-based tools were developed to give craft access to the DSL multiplexer so the state of the line could be read directly without disconnecting the customer modem. Because the tools are web-based, craft can use a laptop, tablet, or smart phone to quickly gain access and retrieve near real time measurements. Knowledge based tools were added to tell craft if a bridge tap or other impairment was present.

Other systems and tools were developed for premises technicians to assist in installing Customer Premises Equipment (CPE) and diagnosing problems in the home network. Where possible, systems engineers work with CPE suppliers to design diagnostics for installation and remote service monitoring to aid customer care as well.

In SBC and subsequently AT&T these tools were developed and patented by the Network Systems Engineering organization. They worked with ASSIA, a company founded by Dr John Cioffi, a DSL pioneer and founder of Amati Communications Corporation [8], to make them available to the industry. The Broadband Forum developed a protocol for CPE Wide Area Network (WAN) Management which is functionally similar to the systems developed at AT&T [9]. The experience illustrates that equipment suppliers are experts at delivering network systems, but carriers often need to develop complex systems to support Engineering and Operations in the introduction of the technology into the network.

5.1.6 DSL Objectives, Metrics, and Line Management

DSL line management starts with the services provided over the line. If VoIP is provided as a supported offering, allowable error rates must be set lower than those needed for Internet service. VoIP is a User Datagram Protocol (UDP) service with no retransmission so uncorrected errors affect the service quality quite directly. IPTV should have error correction mechanisms within the service design, such as packet retransmission, but most of those mechanisms break down at some critical line error rate and when they do the result is usually pronounced. DSL metrics that give a direct measure of line performance are listed below.

5.1.6.1 CPE Initialization

Excessive CPE initialization is an important indicator of customer service. Lines with high variability may synchronize on power up, but fail if line conditions deteriorate, causing an initialization. Four reasons for loss of synchronization are reported by ADSL systems.

- LOF – Loss of framing.
- LOL – Loss of link.
- LOS – Loss of signal.
- LPR – Loss of power to modem.

Some customers regularly turn their modems on and off, and so LPR initializations should be disregarded. The other causes of initialization generally point to a line stability problem.

5.1.6.2 Line Code Violations

Line code violations are detected bit errors. By collecting this measurement every 15 minutes and examining the mean and standard deviation a measure of conditional variability can be derived. The higher the conditional variability, the more likely a higher target noise margin (TNMR) is needed.

5.1.6.3 Current Noise Margin

Monitoring Current Noise Margin (CURNMR) over time is another important indicator. As CURNMR drops below TNMR it signals that the line is in danger of becoming unstable since the line may fail to initialize at the current speed if there is a loss of signal or framing.

Two of the primary tools for managing ADSL service are listed in the following.

5.1.6.4 Downstream Target Noise Margin (TNMR)

TNMR is the minimum noise margin a line must achieve during modem initialization. Typical values are 6 dB for Internet service and 9 dB for VoIP, IPTV, or lines that demonstrate high conditional variability. The penalty for setting the value too low is service degradation, such as lower voice quality, video pixelation, or repeated initialization by the modem as it drops below the minimum margin needed to sustain the session. In extreme cases of high variability 12 dB may be needed. The price of increasing the TNMR is generally a reduction in line speed. That's often a reasonable price to pay for line stability.

5.1.6.5 Fast Channel or Interleaved Operation

Fast channel is a non-interleaved channel. Interleaving is helpful if noise is bursty and the duration of the bursts are of short enough duration to respond to interleaving. Forward error correction is normally used in conjunction with interleaving. If the line does not respond to interleaved operation, raising TNMR may be the only obvious choice for remediation, which is likely to lower the line speed.

5.1.7 ADSL Aggregation Networks

ADSL1 aggregation networks are ATM based. Options for interconnection are prescribed by the ADSL Forum working with ITU [10]. They are generally Point to Point Protocol (PPP) over ATM (PPPoA) or Point to Point Protocol over Ethernet over ATM (PPPoE). PPPoE supports multiple sessions making management more straightforward. As shown in Figure 5.9, these networks use an ATM virtual circuit from the subscriber modem or home gateway to the Internet Services Provider's (ISP's) Broadband Remote Access Server (BRAS) where the ATM VP/VC (Virtual Path/Virtual Circuit) is terminated. VP routing rather than VC routing is usually implemented to reduce administration. Each ISP is assigned a distinct VP and the carrier provisioning systems establishes an automatic VC from the subscriber gateway to the ISP BRAS).

ATM aggregation network capacity is managed by monitoring the physical circuits, typically DS3s, OC3s, or OC12s, and the VP provisioned capacity. ADSL does not have inherent QoS, so only best effort service is provisioned.

5.1.8 ADSL2+ and VDSL Aggregation Networks

ADSL1 evolved at a time when ATM was considered the network transport of the future because it carried voice and data with QoS. The thinking at the time was that Ethernet was a poor transport for voice because of the overhead and QoS was not well developed on Ethernet. By the end of the 1990s it was clear Ethernet is the layer 2 transport of choice. While ADSL2+ and VDSL retain backward compatibility with ADSL, both implement Ethernet framing which is superior to ADSL's ATM framing. With Ethernet compatibility the design of the aggregation network is quite flexible. Figure 5.10 represents one choice. I've omitted redundant equipment in the diagram for simplicity.

Virtual Local Area Networks (VLANs) are a logical choice for service management. They extend into the home and terminate on a layer 2/3 switch in the network. Service provisioning is achieved by interconnecting subscriber VLANs emanating in the home with services VLANs originating at service routers. Each VLAN is marked for the appropriate QoS; traffic policing and admission control are administered at the VLAN level. Service measurements can be obtained directly from VLAN MIBs (Management Information Bases) [11].

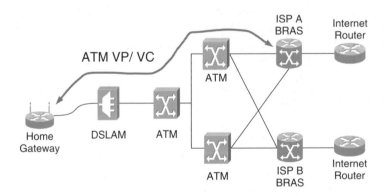

Figure 5.9   ADSL aggregation network.

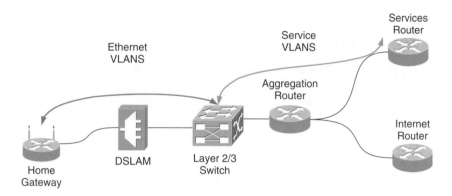

Figure 5.10   ADSL2+ and VDSL aggregation network.

Multicast services can also be enabled at various points in this example, with the Digital Subscriber Line Access Multiplexer (DSLAM) and layer 2/3 switch being logical candidates. An IP multicast tree is established in two stages from the services hosts to any IP enabled CPE requesting a service, such as broadcast video. Static multicast service points are set up using Protocol Independent Multicast (PIM) between the services host and the designated multicast join point, such as the layer 2/3 switch. At the join point, an IGMPv2 group is initiated to serve CPE. The services design must map user services to the multicast group addresses.

5.1.9   Fiber to the Home (FTTH)

FTTH distribution technology has gained in acceptance in many parts of the world as the technology direction for distribution networks. Legacy telephone and cable networks still have the ability to deliver today's services and for a time technology can be used to extend their service life. But FTTH is attractive for greenfield starts (new builds) and selected overbuild or rehab situations. In the US about 21 million homes are passed with FTTH networks and about 7 million homes are connected as of March of 2011 [12]. There are over 40 million homes connected outside of the US, with Asia Pacific leading the way. Fiber to the Building (FTTB) is generally included in FTTH statistics, but the designs can be quite different. Nearly all are based on a PONs for service delivery to the customer premises.

5.1.9.1   Passive Optical Networks (PONs)

PONs are the common denominator for a range of electronic technologies that deliver high speed services to the home. In one sense PONs are attractive for the same reason as legacy copper telephone networks; neither has any outside plant electronics. Cable and DSL distribution networks have significant deployments of electronic terminals that require power and maintenance. PON and legacy copper have no electronics between the CO and the home. Looking at the networks a different way, PON is like cable in that it is a narrowcast network, closer to cable's shared broadcast design than it is to DSL technologies which are dedicated between the CO and the home. Figure 5.11 illustrates a typical PON deployed in a FTTH design.

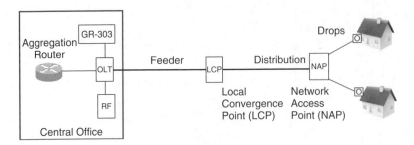

Figure 5.11   Passive Optical Network (PON).

In consumer networks PONs are almost exclusively point to multi-point designs. In the example fiber cable leaves the Optical Line Terminal (OLT) in the CO and is split at two points, a Local Convergence Point (LCP) and a Network Access Point (NAP). Splitters are passive optical devices, requiring no electronics or power. Lightwaves toward the home are split into exact replicas and upstream lightwaves are combined. Fibers are terminated at the side of the home at an Optical Network Terminal (ONT). Different ratios of splits can be made at the two points (LCP and NAP) to achieve an overall desired ratio. Ratios of $1:16$, $1:32$, and $1:64$ are common; $1:128$ is possible. As an example a ratio of $1:32$ can be achieved with a four way split at the LCP and an eight way split at the NAP. The end-to-end concentration is determined by several considerations.

- The optical budget may restrict concentration to $1:16$ if the subscribers are remote from the CO. Higher concentrations are possible when the feeder is shorter.
- Higher concentrations are cheaper. Additional fiber costs are low once you decide to pay for cable placement. But the costs of OLT terminations can raise the cost of deployment and on-going maintenance since the number of OLTs will be higher if the port count exhausts because of low concentration ratios.
- Shared bandwidth is still a consideration, although it is less of an issue with Gigabit Passive Optical Network (GPON) (G.984) than it is with the earlier Broadband Passive Optical Network (BPON) (G.983) technology.
- The size of the NAP also plays into the design. In subdivisions that are populated by multi-family dwellings NAPs may serve far more than 32 subscribers, so there is no need for an intermediate split point.
- Some care is needed to manage the shortest optical path and the longest optical path on a PON. GPON is capable of operating at 60 km, but transmitter output power and receiver ranges must be matched across the PON. Padding can help balance differences but measuring and installing optical pads is an added expense.

Operators may change the split ratio as services become more popular to conserve OLT port costs. Consider a subdivision with 32 ports at each NAP. Early in deployment a carrier may average only four customers at each terminal, for a 12% penetration ratio. At that point a four way split at the LCP and an eight way split at the NAP is ideal. Eight terminals with four subscribers each fully load an OLT port for a $1:32$ way overall split. Over time if 50% subscriber penetration is achieved, a $1:32$ may be dedicated at the serving terminal

Table 5.2 PON technologies

Technology	Down BW	Up BW	Deployment
BPON	622 Mbps	155 Mbps	In the US deployed most widely in Verizon territory beginning in 2004. Well suited for single family units and multifamily units
GPON	2.4 Gbps	1.2 Gbps	Displaced BPON as the standard in 2007 in FTTH deployments. Well suited for single family units and multifamily units
EPON	1.25 Gbps	1.25 Gbps	Widely deployed in Asia and well suited for businesses and high rise apartment buildings
10G EPON	10.0 Gbps	10.0 Gbps	Displacing EPON as a symmetrical standard. First deployments were in 2010
XG-PON	10.0 Gbps	2.5 Gbps	Field trials are underway in 2011. Commercial deployment may begin in 2012

with no split at the area terminal. In practice carriers won't want to change the split ratio more than once since it disrupts service.

The choice of electronics to use over a PON is almost entirely independent from the physical design. Operators have several choices of electronic technology.

BPON – This pioneering standard was developed in the Full Service Access Network (FSAN) working group and submitted to ITU for approval, eventually adopted as ITU G.983.

GPON – Like BPON, this standard was developed in the FSAN working group and submitted to ITU for approval, eventually adopted as ITU G.984.

Ethernet Passive Optical Network (EPON) – A set of IEEE Ethernet standards that evolved from the Ethernet in the first mile (EFM) standard, IEEE 802.3ah. EPON has seen wide deployment in Hong Kong and South Korea.

10G Ethernet Passive Optical Network (10G EPON) – An IEEE Ethernet standard, 802.3av, that supersedes EPON, IEEE 802.3ah.

10 Gigabit Passive Optical Network (XG-PON) – The next in the ITU series, G.987, is not available for commercial deployment at this time, but is well into development.

Table 5.2 summarizes some of the capabilities of these networks.

5.1.9.2 Optical Design

Optical specifications for the different technologies are similar. All use a single mode fiber terminating at the premises and wave division multiplexing to obtain bidirectional transmission. The most common design uses a class B+ laser with a 28 dB optical budget, enabling a 32 way split at up to 20 km. Downstream wavelengths are in the range of 1480–1500 nm and upstream wavelengths are in the range of 1260–1360 nm [13]. Some PON deployments, like Verizon's FiOS have a third wavelength for video at 1550 nm. A video wavelength can carry a full 750–1000 MHz cable television (CATV) passband, making it suitable for greenfield cable deployments, overbuilds, or rehabs.

5.1.9.3   Optical Network Terminal (ONT)

ONTs have multiple designs for different applications. Designs vary significantly, partly because of carrier service differences but also because dwelling units vary substantially. Differences in Single Family Units (SFUs) like the one shown in Figure 5.12 have their genesis in the set of services the carrier chooses to offer, the way they want to distribute Ethernet into the home, and the method of powering the unit. Since SFU ONTs sit on the side of the home and have no metallic connection to the CO, they are home powered.

A typical SFU ONT has two POTS lines (RJ-11) and a 10/100 Base-T Ethernet port (RJ-45). Optionally it may have a coaxial cable video termination (F connector) for digital TV (DTV) service. If video service is provided, the coaxial cable may also carry Ethernet services using Media Over Coaxial Alliance (MoCA)®[3] or Home Phone line Networking Alliance (HPNA)[4] layer 1/2 protocols (see Figure 5.13).

Multiple Dwelling Unit (MDU) Optical Network Units (ONUs) perform the same function as the SFU ONT, but are designed for apartment and condominium deployments. Garden-style MDUs may have 4–8 living units per building where high rises have hundreds. MDU ONUs are placed in a communication space or terminal closet at the MDU in proximity to

Figure 5.12   Single Family Unit (SFU) Optical Network Terminal (ONT).[5]

---

[3] MoCA is a consortium of 60 suppliers supporting the standard.
[4] HPNA, originally the Home Phone Networking Alliance is a consortium supporting a technology similar to MoCA.
[5] Zhone's ZNID-GPON-4220 pictured.

Figure 5.13   Multiple Dwelling Unit (MDU) Optical Network Unit (ONU).[6]

local power, the building's structural wiring and the point of entry for fiber termination. The unit in Figure 5.13 offers 24 10/100 Base-T Ethernet ports and 24 POTS lines, terminated on an Amphenol D type connector.

5.1.10   Fiber to the Curb (FTTC)

FTTC is usually based on BPON or GPON technology, but the ONU terminates the optical path at the curb and is designed as a multiservice terminal for 6–24 subscribers. The copper drop to the home is reused by delivering VDSL2 to the home. POTS can be delivered over the same copper pair or VoIP can replace POTS as the wireline service. The advantage of FTTC over FTTH is that of drop reuse and network powering for FTTC. However, network powering is also an engineering and maintenance challenge. Reliable power must be brought to all of the terminals.

5.1.11   Fiber to the Node (FTTN)

Sometimes called fiber to the cabinet, which invites confusion with FTTC, fiber to the node (FTTN) does not use PON technology. The designation FTTN generally applies to VDSL2 and ADSL2+ DSLAMs deployed as remote terminals or at the SAI.

5.1.12   FTTH Design and Engineering

FTTH design decisions are dominated by capital cost considerations more than technology. So the following sections address specific applications in a preferential order. Deployments that offer the best return on investment, or reduced capital outlays are covered first. Fiber placement costs are generally 70% or more of the total projects costs, not the electronics. Trenching does not follow Moore's Law; costs go up with inflation and productivity improvements are hard to come by. Network and outside plant engineers can reduce those costs significantly by making better design choices.

5.1.12.1   Greenfield Deployments

New subdivisions and new apartment complexes are well suited to FTTH technologies simply because trenches have to be dug, conduit has to be placed. Putting in fiber makes

---

[6] Zhone's ZNID-GPON-8324 Indoor 24 port POTS and Ethernet ONU.

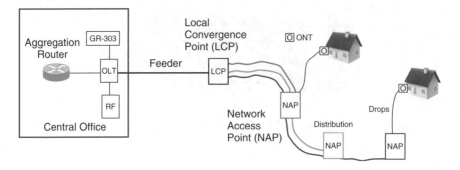

Figure 5.14   Pre-engineered optical distribution.

the design future proof for bandwidth demand and is operationally much cheaper than copper or coax. One constraint is the availability of feeder fiber to the site. If the number of living units is relatively small in a particular build, economics may not support building fiber feeder out to the distribution point. Outside plant engineering economics in carriers tends to be governed by that year's budget and so it is short-term focused. That can lead to a situation where city planners and citizens can see an area is going to experience high growth over the next decade but the carrier makes infrastructure decisions in an incremental way, and no one build can justify the cost of the necessary fiber infrastructure, so copper infrastructure is increased incrementally over time (see Figure 5.14).

Greenfield deployments have another advantage, an opportunity for pre-engineering that is uniform and cost effective. Engineering fiber placement shares some similarities to engineering copper placement. Like copper there are interconnection points and splice points, and finally a drop serving the home. Unlike copper, fiber connections have tight tolerances for loss and return loss. Splicing technique and execution are critical. The two general categories of splices are mechanical connector splices and fusion splices. Fusion splicing is generally performed on mass fiber that is in ribbon form, typically 12 fibers to a ribbon and multiple ribbons in each cable with a strength member. Feeder cable will generally require splicing at some point(s) because outside plant construction is rarely able to place the entire length with one pull, meaning it has to be spliced at meeting points. Distribution cable, between the LCP and NAP can often be pre-engineered with mechanical connectors at each end, and possibly intermediate connectors. The diagram above illustrates three sections of pre-engineered distribution. Pre-engineering is cost effective because it is done in a factory under controlled quality conditions instead of in the field in sometimes difficult conditions. It is also consistent, lowering structural variability.

Completely connectorized drops can be used as well. Where conduit is placed in new builds the drop can be added later or placed and protected. Even in those greenfield subdivisions where the operator does not place fiber in the feeder or distribution there's an opportunity to place conduit and fiber in the drop. If direct buried drops are used, composite cable, sometimes called hybrid or Siamese cable, having both copper and fiber cable is a good choice. It makes the drop future proof. When the carrier decides to deploy FTTH, they will not have to dig up every lawn to place the drop. Cable providers also have the option to place composite coax, with a companion fiber cable.

5.1.12.2 Overbuilds and Rehabs

In an overbuild a carrier places new outside plant in an area served by another carrier. Replacing aging cable in an existing service area is a rehab. Overbuilding other carriers is generally not a winning economic proposition. There isn't a community that has two or three water companies or sewer companies running structure to serve homes to compete simply because the economics make those services a natural monopoly. Communication services aren't quite as closed a case, but they are close. Cable Multiple System Operators (MSOs) and telephone carriers started out in different businesses, but have ended up with similar service offerings. They have made a business of it by opening new services and providing bundles, increasing the Average Revenue Per Unit (ARPU). But cable and telco carriers were able to do that because they made incremental investments to an existing infrastructure to support the new services. Video and Internet access account for the majority of the revenues from these networks today. When satellite video service is taken into account, a penetration rate of 40% of subscribers in an area where cable, telco, and satellite are in competition is a respectable figure. Placing new infrastructure to garner a 40% subscriber penetration, after some years, is a difficult economic case.

Opportunities for rehabs are greater because carriers begin with a customer base and are faced with a need to make some investment to keep subscribers and control operational costs. The opportunity is greatest in areas with high concentrations of multi-dwelling units and aerial cable. MDUs served by ADSL1 from the CO or a distant remote terminal are likely to have data rates below similarly placed SFUs. If fiber can be fed to the building, as described in the following, a substantial improvement in service and the service mix can be achieved. In areas served by aerial cable, particularly those with aerial drops, over-lashing of optical cable onto the existing aerial plant is a practical way to provide relief to the area rather than augmenting or replacing the existing copper.

5.1.12.3 Multiple Dwelling Units (MDUs)

MDUs fall into two broad categories, garden-style MDUs and high rises. Garden style MDUs have one or more buildings on a property, usually connected by conduit and local wiring between the units. Depending on local regulations, access to the buildings wiring may be at a Local Loop Demarcation Point (LLDP) in each building or carriers may gain access at only one building, where the LLDP has been moved to a Minimum Point of Entry (MPOE).[7] In the latter case the buildings are connected by private wiring, generally of unknown quality, perhaps not even twisted. In some cases the local wiring is owned by another party through a contractual agreement.

If the local wiring is CAT3 or better, and is twisted, then installing an MDU ONU or a VDSL2 DSLAM at the MPOE units is a reasonable engineering and economic choice. VDSL2 can be employed at the MPOE to reach the units. One and two bedroom apartments have lower bandwidth requirements than large single family homes, so the 50 Mbps bandwidth provided by a cable length of less than 1500 ft should serve these customers well. If the local distribution cable is not twisted or does not meet at least CAT3 standards, crosstalk, impedance, and noise susceptibility may make VDSL2 problematic. Shared infrastructure

---

[7] See California Public Utility Commission (PUC) Rulemaking 95-04-043 for an example of state rules.

with other carriers can also cause interference issues if they have an incompatible service beyond the MPOE.

High rises feed upper floors via riser cables. Risers terminate in closets on each floor. If the risers and the cabling to the individual units are CAT3 or higher quality, a VDSL2 DSLAM can serve the building. If not, an in building set of PONs can be created by dropping fiber riser cables and placing MDU ONUs on the individual floors, or possibly every other floor.

5.1.13 FTTH Operations

FTTH craft titles are similar to DSL craft titles. There are premises technicians and outside plant technicians. Premises technicians' tasks are similar to VDSL technicians where IPTV services are provided. Outside plant technicians are quite different. FTTH deployments are intensive in the installation phase, placing cable, and bringing service to the NAP, or serving terminal. But once that is accomplished there is little maintenance or engineering needed for the feeder or distribution network. In new builds the carrier waits for service requests and then an outside plant technician is dispatched to install the drop and the ONT. After service is verified to the ONT, a premises technician installs CPE inside the home. Some carriers may combine these titles and have one installer do both jobs. Once an ONT is placed at a home, it is not removed if service is disconnected, nor is the drop removed. Consequently the first installation at a home can be quite lengthy because drop placement can be time consuming and premises work can only go so far before a signal is needed to confirm the work. Subsequent installations at the same house can go much more quickly because the ONT, drop and probably the inside wire can be reused.

Support systems for FTTH have a different emphasis than those that support DSL. As a fiber based network, performance monitoring is much simpler. The structural and particularly the conditional variability of PON are extremely low compared to copper networks. Premises and customer Local Area Network (LAN) monitoring requirements are similar. An additional monitoring requirement for FTTH is customer power. The ONT is powered from the home with a local battery backup built into the power source. Batteries charge in normal operation and power the ONT during commercial power outages. But these batteries have a fixed life, on the order of a few years. The in home converters are designed to alert the customer with a visual indicator, such as a red blinking LED and a low audible alarm when the battery needs replacing. They may also send a notification to the carrier, after which customer care will also inform the customer of the need for a replacement. Battery replacements are usually the responsibility of the customer.

Inventory and engineering systems need to keep track of how many and which ONTs are on which PONs, but the network elements report that automatically, assuming the ONT is in-service. If the customer disconnected and powered down the ONT the inventory needs to be maintained based on the last working assignment. If intermediate splits are used the accounting is a bit more complex but still quite manageable.

FTTH networks that use a 1550 nm video overlay, such as Verizon's FiOS, have an added operational complexity but have significant broadcast bandwidth. Between 750 and 1000 MHz of high quality spectrum can be broadcast on the video overlay. Assuming 256 QAM (Quadrature Amplitude Modulation), the additional wavelength can deliver 4–6 Gbps of downstream bandwidth. That bandwidth is in addition to the shared unicast stream of

2.488 Gbps available on the 1490 nm GPON wavelength. Overlay video spectrum can be split between true broadcast and shared unicast. Broadcast spectrum can be used for analog video, National Television Systems Committee (NTSC), Phase Alternating Line (PAL), or Sequential Color with Memory (SECAM), or digital video, Advanced Television Systems Committee (ATSC), QAM, Digital Video Broadcast (DVB), or other formats. FTTH video overlay designs have some of the same characteristics of a downstream HFC design, but differ in two important ways. Signal quality is higher in FTTH systems because fiber extends to the home, avoiding a coaxial run from a neighborhood node to the home, and one to three distribution amplifiers. FTTH designs do not use an upstream RF spectrum, unlike HFC designs, which means they have no RF noise in the home being transmitted into the common distribution network. The 1310 nm upstream wavelength carries signaling for video on demand (VOD) or other services delivered on the shared video overlay.

5.1.14 FTTH Aggregation Networks

FTTH networks are very flexible, particularly those using a video overlay, so there is no one model that correctly captures all designs. Figure 5.15 illustrates key elements, but designs need not include all the elements shown. For example, both a GR-303 gateway and VoIP Session Border Controller (SBC) are not needed in most deployments.

5.1.14.1 Voice

Voice services in BPON and early GPON deployments were delivered through the voice TDM network using a GR-303 voice gateway to the class 5 switching system. GR-303 extends switched digital voice services to lines on the ONT, leveraging existing provisioning and telco support systems and services. Today GPON and future PON deployments are more likely to use VoIP to deliver consumer voice services, bypassing the class 5 TDM

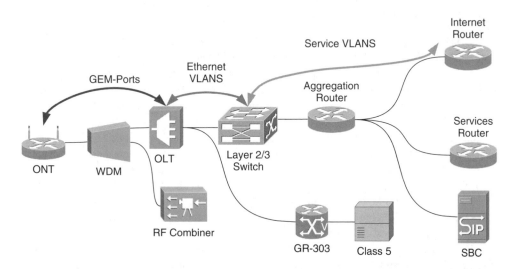

Figure 5.15 FTTH aggregation network.

infrastructure. VoIP offers greater feature flexibility and the opportunity to integrate voice with other IP based services. VoIP Analog Terminal Adaptors (ATAs) are embedded in the ONT to enable traditional POTS GR-57 interfaces [14]. Feature servers, soft switches, and SBCs are deployed within the carriers' voice network infrastructures centrally to scale, simplifying aggregation networks.

5.1.14.2   Video

Video is delivered via two paths. RF video is delivered via a combiner network positioned at the CO or higher in the network. Broadcast video, switched digital video (SDV), and VOD can be combined in the RF spectrum using QAM encoded digital carriers. Analog video can also be combined, but that is increasing unlikely as fewer NTSC/PAL/SECAM sets are produced. Six megahertz channels are dedicated to either broadcast or unicast (SDV and VOD) services. If unicast video is delivered via IPTV on the 1490 nm wavelength, the entire RF spectrum can be devoted to broadcast. Pay per view can also be delivered in the RF spectrum, as a multicast service. Video is combined with digital services coming from the OLT using an optical Wave Division Multiplex (WDM) combiner. Out-of-band signaling on a dedicated VLAN controls video services.

5.1.14.3   Internet Access

IP services, including Internet access can be provided via VLANs and dedicated services routers in a manner similar to FTTN with VDSL as described earlier. GPON ONTs map physical ports to virtual ports using a standard called GPON Encapsulation Method, or GEM. Within each logical port up to eight different priorities can be set using 802.1 p bit marking. GEM-ports with p bit marking enable QoS, service segregation, security and routing segregation.

5.2   Hybrid Fiber Coax (HFC) Networks

Cable MSOs began operation to fulfill a need to extend broadcast TV services to outlying communities using coaxial networks with a coaxial trunk and distribution amplified RF designs. Originally the networks were all coaxial, one way analog broadcast designs. But in the early 1990s lasers and fiber optics were adopted for MSO trunk designs, moving the launch point of the RF amplifiers to nodes in the neighborhood. That design became known as a hybrid fiber coax network. Node sizes were originally well over 1000 homes, but they eliminated a dozen or more amplifiers on coaxial trunk routes. Elimination of long amplifier cascades meant that intermodulation interference and thermal noise were both significantly reduced. Along with improved amplifier technology, that translated into expanded network bandwidth and an opportunity to design a return path onto the network. Prior to HFC, long trunk amplifier networks prohibited return paths because of the collective noise of such large subscriber counts. Telephone networks were used for return paths before the introduction of HFC. Figure 5.16 is a simplified view of a HFC node.

Fiber nodes (FNs), like the one in the diagram serve from 250 to 2000 homes. Video content is taken from satellites, received, decoded, and re-encoded at a super headend, not shown on the diagram. Encrypted content is then multicast from the super headend to

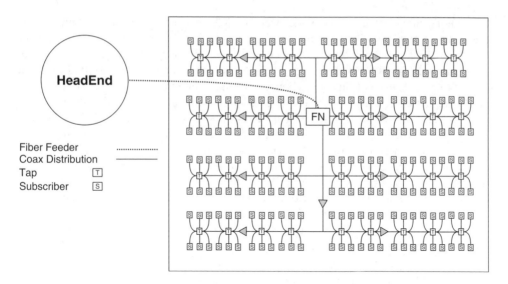

Figure 5.16   HFC fiber node design.

distribution headends, or hubs. Local off air broadcast and Public Education and Government (PEG) channels are added to the cable channels. Video content is RF modulated as either 64 or 256 QAM, depending on network quality. The modulated signal is multicast over fiber to the node where a straightforward O/E conversion is performed. The RF video spectrum is then combined with Data Over Cable Service Interface Specification (DOCSIS) broadband modulated carriers and launched onto coaxial distribution networks. Small nodes in the range of 250–500 homes need only two or three distribution amplifiers in cascade to serve the longest coaxial run.

5.2.1   Node Design

Small node sizes result in lower intermodulation products which are the primary limitation in transmission quality. More aggressive modulation can be used in both directions. In downstream transmission 256 QAM can be used instead of 64 QAM, improving spectral efficiency by about 40%. Upstream transmission can use 16 QAM instead of QPSK (Quadrature Phase Shift Keying), approximately doubling upstream capacity. Unicast traffic capacity is also improved with smaller node sizes because the demand pool is reduced, amplifying the improvements achieved from more aggressive modulation.

5.2.2   Digital TV

Cable MSOs and the CATV industry have evolved a basic analog TV coaxial trunk design from the 1960s into a capable and flexible distribution network serving video, voice, and data services. In most systems the coaxial distribution infrastructure passing the home is unchanged. Hardline cable, directional taps, and drops serving homes carry a new set of services without modification or replacement. Two waves of change triggered in the aggregation, core, and home networks were brought about by the introduction of fiber optics and the regulatory bodies mandating a move away from analog TV to DTV. As regulatory agencies around the world moved to reclaim broadcast spectrum and mandate digital video standards, TV set manufacturers stopped building analog sets and shifted to digital sets capable of displaying higher resolution, or High Definition Television (HDTV), content. Managing the transition from analog to DTV is taking place over a decade with some MSOs being much more aggressive than others. Converting analog channel spectrum to DTV spectrum enables operators to exchange one analog channel in a 6 MHz slot for up to four DTV channels. However, many subscribers continue to hang on to their analog TV sets so a converter, or set top box (STB) has to be provided to decode the DTV signal into an analog format for viewing. Some MSOs had transmitted basic analog service in the clear and so those analog TV sets had not needed STBs before the conversion to DTV. It quickly becomes clear that network decisions cannot be made in isolation. They are part of a complete design and service plan that includes the home, devices in the home, and new services, specifically voice and Internet access.

5.2.3 DOCSIS

DOCSIS is a standard developed in 1997 by CableLabs®, a research and engineering consortium sponsored by the cable industry. The original standard, DOCSIS 1.0 has gone through multiple revisions leading to the current release, DOCSIS 3.0. DOCSIS unifies digital unicast service delivery in CATV networks, enabling broadband services with QoS, security, IPv6, and traffic policing. DOCSIS release 1.1 added QoS which enabled VoIP service, and release 2.0 improved bandwidth for symmetrical services and added IPv6 support. DOCSIS 3.0 made significant improvements in upstream and downstream bandwidth through channel bonding. DOCSIS is largely dedicated to voice and Internet services. SDV services are still broadcast over separate channels, but return path control for those services can use the DOCSIS infrastructure.

5.2.4 HFC Design and Engineering

HFC networks are flexible and while they are built on standards, there are a wide range of options in how they are designed and deployed. Service mix, bandwidth allocations, node size, home network, and CPE choices abound. An MSO competing directly with satellite is likely to have a different service mix and design that an MSO competing with Verizon's FiOS.

Several design choices seem clear;

- **Node sizes of less than 500 homes** – Nodes of greater than 1000 homes are not good candidates for quality voice service, a key differentiator for full service networks [15].
- **DOCSIS deployment** – Broadband competition from telcos with FTTH and VDSL networks should drive deployment of DOCSIS 3.0.
- **Continued analog displacement by DTV** – Unicast service demand drives smaller nodes and speeds DTV deployment.

Others are more complex,

- H.264 (Motion Pictures Expert Group, MPEG−4) deployment – MPEG-4 yields at least a 50% improvement in compression over the dominant CATV DTV standard, MPEG-2. But it requires a complete change out of an MSO's STB base, an expensive and disruptive proposition.
- Greenfield deployment of FTTH or FTTC – CableLabs® incorporated EPON into their Converged Cable Access Platform (CCAP) architecture [16]. The Society of Cable Television Engineers, SCTE, has published a companion specification for radio frequency over Glass (RFoG) [17]. RFoG has an optical design similar to Verizon's FiOS. A 1550 nm video wavelength overlays GPON or EPON wavelengths in the 1490 downstream and 1310 nm upstream ranges. Cultures and organizations are slow to change, but it seems logical to deploy fiber to the living unit instead of coax in new developments. CableLabs choice of EPON over GPON is instructive. EPON is a symmetrical high speed PON, enabling business services more readily than GPON.
- Passband expansion from 750 to 860 MHz and even to 1 GHz is possible, particularly in areas with small node sizes. Gallium Arsenide line extenders or tap amplifiers, can replace existing 750 MHz amplifiers. The rest of the plant, including taps and couplers must be 1 GHz capable or the upgrade becomes quite expensive. RFoG is compatible with 1 GHz operation as well.
- Native IPv6 – DOCSIS 2.0 lays the foundation for native IPv6, but most home networks use IPv4 with Network Address Translation (NAT). IPv6 offers the promise of push services and improved peering, but moving completely to native IPv6 is not straight forward. DOCSIS 3.0 used IPv6 for performance management, but relies on IPv4 for some provisioning functions.

5.2.5 HFC Operations

Cable MSOs manage services on HFC networks using a concept of service groups. In IP networks all services share a common transport, IP. In HFC networks service share a common RF passband but the spectrum must be segregated for each service group. As an example, a 750 MHz network has about 100 6 MHz channels for downstream services. An operator might allocate 30 channels to analog service,[8] 50 channels for SDV service, 10 channels for VOD and 10 channels for Internet and VoIP services. Once the allocations are made and RF combining networks are configured, changes can be difficult if service specific systems feed RF modulators that need to be repurposed.

Complicating factors in HFC operations are lack of uniformity in the configuration of contiguous properties. Consider a city that has grown and plant has expanded to meet the demand. Newer plant may have smaller node sizes than the legacy network. Allocation of spectrum for unicast services is dependent on demand and the size of the service group. Assuming 4 channels per VOD modulator, 10 channels can serve 40 simultaneous subscribers. In a 500 home node that represents 8% of the homes. In a 2000 home node it is

---

[8] The FCC mandated that analog service be provided until mid-2012. After that channels are likely to be converted to all digital transport.

only 2% of homes, a difference of 400%. If nodes are small enough, a SDV approach is superior to digital broadcast. Broadcast dedicates spectrum to a channel and anyone wanting to view the channel simply tunes to it. SDV dedicates spectrum to homes enabling subscribers to watch anything they want by switching it up in the network to their dedicated channel. SDV is the approach used in IPTV by telcos. The serving node size for PON is 32 homes and for VDSL it is one. Switching an entire MSO from broadcast to SDV is expensive and so it's not likely to happen quickly, but as fiber pushes deeper it will happen in many operations in time.

CableLabs® CCAP [17] architecture recognizes the service group issue and addresses it directly. It is a forward looking architecture that combines the DOCSIS Cable Modem Termination System (CMTS) and QAM edge into a single chassis and system. CMTS supports VoIP, Internet, and IP-based SDV services. QAM edge supports digital broadcast, VOD and QAM-based SDV services. By allowing more flexible sharing of modulators it simplifies RF combining and eliminates manual reconfiguration as service demands and mixes change. Both physical design and software design are incorporated into the concept. Like DOCSIS, CCAP also supports integration of EPON.

While CCAP is emerging, DOCSIS has improved operational support in HFC systems by enhancing end to end management through a standard operations support system interface specification [18]. Administrative and service state management is introduced via RFC 2863, IF-MIB. The specification covers Fault, Configuration, Accounting, Performance, Security (FCAPS) functions for both the cable modem (CM) and the CMTS.

5.3 Wireless Mobile Networks

There are three generations of wireless technologies addressed in this section, GSM/GPRS (Global System for Mobile Communications/General Packet Radio Service), UMTS (Universal Mobile Telephone System), and LTE (Long Term Evolution), all of which are in service. First generation networks, Advanced Mobile Phone Service (AMPS) and Time Division Multiple Access (TDMA) no longer have a service footprint. The three generations discussed are covered under the umbrella of Public Land Mobile Networks (PLMNs) and are based on GSM, the dominant worldwide standard. Other standards such as WCDMA2000® have substantial penetration but those networks are also evolving to LTE. The GSM standard was developed by the European standards organization ETSI in the late 1980s and was adopted worldwide in the early 1990s [19]. GSM is considered a second generation or 2G mobile technology and is primarily designed for voice services. GPRS and Short Messaging Service (SMS) extended the GSM standard to include Internet and text services under the 2G umbrella. UMTS, was developed by the Third Generation Partnership Project (3GPP) as the 3G evolution of the GSM standard. The biggest change with the introduction of UMTS was to the radio network. The modulation scheme changed from Gaussian Minimum Shift Keying (GMSK),[9] to Wide Band Code Division Multiple Access (W-CDMA); radio resource control (RRC) also changed with a new set of network elements. The latest mobile technology, LTE is considered a 4G technology and is also a product of 3GPP [20]. Figure 5.17 highlights key elements of the three technologies, as they are deployed in today's networks. There are certainly other network elements deployed in mobile networks, but the

---

[9] GSM networks that adopted Enhanced Data Rates for Global Evolution (EDGE) had already moved to 8PSK.

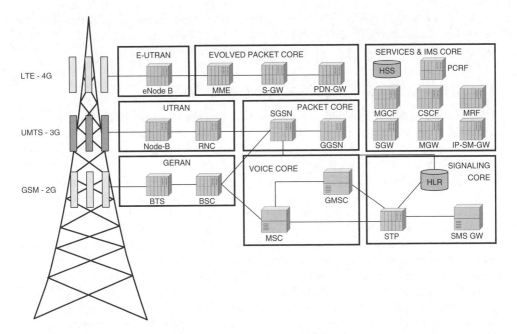

Figure 5.17   Mobile network architecture.

discussion here is limited to those essential elements that highlight design, operations, and engineering issues. Other texts offer a much deeper and more thorough treatment of this subject [21].

5.3.1   GSM

Two important wireline technology transformations were underway when the GSM standard was developed, TDM switching (digital switching) and common channel signaling (SS7).

5.3.1.1   Mobile Switching Centers (MSCs)

GSM voice switching technology is a circuit switched TDM design, based on an extension of the wireline systems in design and production at the time GSM was introduced. Like the wireline network which has two levels in the switching hierarchy for ILECs, GSM also adopted a two level hierarchy for introduction. Mobile Switching Centers (MSCs) are the mobile equivalent of a class 5 switch, with important exceptions, the subscriber database and mobility management. Wireline class 5 switches associate subscribers with line terminations. A subscriber information database in a wireline class 5 switch is indexed with a unique line equipment number (LEN) which indicates where the subscriber's copper pair terminates on the switching system. Included in the subscriber information are the telephone number (TN), type of service, line features (call waiting, call forwarding, etc.), routing privileges, LD carrier, and a range of other information. Wireless network subscribers are mobile; there is no line termination on the switch. Subscriber information is held in a central repository, the Home Location Registry (HLR), and copies of the information are distributed to the network elements currently serving the subscriber. This process is described below for both voice and data services in sections on mobility management.

5.3.1.2 Base Transceiver Station and Base Station Controller

MSCs serve a mobile market and so instead of being anchored to subscriber lines, they serve cell sites. Both MSCs and class 5 switches serve bounded geographical areas, but MSCs' users move in and out of serving areas. The radio network under GSM is called the GSM EDGE Radio Access Network (GERAN), which is an abbreviation for GSM Enhanced Data Rates for Global Evolution (EDGE) Radio Access Network (RAN). EDGE is an advance in radio technology within the GSM standard, moving from GMSK to 8PSK modulation, along with other improvements. In this text GSM refers to GSM/EDGE when describing the air interface. The two major components of the GERAN are the Base Transceiver Station (BTS) and the Base Station Controller (BSC). Cell sites are typically divided into three sectors, of approximately 120° each, although in sparse serving areas one or two sectors are used. Each sector operates on a different set of frequencies, and so three transceivers are needed when operating one set of carriers at a site. If there is sufficient demand, additional RF carriers can be operating at the same site on different frequencies. There is a many to one relationship between BTSs and BSCs. A BTS has the radio head, transmitting and receiving the multiple channels used to exchange signaling and bearer information between the mobile network and a subscriber's mobile device, referred to as a Mobile Station (MS) in pre-UMTS standards and as User Equipment (UE) in UMTS and LTE standards.

5.3.1.3 Radio Resource Control (RRC)

RRC is the active management of radio channels, power levels, and parameters to maintain communication between the mobile network and the MS. RRC is the responsibility of the BSC. The BSC performs all of the logical functions and the BTS acts as the transducer, converting information between RF and electrical energy. Frequencies, timeslots, and power levels are managed by the BSC and it acts as the network interface to the GERAN. Voice transcoding and rate adapting are also under control of the BSC, although they are performed in an adjunct network device, called a Transcoding and Rate Adaption Unit (TRAU). To conserve transport the TRAU is usually located at the MSC.

5.3.1.4 Voice Mobility Management

As users move between sectors or cell sites the relative power levels of sectors change and a handover is performed to connect the MS to the most favorable cell site and sector. Handover between two cells served by a common BSC is the BSC's responsibility. MSs assist in the process by making measurements of different cell sites and sending the results to the BSC. Handover between two cells served by different BSCs is the responsibility of the MSC, if both BSCs home on the same MSC. Other handovers occur between MSCs and even MSCs in different networks when roaming comes into play.

In Figure 5.18 there are three cell sites, labeled 1–3, with different BSCs managing the BTSs at each site. The BTSs are not shown on the diagram but one transceiver is needed

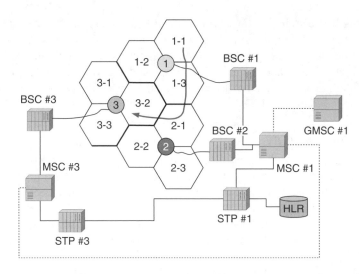

Figure 5.18 GSM voice mobility management.

for each cell. Each cell site has three sectors, numbered sequentially with a prefix of the cell site. In the example, if an MS moves from sector 1-1 to sector 1-3, BSC #1 manages the handover. After the handover the BSC informs the MSC, but does not need the MSC's assistance to execute the handover.

If the MS moves from sector 1-3 to sector 2-1, an inter BSC handover is needed, which is managed by MSC #1. BSCs cannot directly coordinate handovers between themselves. A specification within the SS7 signaling protocol called the Base Station Subsystem Mobile Application Part (BSSMAP) is used for the handover. In mobility networks SS7 messages are exchanged over 64 kbps data links connected via an STP, or directly over IP data links using an adaptation called SIGTRAN [22] (see Figure 5.18).

5.3.1.5 Home Location Register (HLR)

To further the understanding of how mobility management works, the following paragraphs describe a mobile subscriber that receives an incoming call while in sector 1-1 and then travels through sectors 1-3, 2-1 and 3-2. In the diagram the path of the subscriber is shown as a line traversing the sectors.

Mobile Station Registration

As described earlier, subscriber information in mobility networks is maintained in an HLR, not an MSC. When a subscriber switches power on in sector 1-1 the MS is recognized by BSC #1 and it notifies the controlling MSC #1 of the International Mobile Subscriber Identifier (IMSI) of the MS. This serving MSC #1 is called the Visited Mobile Switching Center or V-MSC. An IMSI has three elements, a Mobile Country Code (MCC), Mobile Network Code (MNC), and Mobile Subscriber Identification Number (MSIN). Using the IMSI to determine the home network of the MS, the V-MSC sends an SS7 Mobile Application Part (MAP) request to the HLR in the home network to authenticate the MS; the response is used in a dialog between the V-MSC, MSC #1, and BSC #1 to complete the authentication. Encrypted keys are maintained in the HLR and in a protected area of the Subscriber Identity Module (SIM) to prevent spoofing. Upon successful authentication the V-MSC obtains essential subscriber data from the home HLR when it registers the MS, and places the information into a data store similar to an HLR, called a Visitor Location Register, VLR, which is collocated with the MSC.

Inbound Call from the PSTN

Before a call can be routed the Local Routing Number (LRN) must be found to determine if the call has been ported. Number portability often works differently in mobility networks than in wireline networks. Number translations are not retrieved from an SS7 SCP but rather from a Centralized Database (CDB) which is copied periodically by carriers. Mobile networks consult their copy when forwarding calls or messages. The address of the subscriber's network can also be obtained via a direct query into the HLR of the network identified by the IMSI.

Inbound calls from the Public Switched Telephone Network (PSTN) are routed to the gateway MSC Gateway Mobile Switching Center (GMSC) #1 based on the called number or the routing number. When GMSC #1 receives an SS7 ISDN User Part (ISUP) IAM it consults the home network HLR via an SS7 MAP query to locate the MS in a two-step process. By indexing the HLR with the called number, also called the Mobile Subscriber Integrated Services Digital Network number (MSISDN), an IMSI is returned, which is unique to the SIM card in the MS. The IMSI in turn points to the V-MSC, MSC #1 that currently has the MS registered. The V-MSC can be in the mobile subscriber's home network or in a foreign network if it is roaming. The HLR then sends a MAP query to the V-MSC requesting a temporary routing number for the MS identified by the IMSI. The temporary routing number is assigned by the V-MSC as an identifier for the visiting subscriber. It is forwarded in the MAP response to the GMSC. The GMSC can then route the call to the V-MSC with an IAM containing the unique temporary routing number which identifies the visiting MS. When the call completes, the V-MSC, MSC #1, becomes the *anchor Mobile Switching Center*, or A-MSC.

When the MS moves from sector 1-3 to 2-1, message exchanges are performed between the A-MSC and BSC #1 and BSC #2 using the SS7 BSSMAP protocol. Assuming a call is still in progress, BSC #2 sets up a voice channel to the MS and BSC #1 drops its voice channel.

Transferring to a New MSC

When the MS moves from sector 2-1 to 3-2 the A-MSC, MSC #1 is no longer able to serve the MS directly. Prior to the handover the MS forwards the relative signal strength of received cell sites and it receives an instruction to register with cell site #3, sector 3-2. An identifier of the new serving network is also forwarded to the A-MSC, alerting it that a call transfer is needed. When the MS enters the serving area of BSC #3 a registration process is initiated, like the one described previously for BSC #1. The new visiting MSC, MSC #3 is designated as the *relay Mobile Switching Center*, or R-MSC because it relays the call in progress. Transfer occurs when the A-MSC sends a MAP message to the R-MSC requesting the call, which is hairpinning through the A-MSC, be completed to the MS.

Transfers are performed on a make before break basis, so that new call channels to the MS are established before the old channel is dropped. If the call moves out of the region served by the new R-MSC, MSC #3, the call drops back to the A-MSC, MSC #1, and the call is re-established to the new R-MSC, avoiding successive tandems through multiple R-MSCs.

5.3.1.6 Short Messaging Service (SMS)

SMS surprised network operators and industry analysts with the speed of adoption and broad based popularity it gained in a few short years. Conventional thinking in the 1990s was that mobile subscribers' first choice was to talk live with someone. If that wasn't possible or convenient, they would leave a voice mail or, possibly send an email. But SMS grew out of the GSM [23] circuit-switched network using the SS7 infrastructure. Some might say the service evolved simply because the SS7 network made it possible with little additional investment. While leveraging the SS7 network is cost effective, it also means SMS is constrained to the user payload size of the protocol, 140 octets in the case of SMS. If 8 bit characters are used, that limits text to 140 characters. That limit can be extended to 160 characters by using 7 bit characters. Figure 5.19 illustrates the network elements used to implement SMS.

Mobile Origination

When a subscriber composes and sends a text message from the MS, an SS7 Direct Transfer Application Part (DTAP) message is formatted by the MS and sent to the home Short Messaging Service Center (SMSC). The message includes the message text, destination (MSISDN), the SS7 address of the SMSC, and the DTAP type. The MS determines the SS7 address of the SMSC by consulting its SIM card. Message routing from the MS is via the V-MSC and the SS7 network (MSC-STP-SMSC) since the MS has no direct connectivity to the SS7 network. Networks with sufficient scale use intelligent routing and multiplexing engines, such as Cisco's IP Transfer Point (ITP) to better route and concentrate SMS traffic. The ITP can terminate conventional 64 kbps links from STPs and use SIGTRAN toward SMSCs for traffic efficiency. It performs packet inspection of the DTAP message and is capable of rerouting messages to specific destinations based on user defined rules. The best example of the value of that capability is game show voting contests. When a program such as American Idol® instructs their audience to vote via SMS, it creates a message storm in the mobile SS7 network. An ITP can be configured to route those messages to

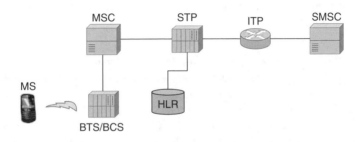

Figure 5.19 SMS architecture.

an application specifically designed to count the votes, avoiding swamping the SMSCs. Messages forwarded to the SMSC for delivery are queued in persistent stores until they can be delivered.

Mobile Termination

The SMSC delivers an SMS message by using the called MSISDN to consult the destination's HLR to determine if the destination MS is attached to a network, and if so, the address of the serving V-MSC. When the SMSC finds the MS is attached, it forwards the SMS message to the V-MSC. Upon successful delivery to the MS the V-MSC notifies the SMSC and the message is deleted from the SMSC's persistent storage. If the MS is not available or not attached, message waiting indicators can be set in the home HLR and in the V-MSC VLR. When the MS becomes attached the originating SMSC is notified and delivery can be reattempted.

Mobile origination and mobile termination are separate procedures. The originating MS receives confirmation from the SMSC when a message is received for delivery. The SMSC receives notification from the terminating network when a message is delivered. But, there is not a formal mechanism for notifying the originating MS when a message is finally delivered.

Over The Air (OTA) Provisioning

The SMS mechanism is used for more than the exchange of user text messages. Carriers use it to provision the SIM card. Services can be activated, user data modified, and firmware updated using the SMS Mobile Termination as a mechanism. The Open Mobile Alliance (OMA) has standardized some of these practices [24].

5.3.1.7   General Packet Radio Services (GPRS)

The move away from circuit switched data, represented by ISDN in the PSTN to packet switched data using the Internet Protocol took place in mobile networks about the same time, the late 1990s. GPRS evolved under GSM after circuit switched voice and so the air interface and BSCs had to be upgraded to support packet switching. Packet switching was added to BSCs in the form of a Packet Control Unit (PCU) and BSCs were modified to allow for the allocation of channels and packet control functions to be delegated to the PCU. The introduction of 8PSK modulation as part of the EDGE initiative was timely, otherwise voice and packet services would have to compete with bandwidth originally designed for voice only using the less efficient GMSK modulation. In addition to modifications to the BSC, new network elements were introduced because circuit switched MSCs were not suited for packet routing and switching.

Gateway General Packet Radio Services Support Node (GGSN)

It should come as little surprise that the general model of control and routing used for the GSM circuit switch model served as the template for packet switching. The Gateway General Packet Radio Services Support Node (GGSN) is the packet equivalent of the GMSC; it serves as the packet gateway to the Internet and to VPNs or other packet networks. An important difference, at least within the confines of IPv4 is that sessions between the MS and the Internet are generally originated by the MS, unlike the GMSC that participates in calls originating from the MS, but also terminates incoming calls from the PSTN.

Another important difference between circuit-switched sessions and packet-switched sessions is how state information is managed. Circuit-switched sessions are defined by connection oriented circuit paths from the MS to the GMSC. Packet-switched sessions are defined by a set of IP addresses that identify the MS, the GGSN and Serving General Packet Radio Services Support Node (SGSN) serving the session, the Internet or other destination address, and points in between. Circuit-switched sessions pivot on an A-MSC. As an MS moves between cells in the GERAN, new connections are established from the A-MSC to the MS through R-MSCs and BSCs. In a general sense the A-MSC manages the circuit-switched session. For GPRS sessions the GGSN acts as a pivot, managing the IP addresses and other key information associated with the session. 3GPP defines a Packet Data Protocol (PDP) Context which contains data elements defining the packet session. The PDP Context is maintained by the GGSN.

A final important difference between packet and circuit switched sessions is that packet sessions do not necessarily share a common path to destination networks. Mobile subscribers invoke different packet applications and the applications are served by different web services, and often by different GGSNs. 3GPP defines Access Point Names [25] (APNs) as a means of identifying mobile packet services, and the servers that host them. Each Mobile Network Operator (MNO) is free to designate their own APNs, depending on their service offerings. New APNs are built when a service uses a unique protocol or when the carrier wants to apply unique policies or charging for a particular service. Two examples of the use of APNs are Wireless Application Protocol (WAP) and the Internet. Early mobile browsers were restricted in the size of the display and the processing power of the mobile device. In response a WAP was created by what is now part of the OMA Forum [26]. An APN is a domain name, such as "wap.att.com" that is used by the MS when activating the service. The name is resolved via a mobility Domain Name System (DNS) to the IP address of the GGSN hosted WAP server, acting as a proxy or gateway to WAP sites on the Internet.

Serving General Packet Radio Services Support Node (SGSN)

As the GGSN functions much like an A-MSC, the SGSN is the packet equivalent of the R-MSC. The SGSN connects to the PCU/BSC downstream, toward the MS, and to the GGSN in the upstream direction. BSCs and PCUs were developed at a time when Ethernet was not commonly deployed in carrier transport networks and so they are generally frame relay or TDM based although Ethernet is available from some suppliers. So SGSNs usually have frame relay connections toward BSCs and optical IP connections toward GGSNs. There is a many to one relationship between BSCs and SGSNs and a many-to-many relationship between SGSNs and GGSNs. Recall different GGSNs may serve different APNs.

5.3.1.8 Data Mobility Management

The mechanics of mobility management for data are similar to voice, but there are differences. Figure 5.20 is the same scenario used in the voice mobility management example. In this example our MS initiates a data session to browse the Internet while in sector 1-1.

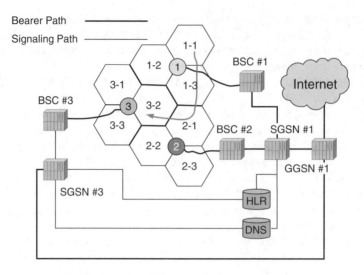

Figure 5.20   GPRS data mobility management.

Authentication and Registration

When the MS activates a service to browse the Internet the BSC notifies the SGSN which registers the MS with the HLR in a similar way to an MSC registration procedure for a GSM call. Upon successful authentication by the HLR and an update of the MS location information, the HLR returns subscriber information to the SGSN, including allowed APNs. The operation is similar to the information transfer of subscriber information to a VLR, only the SGSN is the repository of the subscriber information which is valid for the duration of the session within the scope of the SGSN.

Obtaining a PDP Context

Because different services are provided from different GGSNs, the SGSN must next confirm the requested APN is allowed based on information previously forwarded by the HLR, and then locate the GGSN serving the requested APN. A DNS query is constructed from the APN service name, such as "wap.att.com," the MCC and MNC (Mobile Network Code) contained in the IMSI, and "gprs" as the top level domain name. In the example the Uniform Resource Identifier (URI) is "wap.att.com.150.310.gprs." Including the MCC and MNC are necessary because only the home network of the subscriber offers service compatibility with the APN requested by the MS. No matter where in the world the MS is operating, the PDP context request must be sent to the subscriber's home network so the serving gateway (S-GW) or proxy can be located.

Upon receiving a valid IP address for the GGSN that services the APN, the SGSN formulates a PDP request. The request contains a GPRS Tunnel Protocol (GTP) identifier to enable establishment of session GTPs between the SGSN and GGSN. It also contains subscriber information, including the IMSI, and the identity of the APN to be associated with the context. When the GGSN grants the request it returns a public globally routable IP address to the SGSN that remains active for the duration of the context. The SGSN forwards the newly-obtained IP address and a context establishment notification to the

MS. At that point the MS can initiate the application, such as an Hyper Text Transfer Protocol (HTTP) request using the IP source address and the SGSN – GGSN GTP. When the originating SGSN is a visited SGSN outside the home network, the GTP tunnel can be established through an exchange specifically designed for interconnecting GPRS services among participating MNOs. That exchange is called the GPRS Roaming Exchange (GRX) and was commissioned in 2000.

Transferring to a New SGSN

If the MS moves from sector 2-1 to sector 3-2, it recognizes that an inter SGSN handover is needed. The MS notifies new SGSN, SGSN #3, of the existing packet session. The new SGSN is able to determine the identity of the old SGSN, SGSN #1, from information provided by the MS. Rather than repeating the original authentication process, the new SGSN, SGSN #3, contacts the previous SGSN, SGSN #1, to authenticate the MS and obtain the current subscriber information. Once SGSN #3 has authenticated the MS, SGSN #3 sends a PDP context update request to the serving GGSN to establish a GPRS tunnel. SGSN #3 then sends a location update notification to the HLR. The HLR then sends a release message to SGSN #1. During the change from SGSN #1 to SGSN #3 a tunnel is used between the two SGSNs to insure transient user data is not lost during the handover, although it may be delayed. In SGSN transfer the GGSN acts as the anchor.

5.3.1.9   Multimedia Messaging Service (MMS)

Multimedia Messaging Service (MMS) came out of 3GPP and WAP industry groups and was adopted by the OMA. The popularity of SMS and the revenue stream that accompanied it encouraged carriers to look for ways of expanding mobile messaging services. Expansion of SMS was blocked by the underlying SS7 network which limited the user payload to 140 octets. In an effort to build upon existing technology, MMS uses the following protocols.

- **WAP** – the WAP APN is used to connect mobile MMS originated messages to an MMS server/repository where the message is stored for delivery.
- **SMIL** – Synchronized Multimedia Integration Language (SMIL) – pronounced "smile," SMIL is a markup language resembling Hyper Text Markup Language (HTML) that is used to display MMS messages.
- **MIME** – the Multipurpose Internet Mail Extension (MIME) protocol is used by MMS to include multimedia attachments, such as jpeg, mpeg-4, and gif files.
- **SMS** – SMS is used to notify an addressee of a waiting MMS message.

MMS service is composed of four discreet transactions or processes: composition, transmission, notification and forwarding, and retrieval.

- **Message composition**
  Messages are composed on the MS. Message clients often allow the user to enter text, pictures, sounds, and videos when composing a message. If the message is text only and less than 140 characters, it is sent via SMS. Otherwise MMS is used, including media in a MIME format.

- **Message transmission**

    When the subscriber finishes composition and hits the send button on the MS, a PDP context is established to the WAP APN, and the message is forwarded to the MMS server. Options may allow for delivery or read return notification.

- **Message notification and forwarding**

    When messages are deposited for delivery in an MMS server an SMS notification is sent to the MSISDN address of the recipient. If the addressee is on the originating carrier network and the addressee has not disabled message receipt, the message is forwarded to the MS.

- **Message retrieval**

    If the addressee is on a different network, or a network without an MMS interworking agreement, the MMS server can take one of several actions, depending on the capabilities supported by the carrier. Two options are for the message to be posted to a web server and the addressee notified on how to retrieve it, or for it to be forwarded to the subscriber in an email composed using the MMS contents.

5.3.2   Universal Mobile Telecommunications Systems (UMTS)

UMTS was introduced in the early 2000s as the next evolution of GSM and GPRS/EDGE. While UMTS made few changes to the packet core, it added significant improvements to the air interface, radio resource management, and moved the voice core toward a voice over IP infrastructure but well short of the goal. UMTS advances did not arrive all at once. A series of standards and technology were introduced from the base specification, called Release 99 after the year it was approved, 1999. Subsequent releases of the standard are sequentially numbered, with significant technical advances published in releases 5 and 7.

5.3.2.1   UTRAN

The UMTS Terrestrial Radio Access Network, or UTRAN, standards recognize the ever increasing dominance of data over circuit-switched voice, and implements design requirements to meet that advance. In large part these changes overcome some of the limitations of GSM/GPRS radio interfaces, as described in the following. In keeping with the standards, the acronym MS is used to identify a MS in GSM and UE is the acronym used to identify UE in UMTS.

- **Modulation** – GSM/EDGE uses GMSK and 8PSK modulation in the downlink. GMSK for the uplink. UMTS specifies QPSK for the downlink and QPSK for the uplink in the initial standard, release 99. An extension of UMTS, High Speed Packet Access (HSPA) introduced 16 QAM for the downlink in release 5 [27] (HSDPA) and 64 QAM in release 7. High Speed Uplink Packet Access (HSUPA) introduced 16 QAM for the uplink in release 7 [28].
- **MIMO** – An antenna technology, Multiple-In-Multiple-Out (MIMO) was introduced in UMTS release 7 (HSPA+) as an option. MIMO is effective in environments with multi-path characteristics, such as metropolitan centers with high rises. With MIMO handsets can use signal reflections that are received out of phase to improve reception. Without MIMO out of phase reflections cancel and degrade overall signal quality.

- **Narrow band carriers** – GSM uses 200 kHz carriers that are designed for voice and adopted for data. Two hundred kilohertz carriers significantly restrict the upper end of data speeds in GSM. UMTS uses 5 MHz carriers and W-CDMA to assign bandwidth within the entire 5 MHz carrier, significantly increasing the upper bandwidth available to a single UE in lightly loaded cells.
- **Bandwidth sharing** – GSM uses a combination of Frequency Division Multiple Access (FDMA) and TDMA. MSs are assigned dedicated channels from one of the available 200 kHz carriers and timeslots within that carrier. Eight timeslots are available and an MS can use up to five as a practical limit. UMTS uses W-CDMA over the 5 MHz carrier, significantly improving the bandwidth available to individual UEs in cells experiencing light loading.
- **Channel management** – GPRS packet connections use the same circuit switched model as GSM voice, setting up, and releasing dedicated channels in response to data transfer requests. PDP contexts are persistent for the duration of a session, but connection resources are not. UMTS defines shared and dedicated channels and uses finite state machine models in the UE and network to move sessions between idle, shared, and dedicated channels. This approach reduces the latency inherent in GPRS where dedicated channels are assigned and released as the MS or network senses demand.
- **Mobile station handover** – In GSM/GPRS handover is executed from the MS, not from the network, incurring latency during rehoming. UMTS handovers are controlled by the UTRAN, specifically the Radio Network Controller (RNC). Network control significantly reduces the latency experienced in GSM/GPRS by an MS controlled handover.
- **Transport** – GSM/GPRS transport used Frame Relay. UMTS moved to ATM, with subsequent options for IP over Ethernet, which became the transport preference.

UMTS introduces two new network elements to take the place of the BSC and BTS, the RNC, and the Node-B. A primary goal of UMTS is to more clearly separate radio access management from the core, enabling upgrades in those two domains to be independent in future standard releases.

5.3.2.2   Node-B

UMTS Node-B responsibilities are similar to a GSM BTS, but it uses W-CDMA not FDMA/TDMA; it is responsible for the physical layer protocols used on the air interface. With W-CDMA all UEs in a cell receive and demodulate the same wide band carrier on the same frequency. Channels are differentiated using code multiplexing rather than frequency and time slot differentiation. A Node-B assigns a carrier and coding parameters for each channel based on instructions from the Radio Node Controller (RNC). Power level management is performed by the Node-B with some assistance from the RNC, and it is critical in W-CDMA. All UEs share the same carrier, so one UE's signal is another UE's noise. Unless power levels are carefully balanced, a single UE can adversely affect the bandwidth experienced by other UE's. Frequency reuse in GSM requires contiguous cells to use different frequencies, limiting frequency reuse. With W-CDMA all cells can use the same frequency because code multiplexing is used. Different cells use different scrambling codes for differentiation. A second code, a user data code, is also assigned to the signal for UE data stream differentiation. Traffic loading is still the driver for cell design however.

Unique scrambling and user data coding sequences or vectors are assigned to each UE to be able to uniquely encode and decode that UE's signal on the common carrier. The more UEs assigned to a carrier the longer the vector needs to be to differentiate its signal from other UEs. If too many UEs occupy the same cell, the effective signal to noise ratio (SNR) is degraded because longer vectors are needed for each bit transmitted, lowering the effective bandwidth of the channel for that UE. At the same time more UEs transmitting increase the effective upstream noise, potentially increasing the channel error rate. See Sauter [21] for a thorough description of W-CDMA.

5.3.2.3   Radio Node Controller (RNC)

UMTS adds a level of abstraction at the RNC to packet and to the voice core interfaces. Instead of requesting service specific channels for voice and packet, MSCs and SGSNs request a Radio Access Bearer (RAB) with specific characteristics, such as the bandwidth, QoS, and error rate. Architecturally the objectives are to move all RAN responsibility to the RNC and Node-B, and to move all service knowledge and procedures into the voice and packet cores. Consistent with that philosophy, the formal interface between the MSC and RNC moves the TRAU into the MSC. Recall in GSM the TRAU was logically associated with the BSC, even though the TRAU was physically located at the MSC to conserve transport bandwidth. Besides being a more economic choice, moving the TRAU to the MSC associates voice transcoding with voice switching, not with radio transmission. The other change for voice service is the mandate of the Adaptive MultiRate (AMR) codec [29]. As the name implies, AMR can adapt the bit rate and voice quality dynamically. That characteristic fits well with Code Division Multiple Access (CDMA) modulation. With CDMA as the quality of the channel degrades, the size of the coding factor (vector size) increases and the bitrate of the channel drops to maintain the same overall error rate. AMR can support bit rates as high as 12.2 kbps and as low as 1.8 kbps. Apart from the mandate of AMR and repositioning of the TRAU interface, voice service and mobility management of voice is relatively unchanged when transitioning from GSM to UMTS.

Although the network elements in the packet core, the SGSN and GGSN continue to play their central roles in UMTS as they do in GPRS, there are differences in how the elements function in UMTS. In later releases of UMTS, the SGSN is removed from managing packet bearer traffic by terminating the GPRS tunnel at the RNC, not the SGSN. Recall the tunnel passes data from the APN gateway or proxy at the GGSN to the RNC in this case, where data is transmitted to the UE via the established RAB (see Figure 5.21).

5.3.2.4   Mobility Management

UMTS mobility management improves the transition between cells served by a common controller and cells served across controller boundaries by delegating more responsibility to the RAN and hiding the details of the process from the core network elements, the MSC and SGSN. There are two general categories of handover, soft handovers and hard handovers. Hard handovers operate in a fashion similar to handovers under GSM/GPRS. A new connection is established to the receiving cell; data are buffered by the network, and forwarded, and then the old connection is dropped. That process results in data loss, or at

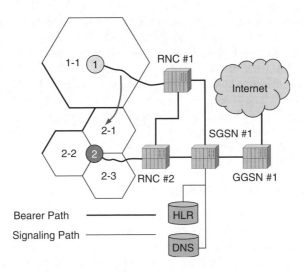

Figure 5.21   UMTS data mobility management.

least a delay in data reception. In soft handovers data are simultaneously transmitted and received by the two cells engaged in the handover.

In the example above a UMTS mobile subscriber moves from sector 1-1 to sector 2-1. Sector 1-1 is a 360° macro cell under the control of RNC #1. Sectors 2-1-2-3 are 120° sectors under the control of RNC #2. Node-Bs are omitted from the diagram for simplicity. All four sectors broadcast on the same CDMA carrier frequency, each sector using a different scrambling code. The two RNCs are directly connected with the Iur interface, specified in UMTS. RNC #1 is the serving controller, the S-RNC, as the subscriber begins their journey. UMTS differs from GPRS in that UE transmission is sent and received from multiple cells simultaneously. The collection of cells communicating simultaneously with a UE is called the Active Set. The Active Set is chosen by the S-RNC and passed to the UE. In our example, 1-1 and 2-1 make up the active set. As the UE moves toward 2-1 relative signal strength is recorded by the S-RNC and by the UE, which reports the measurements to the S-RNC. Data are simultaneously received and transmitted from cells 1-1 and 2-1 while they are members of the Active Set of the UE. The S-RNC, RNC #1, is the arbiter and controller of the session. Data to and from RNC #2 passes over the Iur link from RNC #2 to RNC #1. RNC #1, the S-RNC, decides which set of data to use based on quality and forwards the data to the MSC and SGSN. The core network has no knowledge of the nature of the RAN connection.

When the relative signal strength of cell 2-1 is superior to cell 1-1, the S-RNC instructs the UE to use the carrier and codes associated with cell 2-1, using a soft handover procedure to complete the transfer. Because data are simulcast to both cells, no data are lost during the handover nor is any significant interruption experienced. Once the UE is out of communication with all of the Node-Bs served by RNC #1, the network can be notified of a change in the serving RNC. That procedure is independent from the handover procedure and is called a relocation request. The request is initiated by the S-RNC to the MSC and SGSN, and if accepted, RNC #2 becomes the new S-RNC.

When the UE homes on cell 2-1, the Active Set can also change and in our example could include cells 2-1, 2-2, 2-3, and 1-1. Defining the Active Set for a particular cell has tradeoffs. The more cells included the higher the probability of executing a soft handover instead of a hard handover. But, each cell included in the Active Set adds a duplicate transmission stream, wasting bandwidth. There are timing limitations imposed by the technology, so the relative delay of each data stream reaching the UE and S-RNC has limits that must not be exceeded. There is also flexibility in responding to changing transmission path quality within a given cell. With CDMA the spreading factor, size of the coding vector, can be increased or decreased with a commensurate tradeoff between signal quality and bandwidth.

5.3.2.5 Inter Radio Access Technology (IRAT) Management

Inter Radio Access Technology (IRAT) handover occurs when a mobile subscriber moves from one RAN network technology to another. This became an important consideration with the introduction of UMTS. Technology evolution within a standard, such as the move to HSPA within UMTS, can be orchestrated more easily because the standard can be modified for backward compatibility. As new software loads enter the network, backward compatibility can be managed within a single technology. IRAT is more difficult, because legacy technologies, like GERAN, are more difficult, technically and economically, to modify with the introduction of a different technology like UTRAN. In effect the full burden of compatibility falls upon the new technology with little opportunity to modify the existing legacy technology. There are difficult tradeoffs to be made for IRAT when designing the next mobile technology. Systems engineers, designing the standards and protocols, want the freedom to optimize the next generation and would like to minimize or even ignore previous technologies that constrain and complicate the new designs. Network engineers, responsible for the logistics and economics of deploying the new network want the freedom to deploy the new network technology over time, increasing the likelihood and extent of IRAT interworking and handovers; their preference is for full interworking, giving them more degrees of freedom when developing their roll out plans. Marketing and sales managers insist that the user experience be as seamless as possible so customers don't feel they have to accept compromises to adopt the new technology. At the same time, they want the new technology introduced as quickly as possible to remain competitive. These competing forces inevitably lead to compromises in design and deployment.

To bound the compromises, ETSI has established handover requirements for IRAT [30]. The following scenarios are covered under the IRAT standards:

- Intra network, GERAN to UTRAN
- Intra network, UTRAN to GERAN
- Inter network, GERAN to UTRAN
- Inter network, UTRAN to GERAN.

The permutations of IRAT are daunting. Services affected by IRAT include:

- **Voice service** – circuit switched voice service handover from UTRAN to GERAN and GERAN to UTRAN is required. UTRAN AMR voice codec is not required in GERAN, so codecs may be switched to GSM supported codecs when handing off from UTRAN to GERAN.

- **Facsimile service** – handover not required.
- **Data services**
  - **Circuit switched data services** – handover from UTRAN to GERAN and GERAN to UMTS are permitted but not required.
  - **Packet services** – in general the requirements call for the handover of a PDP context between GERAN and UTRAN. Handing over the context is meant to insure the connection to the Internet or intranet is maintained. That is, a handover should be opaque to the network-side host, avoiding interruptions to established sessions. There are complications because there are not necessarily one to one mappings of QoS levels between services offered on the two networks. In addition, UTRAN devices are far more likely to support multiple RABs, so in a handover to GERAN a decision has to be made about which of the PDP contexts should be handed over, if any. The requirements are not definitive in this scenario.
- **Message services** – SMS handover is not required.
- **Supplementary services** – service handover should not be affected.
- **Cell broadcast** – service handover is not required.

In addition to technical requirements, there are business issues that have to be solved to enable the technology. When handing off between networks, the two networks must agree to work with each other and have a mechanism for revenue sharing.

Much of the burden of IRAT handovers falls on the mobile device. GERAN and UTRAN are completely different radio standards; UTRAN capable devices must implement receivers for both technologies or must be able to monitor both simultaneously in a sharing mode. Dual radios are not only expensive to manufacture, they consume battery life at a high rate. The alternative is to share the receiver and have it perform measurements during gaps in UMTS operation. The operation whereby the receiver time shares across two frequencies is called a compressed mode. By measuring available GERAN cells, a controller intersystem handover can be achieved if the UTRAN signal is degraded and a viable GERAN signal is present. There are several ways of implementing the compressed mode [31], which is also used when multiple UTRAN carriers are present but operating on different frequencies.

UMTS voice service uses the GSM network core and leaves the protocols largely intact with a few changes. When a handover is required from UMTS to GSM, the RNC informs the UMTS MSC which in turn sends a handover request to the target GSM BSC [32]. If the BSC is able to establish a connection to the UE, it informs the MSC, which in turn signals the RNC to release the connection to the UE.

5.3.3 Long Term Evolution (LTE)

LTE is a misnomer in that unlike UMTS which *is* an evolution, LTE is by most measures a complete overbuild of existing GSM and UMTS networks. UMTS changed the RAN but generally left the core voice and packet networks intact. Upgrades from GSM to UMTS apart from the RAN, generally required only software upgrades in the core and some additional circuit cards in core network elements. LTE not only changes the air interface and RAN, it changes subscriber management and policy, the packet core, and most significantly moves from the TDM MSC voice design to an all IP IMS/VoIP (IP Multimedia Subsystem) network.

5.3.3.1 Deployment Considerations

Because LTE is such a sweeping change it requires massive investment and is being rolled out over a period of years in phases. There is no single template for how those phases should be structured and different carriers are starting at different places, depending on their current mobile technology, mix of user devices, and geographical coverage. The intermediate and even final architectures can vary significantly; 3GPP identifies a number of configurations and design options for implementing LTE [33]. Key considerations in building a deployment plan include the following.

Existing Wireless Packet Radio Technology

GSM carriers had the ability to ramp subscriber data rates by upgrading the UMTS network to High Speed Downlink Packet Access (HSDPA) (3GPP release 5), HSUPA (3GPP release 6), and HSPA+ (3GPP release 7) in the period 2008–2010. CDMA2000 carriers upgraded their wireless radio technology to 1xEV-DO (Revision 0 and Revision A). Assuming equally well engineered networks, the downlink performance of the HSDPA is higher than 1xEV-DO; uplink performance is similar [34]. CDMA2000 carriers may choose to deploy LTE more broadly and more quickly to match the speed of HSDPA, and to minimize differences in the subscriber experience when they are handed off from LTE to CDMA2000.

Spectrum Availability

GPRS/Edge retirement will lag LTE deployment by several years in many countries, so new spectrum has to be bought or freed up for LTE deployment. In the US, Ultra High Frequency (UHF) spectrum in the 698–806 MHz range recovered from broadcast television was auctioned off by the FCC for wireless services; that spectrum will largely be used for LTE. There is a tradeoff between spectrum investment and network investment. If a carrier has very limited spectrum, reuse and investments in technology such as Multi-Standard Radios (MSRs) can improve the efficiency of available spectrum. If sufficient bandwidth is available a dedicated LTE overlay is likely to cost less in initial capital outlay and be easier to engineer and manage.

Voice Service Transition Plan

As an all IP network design Voice over long term evolution (VoLTE) assumes the use of VoIP, not TDM as the design target for voice services; LTE does not have a dedicated voice channel. IMS is a 3GPP release 5 standard and is the target architecture for voice and for media and messaging in an LTE network. Carriers are in various stages of IMS deployment, but initial deployments of IMS were not necessarily for wireless mobile services. The set of required VoLTE interworking features were written into the 3GPP One Voice standard in late 2009. Even carriers with extensive VoIP networks are not likely to have implemented the One Voice feature set for their IMS networks to be ready to accept LTE integration at launch. Recognizing the dilemma 3GPP developed bridging technologies, with the principal candidate called Circuit Switched Fallback [35] (CSFB) that supports the use of GSM TDM voice in LTE networks. As the name indicates, LTE users fall back to UMTS when they receive or make a call. Apart from the complexity that creates for UMTS and LTE handoff, it also affects deployment. Since handoffs to complete voice calls also affect data sessions, ideally LTE should be deployed in areas with HSDPA service to minimize user perceived differences in their data sessions after the handoff.

IPv6 Deployment

LTE mandates IPv6 for full deployment but acknowledges initial deployment may be done under IPv4. The concession may be moot since the international inventory of IPv4 addresses, International Address Numbering Authority (IANA), is exhausted. LTE deployments need to either be dual stack (IPv4 and IPv6) capable at launch or shortly thereafter. Conversion from IPv4 to IPv6 well into deployment is costly and filled with risks. Carriers have to make considerable investments in their infrastructure to support IPv6 and a good deal of it should be completed before LTE introduction begins.

Backhaul Transport

In the US a nationwide deployment of LTE requires preparation in tens of thousands of cell sites,[10] in and out of a carrier's franchise region. Sufficient transport capacity has to be in place in a region before deploying LTE. That capacity is over and above what is used for existing GSM and UMTS service. In some cases the backhaul may not be suitable for LTE because of bandwidth, QoS, or IPv6 issues.

Regulatory Requirements

Each country has its own regulatory requirements for lawful intercept and emergency services. Lawful intercept requirements are likely to cover voice, SMS, and other forms of data transmission. For E911 service requirements cover not only call routing, but location identification. Location systems that work for UMTS need modification or must be developed and deployed for LTE. Technical compliance with regulatory mandates is required before a service can be launched. Carriers work with suppliers to adapt or develop country specific solutions to integrate LTE into the existing platforms. At times the development of a compliant solution can gate the introduction of a new technology like LTE.

NMS and OSS Systems

A number of new network elements are introduced with LTE requiring modification or at least significant reconfiguration of existing NMS systems. Provisioning systems also need upgrades and radio controllers like the eNode-B have hundreds of new parameters that need field of use and engineering guidelines, and those too must be included in the requirements for Operations Support System (OSS). Subscriber provisioning systems must support the addition of the LTE HSS (Home Subscriber Server) as a subscriber data store and continue to support the HLR. Normally systems of this complexity go through a controlled introduction over many months, but because LTE must have broad coverage, accelerated deployment of support systems is critical.

---

[10] CTIA reported 256 920 cell sites in the US in June 2011.

5.3.3.2 Deployment Strategies

Given the constraints and considerations for deploying LTE described earlier, LTE is likely to be deployed in different phases by different carriers. GSM carriers are likely to choose the three phased approach shown in the following.

- **Phase 1 – LTE data services only** – rollout high speed data services with air cards and eTablets that are LTE capable.
- **Phase 2 – LTE data services and voice using CSFB** – for carriers with broad HSPA coverage offer service for air cards and for LTE and GSM/UMTS capable phones. Use CSFB and the GSM network for voice calls. The differences in data speeds between HSPA or HSPA+ and LTE are perceptible but not annoying. Some additional call set up delay is encountered with CSFB.
- **Phase 3 – LTE data and VoLTE** – when full IMS is implemented and LTE coverage is complete, CSFB can be phased out in time.

A carrier new to mobile services could elect to deploy Phase 3 and bypass earlier wireless technologies altogether. WDCMA2000 carriers may move from Phase 1 directly to Phase 3 since they are motivated to improve data speeds with wide deployments of LTE. Carriers with extensive VoIP deployment on IMS likewise may skip Phase 2 or even skip Phase 1.

LTE brings with it the following changes to mobile networks. Figure 5.22 aids in the discussion.

5.3.3.3 Radio Access Network (RAN)

LTE's air interface moves away from W-CDMA (UMTS carriers) to Orthogonal Frequency Division Multiple Access (OFDMA) in the downlink and Single Carrier Frequency Division Multiple Access (SC-FDMA) in the uplink, a transmission scheme similar to OFDMA.

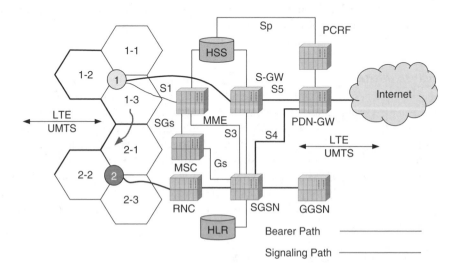

Figure 5.22   LTE and UMTS interworking.

OFDMA is the same modulation technology (OFDM) used by DSL and Wi-Fi (802.11a). In some sense OFDMA adopts the best characteristics of GSM and UMTS. Like GSM, OFDMA uses a combination of frequency and time division multiplexing, and smaller discrete carriers. W-CDMA uses code multiplexing and a wideband carrier. While OFDMA does not adopt code multiplexing, like UMTS it is wideband in design, overcoming the lower data speeds of GSM/EDGE by enabling reception of a large number of carriers simultaneously. LTE passbands are variable; bandwidths of 1.25, 2.5, 5.0, 10.0, and 20.0 MHz are defined in the standard. Within each of these bands are individual carriers that are 15 kHz wide and can be modulated using QPSK, 16 or 64 QAM in the downlink and QPSK or 16 QAM in the uplink. The modulation method changes for each carrier according to the signal and noise experienced in that part of the spectrum. LTE also makes MIMO a virtual requirement instead of an advanced option as it is in UMTS. As a result of these advancements in transmission technology OFDMA has an estimated transmission efficiency improvement of 300–400% in the downlink and 200–300% in the uplink over the closest competitor, HSPA [36].

The LTE RAN is called the E-UTRAN, for evolved UTRAN. GSM and UMTS have separate network elements for the base station and controller. LTE combines them into a single network element, the eNode-B, again using evolved as a prefix. The eNode-B is responsible for transmission and reception, and for RRC, combining the functions of the BTS and BSC in GSM, and the Node-B and RNC in UMTS. In combining the two 3GPP recognized the need for more processing power closer to the radio, making for a more distributed design. In UMTS the RNC manages handover among Node-Bs under its span of control. It also coordinates handovers via a direct link to other RNCs, via the Iur interface. That idea is extended in LTE. eNode-Bs have the ability to coordinate radio handovers with the UE directly, without calling on the voice or packet cores; handovers are possible with configured eNode-B neighbors via the X2 interface. eNode-B neighbors can be declared by the network operator or established through a self-discovery process using UEs.

5.3.3.4   Mobility Management Entity (MME)

Mobility and session management were shared responsibilities prior to LTE. LTE introduces the Mobility Management Entity (MME). In GSM and UMTS the primary responsibility for mobility management rests with the SGSN for packet services and the MSC for voice, both core network elements. With the introduction of UMTS the RNC, or RAN, assumed a larger role in mobility management. In LTE a new network element, the MME has mobility and session management responsibility; unlike earlier architectures the MME has no bearer path responsibilities. In effect it manages sessions, not bearer path traffic. It has the session management responsibilities of the SGSN but none of the bearer path routing and transport responsibilities.

5.3.3.5   Serving Gateway (S-GW)

Bearer paths go from the eNode-B to the S-GW to the Packet Data Network Gateway (PDN-GW), flattening the hierarchy of the bearer network by eliminating the RNC. GTP tunneling is used from the eNode-B to the S-GW (S1 GTP) and from the S-GW to the PDN-GW (S5 GTP), relegating the S-GW to simple IP routing or interconnecting GTP tunnels.

In this way the S-GW serves as an anchor point to the PDN-GW. Mobility handovers are managed either by the eNode-Bs directly and the MME is notified, or they are managed by the MME when the source and target eNode-Bs are not part of the same neighborhood; there is an S1-MME signaling interface between the eNode-Bs and the MME. In both cases the MME instructs the S-GW to terminate a new S1 GTP to the new eNode-B. Multiple bearers may be established by a UE to different APNs and therefore possibly different PDN-GWs. The introduction of IPv6 also means that services may require simultaneous IPv4 and IPv6 bearer paths. The MME manages those connections but the S-GW implements the GTP cross connects.

5.3.3.6   Policy Charging and Rules Function (PCRF)

Defined by 3GPP as part of IMS, Policy Charging and Rules Function (PCRF) has a dual role, managing QoS and setting charging policy for enforcement by a closely related function PERF (Policy Enforcement Rules Function). Policy refers to the ability to manage access to services and QoS of applications. Access to a service is dependent on subscription privileges and in some cases current usage or credits. Authorization for services is governed by the user profile information in the HSS, accessed by the PCRF via the Sp interface. PCRF's role is to apply policies for metered services and restrict usage if current account balances reach or exceed subscription limits. Prepaid services are a simple example. Subscribers using prepaid service have usage information gathered from the MME and other elements and recorded against their credit limit. Usage information is exchanged between the usage measurement system and the PCRF via a Diameter interface, not shown in Figure 5.22. If subscriber usage crosses a threshold the PCRF notifies the PERF to restrict or block access to the relevant services for that subscriber. Carriers have either regulatory or self-imposed rules on how to notify subscribers in advance and give them status of their accounts in near real time to avoid being unexpectedly restricted from services. While the PCRF is responsible for constructing and communicating specific rules for specific subscribers and applications, the actual filtering and application of those rules is performed by the Policy Charging Enforcement Function (PCEF). Unlike the PCRF the PCEF is in line with the PDN-GW or is incorporated into the PDN-GW. The figure assumes the PCEF is incorporated into the PDN-GW.

5.3.3.7   Packet Data Network Gateway (PDN-GW)

The PDN-GW in LTE performs a function similar to the GGSN in UMTS and GPRS. It is the anchor to external packet data networks, typically the Internet, but the PDN-GW can also serve as access to VPNs. The S-GW is the anchor to the RAN and the MME is responsible for managing connectivity between the S-GW and PDN-GW by instructing the S-GW to establish and connect the requisite S1-GTP and S5-GTP tunnels. IP address assignment rests with the PDN-GW in LTE as it does with the GGSN in UMTS, but with two important differences. In GPRS and UMTS addresses are assigned using Dynamic Host Control Protocol (DHCP) and a NAT/PAT (Port Address Translation) translation as part of the PDP context formation process. When a UE requests an authorized service and is directed to an APN a public IPv4 address and outgoing port are assigned by the GGSN during the establishment of the PDP context. Until the first service is requested the UE does not have a routable public IPv4 address. In LTE UEs are assigned routable IPv6 addresses and private IPv4 addresses during network attachment. These are the addresses for default IPv6 and IPv4 bearers between the UE and the PDN-GW. Recognize that there are no private IPv6 addresses; they aren't needed because the address space is so large, 10^{128}. The general issues of IPv6 implications for mobility are discussed further in Chapter 15.

Packet service routing is virtually the same in LTE as it is in UMTS. A service request to an APN traverses from the UE through the eNode-B to the MME. The MME confirms the users request with the HSS and requests a PDN context from the PDN-GW. When users roam in a visited network the requested APN is resolved by the SGSN for UMTS or MME for LTE and a GTP tunnel is established to the PDN-GW in the home network via the GRX. As in UMTS this means that every packet service, including basic Internet access is routed from the UE location through the PDN-GW in the home network. So if a user is in Shanghai accessing a web site in the same city, the request may go through a PDN-GW in Chicago Illinois. The 3GPP standard does have an option called Local Breakout which allows the visited PDN-GW to serve designated APNs. At this point it is not in general use and requires an exchange of policies between the home and visited PCRFs prior to service requests [37].

5.3.3.8   Home Subscriber Server (HSS)

LTE's HSS is similar to the UMTS/GSM HLR. Unlike the HLR, HSS was designed for IMS and so in addition to subscriber information it has information about IP media access privileges, authentication, encryption, and services. It also serves as the authentication center for LTE which is spelled out as a separate function in GSM but often included within the HLR. LTE standards did not adopt the SS7 MAP or Remote Authentication Dial In User Service (RADIUS) protocols used by GSM and instead chose the Diameter protocol. In deployment some HLRs can be upgraded to support both, the HLR functions and interfaces for legacy GSM and UMTS networks, and HSS functions and interfaces for LTE. An upgrade strategy reduces the need for synchronization between HLRs and HSSs by support systems as well as reducing operational costs.

5.3.3.9   Circuit Switched Fallback (CSFB)

As stated earlier LTE is an all IP network that does not have a dedicated architecture for voice. It adopts 3GPP IMS for VoLTE. IMS is not strictly a mobility standard but is instead a multi-media standard based on Session Initiation Protocol (SIP) that supports voice. The SIP [38] has evolved as the *de facto* standard for voice over IP. IMS was standardized in 3GPP Release 5 but has not been widely adopted. Recognizing IMS deployment could delay LTE, 3GPP standardized CSFB as an interim solution in 3GPP Release 8 with some enhancements in Release 9. CSFB is needed when a carrier chooses to offer LTE service to smart and feature phones prior to full deployment of VoLTE, described as a Phase 2 deployment above. CSFB is also needed when a subscriber roams out of LTE coverage but can use UMTS coverage for service. In both cases the MME assumes the role of a serving UMTS SGSN that is handing over service to a target SGSN and MSC. The same procedures described under the sections on GSM Voice Mobility Management and UMTS Data Mobility Management are used here, except the source SGSN is the LTE MME. Coordination with the UE, the source eNode-B, and the target RNC are required to select the best Node-B; that exchange and other overhead can increase call setup delay by 2-6 seconds. Another complication is that the MME must facilitate joint registration in LTE and UMTS networks for the UE when it powers on to prepare for the possibility the UE will need to fall back to UMTS. Without prior joint registration call setup delay would extend even further.

If a data session is active during fallback, both the data session and the voice session must be moved to the UMTS network or terminated because the UE cannot operate on two different RANs at the same time. Complications in the data handover occur if the UE has an IPv6 session in progress or an LTE IPv4 service that is not supported on the UMTS network.

5.3.3.10   Short Messaging Service (SMS)

SMS works the same way in LTE as it does in UMTS, at least until IMS is fully deployed. Again the MME emulates an SGSN via the SGs (Serving Gateway) interface which acts like the UMTS Gs interface. Unlike voice, SMS originated and terminated messages work without CSFB being invoked. SMS messages are passed between the MSC and MME. The MME forwards them to the UE via the eNode-B and the MSC forwards them to the SMSC.

5.4   Wireless Design and Engineering

Design of wireless networks requires all of the skills and rigor of wireline networks and then some. Factors complicating these designs are the air interface, mobility, IRAT, and device behavior. Volumes have been written on these topics [39]; the following paragraphs touch upon some of the important considerations that face wireless engineers.

5.4.1   Air Interface

Wireless network design is most critical at the air interface. The degree of control of the actual performance of the RAN is far less than the rest of the network and it is far more variable. Topology, weather, and a range of other factors outside our control affect data rates and error performance. The dominant technology considerations are described below along with some general guidelines.

5.4.1.1   Operating Spectrum

The region of the RF spectrum in which the wireless carrier operates is determined by regulatory bodies and licenses obtained by the carrier. There are also historical and technical reasons that each technology ends up operating with certain regions. A global listing of carriers, technologies, bands, and channels is beyond the scope of this text, but Table 5.3 lists some bands commonly used by wireless carriers in North America along with some of the technologies generally associated with the bands. Comparable lists can be prepared for other parts of the world but the principles of how they affect design are the same.

Loss computations are only relevant for the sector that is being engineered. Topology, buildings, and interference sources all play into a specific budget and design for that sector. Computerized tools that use terrain maps, RF readings, and detailed characteristics of the

Table 5.3 US 3GPP wireless spectrum

Service	Band	Up link (MHz)		Down link (MHz)		Mode
		Low	High	Low	High	
LTE	LTE-12	699	716	729	746	FDD
	LTE-13	777	787	746	756	FDD
	LTE-14	788	798	758	768	FDD
	LTE-17	704	716	734	746	FDD
GPRS/EDGE	PCS	824	849	869	894	FDD
UMTS	AWS	1710	1755	2110	2155	FDD

equipment and technology used are employed to develop specific antenna recommendations and coverage estimates. However, for a general discussion of the factors that affect technology and engineering decisions it is useful to begin with an understanding of free space loss, the loss of RF energy in space where no buildings or topology come into play. Free space loss is given by the equation:

$$\text{Loss} = 20 \log_{10}(f) + 20 \log_{10}(d) + 32.45$$

where,

- Loss is the free space path loss in decibels, f is the frequency in megahertz, and d is the distance in kilometers.
- Computing the free space loss for a selected set of bands from Table 5.3 we have the following results presented in Table 5.4.

LTE operating in 700 MHz spectrum has a 10 dB advantage over UMTS operating in the Advanced Wireless Services (AWS) spectrum. The 10 dB difference is equivalent to 7 km. LTE path loss at 10 km is approximately the same as AWS path loss at 3 km.

Table 5.4 Free space loss for selected US wireless bands

			Free space loss (dB)			
Band	Link	Frequency	Distance (km)			
		(MHz)	1	3	5	10
LTE-12	Up	716	89.5	99.1	103.5	109.5
	Down	746	89.9	99.4	103.9	109.9
PCS	Up	849	91.0	100.6	105.0	111.0
	Down	894	91.5	101.0	105.5	111.5
AWS	Up	1755	97.3	106.9	111.3	117.3
	Down	2155	99.1	108.7	113.1	119.1

5.4.1.2 Modulation and Coding

UMTS and LTE have adaptive modulation and coding schemes. GMSK, used in GSM is not adaptive and transmits 1 bit for each symbol. With GMSK a device has sufficient noise margin to receive and decode a signal or it doesn't. HSPA, an extension of UMTS, has a range of supported modulation and coding formats that can be used depending on the measured signal strength and SNR. UMTS was originally introduced with QPSK which encodes 2 bits for each symbol transmitted. HSPA added 16 QAM with 4 bits per symbol and 64 QAM with 6 bits per symbol. LTE's OFDM modulation scheme uses up to 1200 subcarriers for a 20 MHz band with 15 kHz spacing. Each subcarrier can adapt independently and use QPSK, 16 or 64 QAM depending on the SNR experienced by the subcarrier.

Table 5.5 lists mobile wireless coding schemes and their relative efficiency [40]. Higher data rates achieved by more aggressive encoding come with a price. Because symbols are more closely packed channels must have less noise and interference. E_b/N_0 is a measure of the quality of the channel. Higher data rates require a higher floor for E_b/N_0.

For an ideal, cell coverage is a set of concentric rings with more robust modulation schemes in the outer rings and more aggressive higher throughput schemes in the inner rings. Figure 5.23 is a hypothetical cell with adaptive modulation.

E_b degrades as we move away from the transmitter because of attenuation, including free space loss. Other factors such as foliage and buildings add to free space loss. Hills create

Table 5.5 Modulation and coding efficiency

Coding	GMSK	QPSK	16 QAM	64 QAM
Efficiency (bits/s/Hz)	1.0	2.0	4.0	6.0
E_b/N_0 (dB)[a]	10	10	18	28

[a]E_b/N_0 is a measure of signal energy per bit, divided by noise measured over 1 Hz.

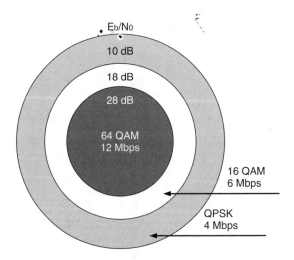

Figure 5.23 Adaptive modulation example.

shadows and further degrade transmission. An actual RF map of expected transmission performance looks more like the elevation contours of a topological map of a hill than they do the symmetrical bulls-eye shown in Figure 5.23. N_0 can increase in the uplink as the device moves away from the cell site because device transmit power is increased to overcome increased path loss. Other devices in the area collectively add noise creating an increase in the denominator.

From the earlier discussion the difference between operating in the 700 MHz spectrum and AWS spectrum was 10 dB for a given distance. In the hypothetical diagram in Figure 5.23 that translates to a difference of 6 Mbps if a device operates with 16 QAM instead of 64 QAM. The example also illustrates the classic relationship between information rates and signal quality formalized by Claude Shannon. Although it is a simplification of how networks behave the effects shown are representative.

5.4.1.3   Traffic Engineering

Physical characteristics of the cell site, available spectrum, and radio technology set limits on channels that can be used to serve subscribers. Classic traffic engineering, based on queuing theory has to be integrated with radio engineering to arrive at a design that best fits the particular cell. Service quality at a cell site can be limited by RF performance, available channels, or both. Limitations manifest themselves and have to be managed according to the radio technology used.

In GSM/GPRS timeslots are a fixed resource at a given site and are shared between GSM for voice and GPRS for data. Operators can engineer dedicated serving groups for GSM, GPRS, or both. Dedicating a pool for GSM is a way of giving a level of service protection to voice service. If voice demand exceeds the size of the service protection pool, voice channels can still be assigned from the available channels if they are not being used for GPRS.

UMTS uses a dedicated channel for voice in a fashion similar to GSM but can use shared channels for packet service. The RRC state machines in the device and the network determine how and when each type of channel is assigned. Because UMTS uses W-CDMA another layer of complexity is added, code assignments. In W-CDMA scrambling codes differentiate channels, not frequency or timeslots. Codes are a fixed resource and have to be engineered just as carriers are in GSM. As more devices are added to a particular carrier the codes have to become longer to maintain orthogonality, reducing the bandwidth available to any one device. When code management is added to RF management UMTS traffic engineering becomes far more complex than GSM.

LTE in some sense should simplify engineering by the addition of an all packet transport layer with QoS. Global carriers have been designing VoIP networks using QoS for about a decade; as mobile wireless moves to VoLTE engineering end to end quality should become much more uniform.

5.4.2   Mobility

Many of the same engineering issues appear in fixed wireline networks and mobile wireless ones allowing carriers to apply tried and proven solutions. But the changing technologies and user patterns found in mobile networks demand that new models and approaches be developed as these networks and services evolve. In particular the scope of the engineering space differs between wireline and wireless networks. Planning and engineering in a wireline network start with a wire center with a CO at the center. Demand and services are defined by the boundary of the copper and fiber that emanate from the office. Engineers focus on designing transmission and traffic capacity within that scope. Mobile networks also have serving areas, cell towers, and sectors, but the greatest engineering challenges are not within the cells but at the intersections of cells. Traffic capacity and service within a sector follows some of the same engineering principles as the wireline network, as described in the previous section. But as devices cross sector boundaries the problems of RF and traffic engineering become much more dynamic. For engineers to understand how to allocate resources within a cell they must be able to predict demand within some bounds. Cell sites then need to be planned and engineered in clusters rather than as individual sites because demand moves across the clusters.

Planning and engineering start by gathering demand data from the network. Data include cell demand by time of day and transition data that shows how demand moves between cells. The number of handovers between cells is a measure of the associativity. Cells that are highly associated should be homed on the same radio controller and voice and packet core elements to promote soft handovers rather than relocation. Cells with high associativity that also have higher dropped calls and dropped packet rates are an indication of an edge problem in the RAN that affects handover or possibly congestion in direct links such as an Iur interface that carries handover traffic. Understanding associativity is also a basis for engineering antenna coverage. Directionalizing antennas at the ends of a corridor may allow the elimination of an intervening cell, reducing the number of handoffs. Cell boundaries, particularly UMTS boundaries change as a function of traffic. More devices in a cell create more uplink noise and can shrink the cell boundaries creating gaps or cutoffs as traffic moves into the cell. The effect can be dramatic because devices are instructed to increase output power when the quality of the uplink degrades, causing a widespread rise in the level of noise.

5.4.3   Inter-Radio Access Technology (IRAT)

As UMTS coverage expanded, handover between GSM and UMTS diminished. With the introduction of LTE the cycle starts again and will grow for several years. Handover from LTE to UMTS occurs when LTE a device moves out of LTE coverage and when CSFB is invoked because of a user originated or incoming voice call, prior to the implementation of VoLTE. GSM and UMTS share voice and packet cores so handover is relatively straightforward. LTE's packet core is different than UMTS; the LTE MME emulates a UMTS SGSN to affect the handover but PDP contexts must be re-established or must be duplicated in advance to affect packet session handover. The process may be further complicated by the reliance on IPv4 in the UMTS core and IPv6 in the LTE core. Messaging and ultimately voice will ride LTE's IMS core, another possible complexity. It's early in the deployment of LTE but management of IRAT is likely to decide the reliability and performance of LTE networks.

5.4.4   Device Behavior

The introduction of smart phones and the avalanche of applications they triggered dramatically alter the volatility of traffic in wireless networks. The iconic pioneer in the market place, the iPhone® has a half million applications available to subscribers, many at no cost. Feature phones, the predecessor to smart phones largely relied on in-house developers or carefully screened third party sources. The iPhone and Android™ by design encourage and evangelize their operating systems' (OS) open software approach and they provide sophisticated software development kits (SDKs) to encourage application development across the industry. Apple® and Google® are thoughtful about the process and put in safeguards to protect the device, users and the network from developer's mistakes. However, the way applications use the network varies from application to application and developers do not always consider how to husband network resources and use them in an effective way so that all parties benefit. Developers are most likely to come from a desktop or server environment where resources and network capacity have generous limits. Examples of poor design decisions are brought to light by carriers when they gather statistics on devices and their relative performance under identical network conditions. Some applications are extremely chatty even when they are running in the background. They contact their servers to update information in the event the user opens the application; that way they won't have to wait for a response. Server connections are also opened and kept alive to allow the server to contact the client if some network application condition is satisfied. Applications can also open scores of connections rather than multiplexing requests and responses. All of these design decisions create network load and are particularly burdensome when handover and IRAT are invoked. Carriers and device OS suppliers need to continue to work with application developers to educate and encourage design practices that take into account the resource limitations of the networks.

5.5 Summary

Access and aggregation networks are the most diverse and complex designs in the carrier's portfolio. They collect and distribute services that range from low speed data and voice through video and complex enterprise hosting and collaboration services. Wireline and wireless networks are both quickly evolving from voice networks to broadband networks, delivering voice and all other services over IP. Like the services they support, access technologies are varied and complex. Copper's ability to deliver broadband continues to advance with VDSL2 and bonding. Fiber pushes ever deeper into consumer and enterprise networks and radio technologies have experienced significant upgrades or entire overbuilds every three to four years. Each access technology has unique design and engineering constraints and rules that have to be addressed by systems engineers, network engineers and operations analysts to deliver reliable and high performance service.

References

1. ANSI (1992) T1.110-1992. *Telecommunications Signaling System No. 7 (SS7) General Information*, American National Standards Institute, New York.
2. Bellcore (1990) *Telecommunications Transmission Engineering*, Facilities, Vol. **2**, Bellcore Technical Publications, p. 94.
3. Bellcore (1991) SR-NWT-001937, National ISDN-1.
4. Goralski, W. (1998) *ADSL and DSL Technologies*, McGraw-Hill.
5. ITU Recommendation (2009) G.992.5.
6. ITU Recommendation (2011) G.993.2.

7. Starr, T., Sorbara, M., Cioffi, J.M. *et al.* (2003) *DSL Advances*, Prentice Hall.
8. Starr, T., Cioffi, J.M., and Silverman, P.J. (1999) *Understanding Subscriber Line Technology*, Prentice Hall.
9. TR-069 Amendment 3 (2010) *Protocol Version: 1.2*, Issue 1, November 2010.
10. The Broadband Forum (2001) TR-043 *Protocols at the U Interface for Accessing Data Networks using ATM/DSL*, Issue 1.0, August 2001.
11. IEEE Standard (2011) 802.1ap. *Management Information Base (MIB) definitions for VLAN Bridges*, IEEE.
12. Ross, S.S. (2011) FTTH Deployment Trends: The Bounceback, FTTH Market Report, May/June 2011.
13. ITU Recommendation (2005) G.983.1.
14. Telcordia (2001) Functional Criteria for Digital Loop Carrier (DLC) Systems, GR-57, October 2001.
15. Strater, J. and Nikola, S. (2004) Motorola White Paper, Engineering CMTS and HFC for VoIP with Capital and Operating Expenses in Mind, December 2004.
16. CableLabs (2011) Converged Cable Access Platform Architecture Technical Report, CM-TR-CCAP-V02-110614, June 14, 2011.
17. SCTE (2011) ANSI/SCTE 174. *Radio Frequency over Glass Fiber-to-the-Home Specification*, SCTE.
18. CableLabs (2011) Data-Over-Cable Service Interface Specifications DOCSIS 3.0, Operations Support System Interface Specification, CM-SP-OSSIv3.0-I15-110623.
19. Mouly, M. and Pautet, M.-B. (1992) *The GSM System for Mobile Communications*, Cell & Sys Correspondence.
20. Rumney, M. (2009) *LTE and the Evolution to 4G Wireless, Design and Measurement Challenges*, Agilent Technologies.
21. Sauter, M. (2011) *From GSM to LTE: An Introduction to Mobile Networks and Mobile Broadband*, John Wiley & Sons, Ltd, Chichester.
22. Ong, L. Rytina, I., Garcia, M. *et al.* (1999) IETF, RFC 2719, *Framework Architecture for Signaling Transport*, October 1999.
23. 3GPP TS 23.040 (2011) *Technical Realization of the Short Message Service (SMS)*.
24. OMA (2009) Provisioning Architecture Overview, Approved Version 1.1, 28 July 2009.
25. 3GPP TS 03.03, v7.8.0 (2003–09) *3GPP Technical Specification Group Core Network, Numbering*, addressing and identification (Release 1998).
26. WAP Architecture (2001) Wireless Application Protocol Architecture Specification, WAP-210-WAPArch-20010712, Version 12-July-2001.
27. Overview of 3GPP Release 5 V0.1.1 (2010–02) (2010).
28. Overview of 3GPP Release 7 V0.9.15 (2011–09) (2011).
29. 3GPP TS 26.090 (2004) *Mandatory Speech Codec Speech Processing Functions, Adaptive Multi-Rate (AMR) Speech Codec*.
30. ETSI TS 122 129 V10.0.0 (2011–05) (2005) *Handover Requirements between UTRAN and GERAN or other Radio Systems*.
31. ETSI TS 122 331 V8.1.0 (2008) *RRC Protocol Specification*.
32. ETS 3GPP ETSI TS 123 009 V10.0.0 (2011–04) (2011) *Handover Procedures*.
33. Lte (2010) GSM-UMTS Network Migration to LTE: LTE and 2G-3G Interworking Functions, 3 Americas Whitepaper, February 2010.
34. Sauter, M. (2011) *From GSM to LTE: An Introduction to Mobile Networks and Mobile Broadband*, John Wiley & Sons, Ltd, Chichester, p. 203.
35. 3GPP Standard TS 23.272(2012).
36. Motorola (2007) Overview of the LTE Air Interface, Whitepaper, undated.
37. Punz, G. (2010) *Evolution of 3G Networks: The Concept, Architecture and Realisation of Mobile Networks beyond UMTS*, Springer Wien New York Press.
38. IETF RFC (2002) SIP: Session Initiation Protocol, J. Rosenberg, dynamicsoft, H. Schulzrinne, Columbia U., G. Camarillo, Ericsson, A. Johnston, WorldCom, J. Peterson, Neustar, R. Sparks, dynamicsoft, M. Han, ICIR, E. Schooler, AT&T, June 2002.
39. Agbinya, J.K. (2009) *Planning and Optimisation of 3G and 4G Wireless Networks*, River Publishers.
40. Sklar, B. (1988) *Digital Communications Fundamentals and Applications*, Prentice Hall.

6

Backbone Networks

Backbone networks are the interstate highways of the communication network. Access networks serve as on and off ramps and are responsible for functionality, features, and service management. Backbone networks have scale, reliability, and connectivity. They are there to serve all traffic generated by access and services and do it in the most efficient and reliable way possible. Backbone networks have two important dimensions, a horizontal or regional view, and a vertical or functional view. Figure 6.1 illustrates the regional view at a high level.

Incumbent Local Exchange Carriers (ILECs) tend to have their highest traffic volumes in franchise regions. The breadth of services in region generates higher traffic and connectivity demands and substantial traffic from peering partners. Out of region backbones are built to service multi-region or multi-national customers, to extend consumer services such as mobility, and to extend global connectivity. Out of region networks have fewer and smaller access networks, because of regulatory and economic constraints. For many consumer services, apart from mobility, it is difficult to make an economic argument for building an out of region access network. That generally means overbuilding a national incumbent with advantages of market recognition, an existing infrastructure, and substantial local operations. Far more common is interconnection with the local ILEC. ILEC interconnection occurs at many levels for technical and business reasons. Transport, Internet Protocol (IP), and voice interconnection are examples of the different levels common in out of region interconnection.

As the diagram illustrates, carriers segregate backbones horizontally for several reasons.

- Backbones are high connectivity designs. The number of edges in a backbone expands as the geography, traffic mix, and traffic volumes grow. Maintaining full connectivity, every node directly connected to every other node, means the number of links increases exponentially by $N \times (N - 1)/2$, with N being the number of nodes.
- Control plane routing tables grow arithmetically rather than geometrically with backbone growth, but control plane communication, route reconfiguration, and reconvergence in particular, become more difficult to manage within acceptable service times as backbones grow.[1]

---

[1] OSPF grows with the growth in backbone topology. BGP grows as a function of external network reachability.

Global Networks: Engineering, Operations and Design, First Edition. G. Keith Cambron.
© 2013 John Wiley & Sons, Ltd. Published 2013 by John Wiley & Sons, Ltd.

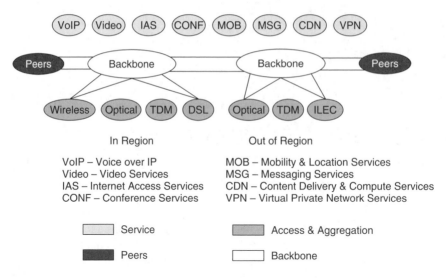

Figure 6.1   Backbone network regional view.

- Homologation in telecommunications is the process of adopting a technology to comply with local standards. As backbone connectivity extends into countries with different standards, select meeting points are used to convert between differing national or regional standards. Synchronous Optical Network (SONET) [1] and Synchronous Digital Hierarchy (SDH) [2] are two synchronous digital standards used for transport systems in different parts of the world, as one example. But homologation extends to services as well. μ-Law digital voice commanding is the standard used in North America and Japan, but A-law is used in Europe. Bridging at select gateways limits the costs of conversion by using economic designs that scale at those intersections.
- Risk containment is a less obvious but important advantage of backbone segmentation. There are very rare, but finite risks of propagating failures disabling entire networks. In spite of all the care and effort carriers and suppliers take in designing and developing network technology, these black swan events occur every few years. Because they propagate using the same mechanisms networks use to maintain routing, the more nodes there are in a network the more difficult it is to place a quarantine around affected nodes. Reducing connectivity, and thereby dependency, to specific bridge points improves the ability to isolate propagating failures. As important, it means networks can be isolated and brought back on line independently.

In the functional or vertical dimension backbone networks and subtending regional and metro networks are designed and organized in layers, like the Open Systems Interconnection (OSI) model. Transport networks are the underpinning of upper layers and so they support all connectivity demands, from legacy Time Division Multiplex (TDM) services to IP, Voice over Internet Protocol (VoIP), and video services. Not only must legacy services be supported, legacy transport equipment has to be integrated and managed along with newer technology. Transport networks have multiple design objectives.

- **Economical and reliable transport** – Bandwidth, error rate, and reach are key considerations for transport systems.
- **Compatibility** – A range of standard interfaces that have evolved over decades, organized by physical interface and data rate, are used in carrier networks. Transport systems in metro networks in particular need to be able to support legacy and new interfaces.
- **Grooming** – Metro and regional networks aggregate, transport, and distribute traffic. But not all traffic originates or terminates at the same points. Transport systems that can add and drop tributary traffic without incurring prohibitive costs are difficult to design, particularly when the traffic is of mixed format and data rates. Historically, grooming is a multi-tiered affair with lower speed systems at the bottom of the hierarchy, grooming, and aggregating traffic for the next tier up until the high speed long haul system takes the traffic for delivery.
- **Route protection** – Once a transport facility leaves the central office it is subject to a long list of potential misfortunes. Cable disruptions (backhoe events), floods, earthquakes, hurricanes, tornados, and automobile accidents are just a few of the calamities that interrupt facility routes. Many transport systems, particularly those that are SONET/SDH based, have automatic fault detection and switching mechanisms that restore connectivity in sub-1 second times.

Figure 6.2 is a simplified view of transport systems and the associated IP routing infrastructure.

6.1   Transport

To lay the foundation for a discussion of the backbone a digression is needed on transport hierarchy. Transport technology evolved from lower speeds to higher speeds just as the network evolved from voice services to data services. Transport systems evolved to carry, groom, and manage units of bandwidth consistent with the service mix and demand at the time. Digital and optical technologies opened the door to economic transport over wide distances at a rate of growth that approached, but did not meet that of Moore's law. Table 6.1 summarizes transport speeds and systems.

The system mappings are normative; line and trunk interfaces vary among individual suppliers. Although an interface is supported on a system, it may not be economical when compared with alternatives. The systems in Table 6.2 are representative of systems in the class.

6.1.1   Transport Services

There are four broad categories of transport services engineered into the transport layer, each of which is described in the following.

6.1.1.1   Enterprise Private Line – Point to Point

Enterprises grew their networks organically in the 1980s and 1990s, often purchasing TDM transport from carriers and designing and operating their own dedicated networks using

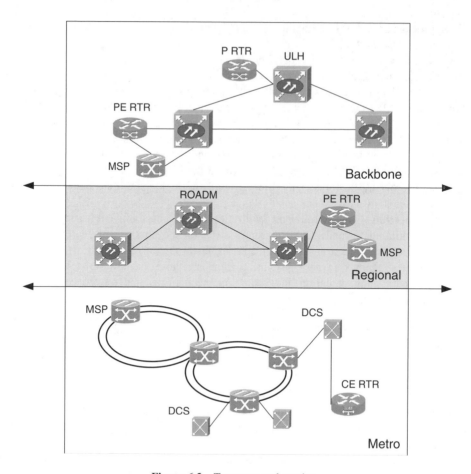

Figure 6.2   Transport and routing.

private line, Asynchronous Transfer Mode (ATM), and frame relay services. These networks, usually hub and spoke, used TDM access from branch locations to home on a central hub, or a few hubs which were interconnected with backbone transport. Many of the branch locations used channelized T1 for NxDS0 frame relay service. Connections between hubs were T1 or channelized DS3s. IP based Virtual Private Network (VPN) services have rapidly displaced these networks, offering any-to-any connectivity, easier management, and higher reliability. However, some of these dedicated networks remain in place where they support relatively low speed services such as point of sale networks for smaller retail outlets. Other networks, such as lower speed data circuits, private voice, and closed circuit video circuits will persist as long as it is not economical to replace the legacy technology. Eventually these applications will move to IP transport and at that time many of the remaining circuits will migrate to Ethernet or other IP-based transport.

6.1.1.2   Enterprise Private Line – Dedicated SONET/SDH Rings

Some Fortune 1000 enterprises and government agencies take a more comprehensive and organic approach to their network services by purchasing complete SONET networks and

Table 6.1 Transport hierarchy

System		D4		NB DCS		BB DCS		MSPP		ROADM		ULH	
Type		DCS		DCS		DCS		ADM		DWDM ADM		DWDM ADM	
TSI grooming		64 kbps		64 kbps		1.5 Mbps		51.8 Mbps		Per lambda		Per lambda	
Trunk bandwidth		1.5 Mbps		45 Mbps		2.5 Gbps		10 Gbps		1.6 Tbps		3.2 Tbps	
Protection		None		DS3		APS UPSR		APS BLSR UPSR		APS O-UPSR		–	
Bandwidth	**Standard**	L	T	L	T	L	T	L	T	L	T	L	T
64 kbps	DS0/E0	D4		NB	NB								
1.5/2.0 Mbps	DS1/E1		D4	NB	NB								
10 Mbps	Ethernet			NB									
45/34 Mbps	DS3/E3			NB		WB							
51.8 Mbps	STS-1/VC-3					WB		MS					
100 Mbps	Ethernet					WB		MS					
155.5 Mbps	OC-3/STM-1					WB	WB	MS		RM			
622 Mbps	OC-12/STM-3					WB	WB	MS	MS	RM			
1 Gbps	Ethernet/ODU-0							MS		RM			
2.5 Gbps	OC-48/STM-16/ODU-1						WB	MS	MS	RM	RM	LH	
10 Gbps	OC-192/STM-48/ODU-2								MS	RM	RM	LH	LH
40 Gbps	OC-768/STM-192/ODU-3									RM	RM	LH	LH
100 Gbps	ODU-4											LH	LH

NB – Narrowband, BB – Broadband, DCS – Digital Cross Connection System, MSPP – Multiservice Provisioning Platform, ROADM – Reconfigurable Optical Add Drop Multiplexer, ULH – Ultra High Frequency, APS – Automatic Protection Switching, TSI – Time Slot Interchange, ODU – Optical Data Unit.

Table 6.2 Representative transport systems

System	Key	Manufacturer	Model
Channel Bank	D4	Lucent	D4
NB-DCS	NB	Adtran	ATLAS 890
WB-DCS	WB	Adtran	OPTI-6100 MX
MSP	MS	Cisco	MSPP 15454
ROADM	RM	Fujitsu	Flashwave 7500
ULH	LH	Nokia-Siemens	Surpass hiT 7500

designing their own services on top of dedicated resilient SONET rings. Companies with a large campus presence in a metropolitan area can take advantage of this approach. It enables the enterprise to customize and control services like Fiber Channel and metropolitan Ethernet. An added advantage is that Internet access can be centralized, reducing the effect of a Denial of Service (DoS) attack to a single set of portals. SONETs reliability and flexibility fit these metro networks well.

6.1.1.3 Enterprise Access

TDM and Ethernet transport are used for access to managed Internet Access Services (IASs) and VPNs. A range of transport options are available, depending on the desired speed of access and the availability of facilities. Lower speeds can be served by T1 or multiple T1s using inverse multiplexing or multi-link Point to Point Protocol (PPP) (NxT1). Higher speeds are served by DS3s or NxDS3, by Ethernet over SONET, or Dense Wave Division Multiplexing (DWDM) at speeds of 1 and 10 GigE, native Ethernet, or Ethernet over microwave.

6.1.1.4 Carrier Infrastructure

Transport is provisioned within a carrier network as "official services" in support of the wide range of higher level offerings. Transport for the voice network and backhaul for radio cell sites are common examples. There are about 260 000 cell sites in the US with multiple carriers needing backhaul to their radio controllers and mobility aggregation networks. The incumbent carrier provisions backhaul as an infrastructure service and sells transport service to the non-franchise carriers. In a similar fashion the entire wireline voice network and shared IP services network are provisioned.

6.1.2 Transport Resiliency and Protection

SONET and SDH were developed in the 1980s and widely adopted in the 1990s as bandwidth demands and economics drove carriers to move from copper transport to optical transport. Prior to the introduction of SONET and SDH, reliability was largely achieved through distributed redundancy inherent in the voice networks. With multiple T1 copper circuits in a metropolitan area between any two tandem switching offices the loss of a single cable would force a trunk group, or a substantial number of trunks, out of service. But as those trunks were busied traffic would automatically reroute to alternate members and trunk groups. Prior to SONET/SDH reliability was achieved through switching, not through resilient transport.

6.1.2.1   SONET/SDH

The economics of SONET and SDH were compelling, enabling hundreds of T1 and E1 circuits to move onto a single fiber pair. An OC-12 SONET system carrying 336 T1 equivalents or 8064 DS0 voice circuits constitutes a large risk, considered too large to leave entirely to the switching network to defray. Architects have a tendency to view networks as shown in Figure 6.3 which does not portray the real jeopardy.

This view's lack of real topographical information doesn't convey the vulnerability of the transport network. Figure 6.4 is representative of facility routes in metropolitan areas,

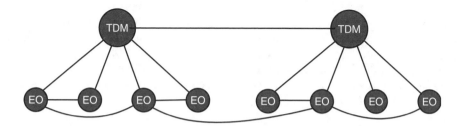

Figure 6.3   Architect's network view.

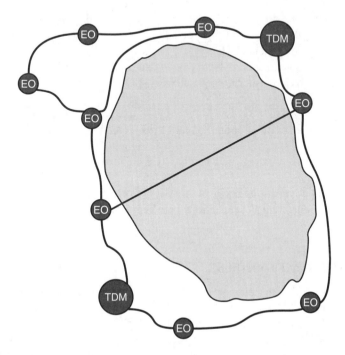

Figure 6.4   Topological network view.

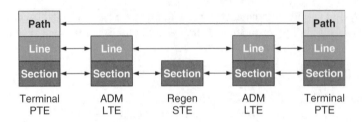

Figure 6.5   SONET nodes.

although the diagram is fictitious. It illustrates how natural boundaries, such as bays, rivers, and hills determine how metropolitan infrastructure and rights of way evolve. Facility routes follow rights of way, concentrating interoffice ducting and fiber into narrow corridors.

In this simple example there are six distinct loops that are candidates for transport rings. SONET rings are configured as either Uni-directional Path Switched Rings (UPSRs) or Bi-directional Line Switched Rings (BLSRs), with BLSRs being more common. Designing rings for such a metro area has to take into account many factors, starting with communities of interests and traffic volumes (see Figure 6.5).

Three types of elements are used in SONET/SDH designs: terminals, Add-Drop Multiplexer (ADM)s, and regenerators, shown in Figure 6.5.

- **Terminals** – a single ended SONET node that terminates signals and multiplexes them into higher speed TDM or Ethernet facilities. Terminals are SONET Path Terminating Equipment (PTE).
- **ADMs** – a double ended SONET node that sits on a ring, capable of passing traffic through the node, dropping off traffic or adding traffic to the ring. ADMs are SONET Line Terminating Equipment (LTE).
- **Regenerators** – a double ended SONET node that passes signals by performing an O-E-O conversion, acting as a repeater. Regenerators are SONET Section Terminating Equipment (STE).

SONET/SDH connectivity is organized into three layers;

- Sections are network segments between any two adjacent SONET compliant nodes, such as repeaters and ADMs.
- Lines are logical connections between ADMs and/or terminals.
- Paths are end to end logical connections between terminals or ADMs functioning as an end point.

6.1.2.2   SONET Ring Failure-BLSR

SONET rings exemplify the elements of design supporting the Virtuous Cycle. When a fiber cut occurs on a SONET ring, shown in Figure 6.6 as a cross, a loss of signal (LOS) is detected at the receiving ends of the span at ADM#1 and ADM#2.

The LOS causes ADM#1 to forward a ring bridge request to ADM#2 via ADM#4 and ADM#3 for a bridge of working traffic onto the protect timeslots on the reverse ring.

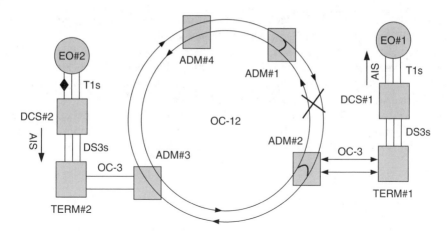

Figure 6.6   SONET ring.

Likewise, ADM#2 sends a ring bridge request to ADM#1 in the clockwise direction for a bridge of working traffic onto the protect timeslots. Traffic that previously traveled ADM#4 → ADM#1 → ADM#2 → TERM#1 now goes ADM#4 → ADM#1 → ADM#4 → ADM#3 → ADM#2 → TERM#1.

Alarm Indication Signals (AISs) can be generated at the section (AIS-S), line (AIS-L), or path level (AIS-P). When an AIS is received at a terminal, upper layers may invoke alternate routing or other recovery procedures. The receiving terminal also informs the originating terminal of the failure by sending a Remote Defect Indication (RDI-P). The originating terminal can then invoke alternate routing or other upper layer recovery procedures. Upper layers should delay invoking recovery procedures until the normal SONET recovery interval, less than 100 ms, has passed.

SONET standards [3] require the LOS be detected within 10 ms and that protection switching occur within 50 ms. These requirements originate from voice networks and voice services. Loss of bearer or signaling path for 60 ms will only affect calls in the addressing state, and even then the call is likely to complete. SONET has a robust fault detection and notification design with signaling channels reserved for section, line, and path layers.

In the Virtuous Cycle we can expect the following sequence from the failure.

- **Simplex operation** – Within 60 ms of the span failure we expect to be in simplex operation, bridging traffic at ADM#1 and ADM#2. Working traffic at ADM#3 and ADM#4 could experience a 60 ms outage, but not more than that.
- **Notification** – Major alarms are generated at ADM#1 and ADM#2 indicating LOS in the direction of the failed span. Notifications are also generated by the bridging requests from the ADMs. The controlling Element Management System (EMS) may receive other related alarms but it should correlate and suppress secondary notifications.
- **Sectionalization** – Upon receiving the alarms craft logs into ADM#1 and detects the span from ADM#2 is OOS-AU with a cause of LOS. Craft should also note the bridged state of traffic bound for ADM#2. A similar session with ADM#2 mirrors the findings at ADM#1. Having sectionalized the trouble, a trouble ticket is issued to the Line Field Organization (LFO) to locate and repair the fault.

- **Fault isolation** – The LFO isolates the physical fault in the field, using an Optical Time Domain Reflectometer (OTDR) or other techniques.
- **Fault mitigation** – The fiber is repaired and the transport center is notified.
- **Test and verification** – The transport center places the ports on the affected span at ADM#1 and ADM#2 in an OOS-MA state and tests the span remotely to insure signal and bit error rates are within the standard for service restoral.
- **Service restoral** – The transport center places the ports on the repaired span into an In-service-Normal Service State (IS-NR). When alignment is achieved notifications are sent from ADM#1 and ADM#2 to drop the protection bridges and normal duplex service is restored.

6.1.2.3 Tributary Failures

SONET has maintenance visibility into the section, lines, and paths as described above, but it also has another important characteristic, transparency to metallic TDM failures. In the SONET ring diagram in Figure 6.6 previously a failure of a metallic T1 is marked by a diamond. An AIS will be originated at DCS#2. The AIS travels over the SONET ring and is received at end office #1. EO#1 raises a carrier group alarm and marks all trunks associated with the carrier as OOS-AU. Traffic is re-routed to alternate trunks or trunk groups. When the fault is repaired, the AIS signal is removed and the carrier groups can be tested and returned to service.

6.1.2.4 Mesh Network Service Protection and Restoral

SONET service protection is standardized with a few proven options that detect faults and restore service in about 50 ms. Mesh optical networks are almost exclusively DWDM networks. Today's optical switches can switch optical streams with the constraint that the wavelength does not change. So optical paths begin and end with a single wavelength. The endpoints convert the optical signal into electrical streams using either transponders for a single port or muxponders where multiple streams ride a single wavelength. With today's technology transponders must be designated for a particular wavelength.

There are three common protection schemes used with optical cross connects. The implementation of these schemes is supplier dependent.

- **1 + 1 Protection** – two complete paths are provided end to end. The two paths carry identical traffic. Terminals monitor signal quality and select the highest quality path. To be effective, the two paths must be completely independent, with independent routes and electronics.
- **1 : 1 Protection** – two complete paths are provided end to end, a working path and a protection path. Terminals monitor signal continuity and switch from the working to the protected path on a loss of signal. To be effective, the working and protection paths must be completely independent, with independent routes and electronics.
- **1 : N Protection** – similar to 1 : 1 protection, a spare path is provided for N working paths. If any of the N working paths fail, the spare carries the traffic.

While these schemes can provide local protection, they are difficult to implement on a global scale in a carrier network. Global networks have a large number of adds and drops, so a 1 : N scheme is not practical. 1 + 1 or 1 : 1 schemes are very expensive, essentially stranding half of the network capacity. Backbone and regional OADM networks are generally more economically served by relying on Multiprotocol Label Switching (MPLS) traffic engineering and fast re-route for service protection. The responsibility for service protection is assigned to the MPLS/IP layer rather than the optical layer.

6.2   IP Core

Global carriers can take two different approaches in designing and operating backbone networks. One approach is to physically segregate public (Internet) and private (enterprise) traffic onto two separate networks. The second approach is to build one physical network and logically segregate private and public traffic. Motivation for physically separate networks is often founded on security and regulatory concerns.

Separating public and private investments is insurance against regulatory policies that mandate investments in public networks that have little opportunity for a compensatory return. US carriers are still recovering from the 1996 Telecommunications Act that mandated investments in local loops be shared at rates the carriers believed were below costs, with the stated objective of improving competition. The result was a decade of reduced investments in loop technology and a drop in customer service caused by mandated arm's length separation of services within the telco's consumer sales, operation, and care organizations. Segregating backbone networks is a form of insurance against net neutrality or other initiatives that may claim public Internet traffic has an equal right to the same quality of service (QoS) as private enterprise traffic, with no consideration of the significant differences in revenues and return on investment of the two markets.

A second reason for segregating private and public networks is security. Increasingly enterprises are targeted by Internet activists and other cyber vandals with distributed denial of service (DDoS) attacks. Attacks can originate from unsuspecting hosts using botnets, or directly from activists such as the attacks inspired in retaliation for the arrests of WikiLeaks™ founder Julian Assange™.[2] Mechanisms used for these attacks are constantly changing, targeting web hosts, infrastructure such as Domain Name System (DNS) or networks. In response to the threat, defenses are improved to deflect and defeat attacks, but attacks with massive volume targeted to specific sites can flood terminating networks. Aggregation and edge routers are susceptible to massive overloads which can affect all traffic served, not just Internet traffic. Defeating these attacks is best accomplished by characterizing the attacking traffic and instituting controls close to the point of network ingress that filter out malicious traffic before it can enter the network. However, the changing nature of the attacks makes it difficult to identify the threat and institute defenses quickly enough to eliminate the danger.

There are collateral benefits to segregating networks. Internet traffic is far more volatile than enterprise traffic. Social networking or news in general can cause a site to receive a tidal wave of hits because of some event no one forecast. Physical separation protects

---

[2] That's not a mistake. Julian Assange, champion for the freedom of information, applied for a trademark for his name.

enterprise traffic from Internet volatility. Features needed for enterprise services are far more complex that those needed to serve the public Internet. Segregating private and public networks simplifies management by isolating changes and upgrades to the two networks. Each network can evolve on its own path.

For most carriers the economic and operational advantages of a single physical network outweigh other considerations. With the exception of transport services such as legacy ATM and Frame Relay most services use or are moving to IP as their network layer protocol. A result of the universal adoption of IP services by enterprises and consumers is that global IP carrier networks of scale have far more edge routers and subtending customer routers than was envisioned 10 years ago. Open Shortest Path First (OSPF) convergence computations are an N^2 problem, meaning that it expands geometrically with the number of nodes. Three key architectural tools, regionalization, MPLS, and route reflectors enable core networks to scale linearly with the increases in the number of nodes.

6.2.1 Regional IP Backbones

As an example of the N^2 scaling problem, assume we have a network within one region and 100 000 nodes. It has an N^2 figure of complexity of 1×10^{10}. If we break the network the network into 5 equal regions, each with 20 000 nodes, N^2 drops to 2×10^9, an 80% reduction. Scale networks use regional designs for private and public IP services. There are different models for interconnecting regional backbones, one using direct bridging and another using an underlying global backbone for connectivity [4]. Throughout the text the direct bridge model is described. Regions are logical from a topological viewpoint as well, often following political, geographical, and Internet numbering boundaries. As we might expect, traffic communities form around these same boundaries. Carriers draw regional boundaries based on the international regional assignments set by IANA. Conforming to IANA's boundaries makes it straightforward to consolidate routes and significantly reduce the size of routing tables and lower convergence times.

- **Africa** – the African Network Information Centre is the regional registry (AfriNIC).
- **North American (US, Canada, and parts of the Caribbean)** – the American Registry for Internet Numbers (ARIN) is the regional registry.
- **Asia-Pacific (Australia, Asia, New Zealand, and the Pacific Rim)** – Asia-Pacific Network Information Centre (APNIC) is the regional registry.
- **Latin and Central America** – the Latin America and Caribbean Network Information Centre (LACNIC) is the regional registry.
- **Europe, the Middle East, and Central Asia** – the Réseaux IP Européens Network Coordination Centre (RIPE) is the regional registry.

Regions are further divided into autonomous systems (ASes). The number of ASes in a region reflects segregation of private and public domains, the size of the region, and the history of carrier consolidations. When carriers consolidate it may not make sense to combine their domains since changing AS designations in all of the carrier routers, and customer routers is a burden and risk not worth taking. So it is not uncommon to find multiple private and public domains within a region for carriers like AT&T and Verizon that have absorbed other carrier IP networks over time.

6.2.2   Points of Presence (POPs)

Customers and peers interconnect with carrier networks at Points of Presence (POPs). POPs are usually standardized in their design and those designs are tiered by capacity and services.

Mega POPs, the highest tier, are often designated as Tier 1 locations. Tier 1 sites are facility hubs located on long haul fiber routes with ULH DWDM connectivity to other Tier 1 sites. They also have P routers and a family of high performance edge routers for VPN and Internet services. If traffic warrants it, a layer of aggregation routers or aggregation switches is engineered between the edge routers and P routers.

Regional POPs, designated as Tier 2 locations are located at regional facility hubs, from which metro areas are fed. In the US Tier 2 POPs serve one or a few states with large states like California having northern and southern Tier 2 POPs. Tier 2 POPs are likely to be served by ROADM optical switching.

Metro POPs, designated as Tier 3 locations are metro facility hubs, possibly served by a ROADM from Tier 1 and Tier 2 POPs, but likely to use SONET for distribution to customers and connections to the incumbent carrier, if the POP is out of region. Metro POPs may not have a complete service set. A service like VPN multicast may be provisioned by transport back to a Tier 2 or Tier 1 POP.

6.2.3   Multiprotocol Label Switching (MPLS)

MPLS under the name of tag switching was introduced and championed by Cisco Systems and later became an IETF standard [5]. Many IP backbones of scale use MPLS as the primary bearer protocol. Core networks using MPLS can route IPv4, IPv6, and Virtual Local Area Network (VLAN) traffic with equal ease and no special engineering or management [6]. Service specific infrastructure VPNs can be designed and managed for VoIP, managed video conferencing services, Internet Protocol Television (IPTV), and others. Internal and external routing and route advertisement can be confined to the essential domains, improving scalability and security. MPLS has been described as a shim layer protocol, sitting logically between layers 2 and 3. It is compatible with different layer 2 protocols, such as ATM, Frame Relay and Ethernet, and supports IPv4 and IPv6 at layer 3, as well as VPN and Virtual Private LAN Services (VPLS).

6.2.3.1   Architecture

The essential network elements of an IP core, based on MPLS are shown in Figure 6.7. The diagram is functional. A single physical router can serve as both a Provider Edge (PE) router and a P router.

- **CE (customer edge) router** – The CE router acts as a gateway for a customer location, connecting the locations with the service provider network. Internet service and private network services can both be provided, although typically they are provided over separate physical facilities or separate virtual facilities.
- **PE router** – The PE provides access and egress services to the CE. Customer connections terminate in an independent *routing context*. A routing context is in effect a virtual router, maintaining a separate routing database. In a VPN, the routing context is called a Virtual

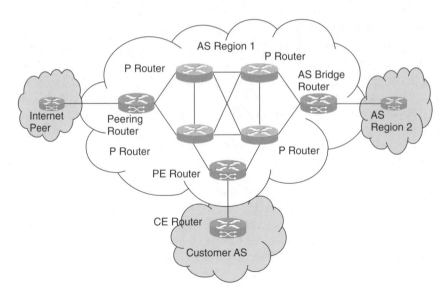

Figure 6.7 IP routing core.

Routing and Forwarding (VRF) instance. A single PE router can provision scores of VRFs each of which acts like a dedicated physical router, isolating routing, and address management, but aggregating traffic for delivery to core P routers.

- **P router** – At the very core of the network, P routers tend to be the largest routers in the network capable of routing terabits of information each second. Except for label distribution, internal routing information, and management packets (Internet Control Message Protocol, ICMP) they use label switching to achieve scale and speed.
- **Peering router** – Internet traffic may be destined for a host on the carrier's network, but more often it is meant for a host on another network. Traffic is passed between networks at peering points. Peering routers differ from PE routers since they do not pass private traffic and are often capable of higher throughput than PE routers.
- **AS border routers** – AS border routers interconnect regions, preserving QoS, and serving as the conduit for inter-AS address management.

6.2.3.2 Packet Forwarding

MPLS has some characteristics of a connection oriented protocol. Packet forwarding under MPLS is deterministic within the core network, or PE to PE. Once an MPLS packet leaves the ingress PE router it follows a Label Switched Path (LSP) that is pre-determined. Unlike IP routing in which each router determines the next hop, MPLS routing is deterministic once the headend router has chosen an LSP. All routers in the core, P routers, are Label Switch Routers (LSRs); PE routers are Label Edge Routers (LERs).

Ingress LER
The ingress LER selects a routing context for each incoming interface. Incoming interfaces can be physically or logically distinct, like a GigE link, a Generic Routing Encapsulation

(GRE) tunnel, or a VLAN. Destination IP prefixes at the ingress LER are listed in a forwarding table and resolved to a Forwarding Equivalent Class (FEC). Packets within a routing context that have destinations that point to the same egress LER, and have the same QoS belong to the same FEC. All packets belonging to an FEC are routed along the same LSP and terminate at the same egress LER. Each FEC is bound to an assigned unique outbound label and an outbound inter-nodal interface.

Once the ingress LER has determined the FEC, the associated label is inserted after the layer 2 header and before the payload as part of the MPLS header. The MPLS shim header includes the following.

- A 20 bit label unique to this LSP on the outbound facility.
- A 3 bit Class of Service (CoS) field.
- A single End of Stack bit.
- A Time to Live (TTL) octet.

Next Hop LSR

The packet is then forwarded via layer 2 to the next hop LSR. There the layer 2 header is removed and the incoming MPLS label is used to index the Label Forwarding Information Base (LFIB). The table is indexed with the received label and returns the outgoing interface and the new label. The received MPLS label is removed (popped) from the header and the new MPLS label is inserted (pushed). The TTL value is decremented and the packet is transmitted on the outgoing interface.

Egress LER

When a packet reaches the egress LER the process is the same, except the forwarding table points to an egress interface instead of an MPLS route. The received MPLS label is removed and the packet is forwarded on the outgoing interface; terminating packets are forwarded using IP routing.

Virtual Private Networks

In the example sited thus far, one ingress context was described. But MPLS labels and MPLS domains can be stacked like Russian dolls. Enterprise VPNs are an example of how label stacking can be used. When a VPN packet arrives at the carrier LER the originating enterprise interface is associated with a routing context associated with a VRF instance. The destination IP address in the received packet is used to index the forwarding table in the VRF, and a label associated with the VPN route is pushed onto the packet first. A separate label associated with the LSP to the egress PE router is also pushed onto the stack in front of the VPN routing label. Routing across the carrier LSP is performed using the outermost label. When the packet reaches the egress carrier LER the outer (LSP) label is popped and the carrier uses the VPN routing label and the IP address, if needed, to select the outgoing interface.

6.2.3.3   Label Distribution

Label distribution is the process of advertising the binding of label mappings to FECs and informing LSRs of the bindings so they can properly interpret the labels on received packets.

Within the carrier's MPLS domain the choice of label distribution method depends on how the carrier decides to manage the MPLS network. LSPs can be automatically constructed and distributed using an Interior Gateway Protocol (IGP), such as OSPF, and the Label Distribution Protocol (LDP). While MPLS routing is more efficient than IP routing, in its basic form it still has a shortcoming of basic IP routing, that is route congestion. Congestion along an LSP and an IP route is no better than the weakest link in the route. If a link along the way becomes congested, traffic will continue to be forwarded from the headend without regard to the link congestion. MPLS has a protocol extension to deal with that problem, Multiprotocol Label Switching Traffic Engineering (MPLS-TE).

6.2.3.4 MPLS – TE

If the carrier opts for a network design that guarantees bandwidth and service quality along the LSP, MPLS-TE is used. With MPLS-TE multiple MPLS tunnels are constructed from the headend PE router using Constraint-Based Routing (CBR). Constraint based routing protocols are used to distribute MPLS labels and reserve bandwidth at each LSR along the way; the Resource Reservation Protocol with Traffic Engineering (RSVP-TE) and the Constraint Routed Label Distribution Protocol (CR-LDP) are choices. An added advantage of MPLS-TE is better fault management. By building multiple LSPs and reserving resources on each path, under fault conditions the headend router or an intermediate LSR can select an alternate LSP tunnel to reach the destination LER, with a guarantee of bandwidth. MPLS fast reroute times are sub-1 second, approaching SONET recovery speeds. Labels are distributed and resources reserved via RSVP-TE or CR-LDP. MPLS-TE algorithms for constructing tunnels that take into account QoS and bandwidth availability are beyond the scope of this text and are treated in depth elsewhere [7].

6.2.3.5 Routing Protocols

Network design has many parallels in software design. In both the ideas of encapsulation or compartmentalization and coupling are central to clean designs. In carrier backbones routing within the core must be independent of enterprise customers, consumers, peering partners, and any external parties using the network. Carriers must be free to make internal network changes without affecting subtending networks. In a similar fashion, customers and partners of the carrier should be able to make changes within their networks with minimal impact to the carrier core. The idea of compartmentalization is extended to the separate regions and ASes chosen by the carrier. Changes within one region should not affect another. Following this principle, routing protocols used within an AS, called Interior Gateway Protocols have little or no impact on external networks. But networks interconnect at their borders and there they must share some minimal amount of information in order to exchange packets and advertise which routes can be reached via that edge. Since MPLS forwarding is used within the core, the responsibility for maintaining border connectivity and routing information falls on the PE and CE routers. The protocol family used in their exchange is the Border Gateway Protocol (BGP) (see Figure 6.8).

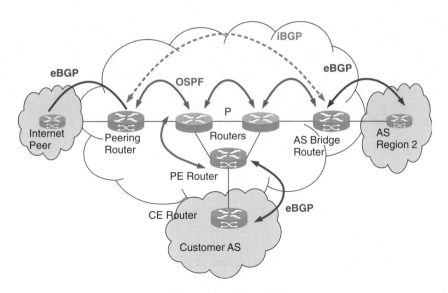

Figure 6.8   Core routing protocols.

Interior Gateway Protocol (IGP)

An IGP is necessary because routers in the MPLS domain need to communicate for a range
of reasons, including forwarding MPLS label information. There are two popular IGPs,
OSPF and Intersystem-Intersystem (IS-IS). While the two have common functions, OSPF
is the one described here at a summary level; a full and formal description of OSPF can be
found in the IETF RFC [8]. In OSPF information is shared among routers using Link State
Advertisements (LSAs). LSAs are records reflecting connectivity and reachability. LSAs
are exchanged to update the Link State Database (LSDB) maintained by each router. OSPF
performs the following functions.

- **Topological mapping** – router adjacencies are discovered and shared.
- **IP reachability** – whether the issuing router can reach nodes within the OSPF domain.
- **Connectivity monitoring** – "hello messages" are exchanged at regular intervals to confirm adjacent routers can be reached. If a set number of consecutive messages fail, LSAs
  are issued reflecting the change.

Using the LSDB, each router computes the "OSPF," or the preferred route to an entry
in the IP forwarding table. Network operators can control how the algorithm computes
preferred routes using link weighting. Weights, or costs, are set by the network operator to
influence the algorithm.

Border Gateway Protocol (BGP)

Because we are discussing how it is used in an MPLS network, we only need to consider
BGP at the edge, or PE routers. P routers have no need to route external packets based on

IP address, so they have no need to implement BGP. When VPN/MPLS was introduced BGP was extended. Originally BGP only allowed one advertisement for a single IP address prefix. That meant BGP could not support VPNs on MPLS because VPNs typically use private address space. With VPN/MPLS private addresses from multiple enterprise overlap at the PE and so a multi-protocol extension was added to BGP. The extension [9] adds an address family attribute, enabling BGP to support not only VPNs, but IPv6 and VPLS. BGP has four characteristics that suit it to this application.

- BGP uses Transmission Control Protocol (TCP) for communication, and so it is reliable at high volume transmission.
- It can handle large updates and so it scales better than other protocols, such as OSPF and Enhanced Interior Gateway Routing Protocol (EIGRP).
- It is not based on adjacency. Sessions can be established across a network between two peers.
- BGP implementations have flexible importing and exporting filters and policy settings that improve routing table efficiency and scalability.

There are two uses of BGP in MPLS, as an interior border gateway protocol (iBGP) and an exterior border gateway protocol (eBGP).

eBGP

As shown in Figure 6.8, eBGP is used to exchange routing information with adjacent edge routers in other ASs or with CE routers. The edge, or border routers from the two ASes using eBGP are known as BGP peers. eBGP peers summarize IP prefixes that they can reach, along with the list of ASes in the path and forward it to their peer. Since reachability can change regularly in large networks, BGP can be somewhat chatty. Timers and policies control the frequency and number of updates allowed as a balance between efficiency and overhead. If the receiving BGP router finds its own AS associated with an address, it disregards the entry to prevent looping. The receiving router can compare competing routes and choose the one that is most direct.

eBGP is also used to bind private addresses to VPN route descriptors configured in the PE. Adjacent PEs at AS borders, shown in Figure 6.8, can use multi-protcol (MP) eBGP to share reachability information for the Internet as well as VPNs. Adjacent domains update the connecting LER using eBGP and iBGP is used to update other LERs within the MPLS domain.

iBGP

Edge routers need up to update address tables for external destinations, but the updates must be shared with the other edge routers in the MPLS domain. Just as eBGP is used to exchange information between ASes, iBGP is used to exchange routing information among edge routers within an AS. In that way every ingress edge router has a full view of reachable destinations and which edge has the best path to them. In turn, the ingress router can share updates received from other LERs with their eBGP peers (CEs).

The iBGP exchange between MPLS domain LERs includes the following information.

- **Address family** – the network address protocol used at the LER for this context.

- **Next hop** – usually the address of the LER, using the same address family.
- **Network Layer Reachability Information (NLRI)** – an IP prefix or other address conforming to the address family.

6.2.3.6 Advantages

Label switching brings several important scale advantages to MPLS.

- The association of a packet to an FEC happens just once, at the ingress PE router. Afterwards, forwarding occurs through switching of the label bound to the FEC. With IP routing each router along the way has to make an IP forwarding decision by analysis of the IP header information.
- Customer VPNs use private virtual routing contexts so address conflicts are avoided, which allows the use of overlapping address space for different VPNs. Without MPLS P routers would need a mechanism for distinguishing among VPNs for packets arriving over the same inter-nodal links.
- MPLS, properly designed, makes it impossible for customers to receive misrouted traffic and provides a degree of protection against BGP flooding by eliminating customer routing information from the core of the network.

6.2.4   Route Reflectors

MPLS makes extensive use of BGP, but BGP is a full mesh protocol; each node establishes sessions with every other node. Once again we are back to the N^2 problem. There may be hundreds of PE routers in an MPLS domain, and thousands of CE routers. Any CE router that experiences a route update will inform the adjacent PE router via eBGP. In a full mesh iBGP design the PE router will update all other PEs. Route reflectors significantly reduce the amount of signaling traffic by acting as a hub or central repository for route updates. PE routers exchange route updates with the route reflector, where they are passed to affected PEs. Instead of an update requiring N^2 advertisements, we need N advertisements using a route reflector. Route reflectors can also be organized into clusters with PEs, either geographically or by address family, further managing signaling. Route reflectors can reside within a router, or function as a standalone network element. Because of their criticality, they should always be deployed in a manner that eliminates single points of failure.

6.3   Backbone Design and Engineering

Without question backbone design is a most demanding and potentially contentious undertaking for carrier engineering teams. In other parts of the network choices revolve around divergent technology paths (e.g., FTTH: Fiber to the Home vs. FTTN: Fiber to the Node) and suppliers, once a path is chosen. With backbones there are multiple design decisions that are interrelated. By its nature backbone design is iterative. Moreover, no design is more critical. All other services ride the backbone, so a failure there can have a massive effect on customer services, and with the largest carriers can propagate into other networks. Here are some of the decisions that go into backbone design.

6.3.1   Location and Size of POPs

Traffic and service analysis begin with an understanding of where demand originates and terminates. Communities of interest are identified by analyses of empirical flow data. Traffic includes transport, layer 2 and layer 3 time variant metrics broken down by service demand. Constraints are the counter to demand, limiting the solutions to a manageable number. Fiber routes, building space, and conditioning and operational suitability limit the alternatives to a figure that can be assessed on the bases of economics and schedule. Service demand and projected shift in services may weigh heavily on the solution. The classic tradeoff between switching and transport, coupled with estimated service demands can swing the results dramatically. For example, mobile networks designed in 2000 used TDM transport for voice service, and so TDM connectivity to Mobile Switching Centers (MSCs) was part of the total demand when assessing backbone networks. In 2012 that demand is moving to VoIP and the traffic is multimedia, not just voice. In a short time MSCs will no longer be part of the service hierarchy; data centers which are far fewer and centralized are the destination of traffic originated at the cell sites for LTE.

6.3.2   Fault Recovery

Fault recovery has to be considered early in the design process. The reason is that it can be managed at different layers of the OSI stack, and the choice affects design choices for traffic management and topology. Applications such as IPTV can only tolerate interruptions in the tens of milliseconds. Applications with such stringent requirements may require dual feed networks at the application or session layer. In the case of IPTV it means multicasting over two separate IP networks and relying on the application to monitor the feeds and choose the best one. That choice affects not only the application and IP layer, it affects the optical layer. Timers used to detect and switch at the optical layer may need to be extended to prevent an oscillating recovery between layers.

VoIP can generally tolerate interruptions of a few seconds. Longer interruptions can cause registration storms and lead to network congestion or even prolonged node outages. VoIP data centers, home to soft switches, and feature servers, can scale to handle far greater volumes of traffic than conventional TDM switching systems. VoIP sites are usually geographically diverse and redundant, but failovers usually require reregistration. With the adoption of VoIP as part of 4G mobility, sites supporting 20M subscribers or even much higher will be common. A backbone transport failure that is resolved via IGP reconvergence can result in a VoIP outage of 10–20 seconds, long enough to force reregistration. The resulting registration storm at the alternate site will increase traffic by an order of magnitude, possibly disabling the site. A preferred solution is to use MPLS-TE on the backbone routes serving VoIP centers. It scales by using a mesh DWDM/MPLS/IP infrastructure and relies on fast reroute for sub-1 second reroute on failure, which is well below the threshold of VoIP registration timers.

There is a gap in standards and practice in the realms of application and network design. That gap is the lack of a clear statement of fault and recovery objectives for both networks and applications. Redundancy is the underpinning of global networks, so by design the overwhelming majority of faults experience a short service interruption while the redundant system assumes the load. Examining these events from the network viewpoint we find a bimodal distribution with a few outages of duration of 10 minutes or longer and thousands of outages of 10 seconds or less. Enterprise telecommunications managers see it quite differently. Interruptions that last more than a few seconds cause their applications to fail, sometimes initiating a complete restart that lasts tens of minutes or even an hour. The lack of an explicit industry objective for automatic recovery in carrier networks via redundancy has not served network operators or applications developers well. Application developers with little experience in networks use SONET recovery times, sub 1 s, as the expected interruption time when designing their systems. However, IP standards and design practices never adopted SONET's stringent requirements, and did not adopt a clear objective for recovery at scale in any case. From a practical viewpoint, the burden of establishing a standard and communicating via service level agreements falls on the carrier. Carriers should be explicit in setting the objective and then communicating it to enterprise customers and to carrier partners, where services are carried across a boundary. Enterprise telecommunications managers should set their application recovery timers above the advertised standard to ride over short outages and prevent unnecessary churn in their networks and applications.

If 1 s is used as an objective for short duration failures, fault recognition, and recovery designs can be standardized for backbone, regional, metro, and access networks. Legacy carriers can use SONET in metro and possibly some regional networks. MPLS-TE works well for carriers of scale in the backbone. Regional networks are likely to have a mixed approach. In addition to SONET, point to point DWDM with 1 : 1 or other fault recovery designs at the optical layer can be used. MPLS-TE can also be extended into regional centers and even metros if traffic volumes and service mixes warrant it. The need to extend the service standard into access networks can be taken on a case basis. Legacy Digital Subscriber Line (DSL) access networks generally do not need sub 1 second recovery, but mobility access and aggregation networks do.

Scaled recovery escalation is a design principle implemented decades ago in stored program control systems. As network designers move up the OSI stack in each of the networks (backbone, metro, and regional) an explicit time-based escalation protocol should be used. As an example, if a regional MPLS LSP travels over a SONET network for a portion of the route, MPLS Fast Reroute (FRR) timers should be set to allow SONET an opportunity to reconfigure before FRR is invoked. That can be accomplished by specifying how many successive Resource Reservation Protocol (RSVP) hello messages must fail and the frequency of the messages [10].

6.3.3   Quality of Service QoS

While these design topics are addressed serially, they are interconnected in complex ways. Backbone design is necessarily an iterative process. QoS is closely linked to fault recovery and traffic analysis. Historically networks enforce QoS at the edge of the network, on ingress and egress. Today IP backbone networks are generously engineered, even in the face of all single fault link and node failures. ATM networks in the 1990s divided traffic into priority classes such as constant bit rate (CBR), variable bit rate (VBR), available bit rate (ABR), and unspecified bit rate (UBR). Complex queuing, admission control, and traffic management algorithms were designed and implemented. Yet my experience with ATM network design and engineering is that there was never a condition in which UBR traffic was discarded and other classes survived. Network engineers always engineered the network well above the measured and projected demand. During many failure conditions traffic was either rerouted and all grades of traffic experienced the same interruption, or traffic was completely cutoff in a significant outage. The complex algorithms implemented by the systems engineers simply had no effect because the network engineers were instructed to engineer the ATM network in the same way they engineered the voice network; insure there is always sufficient capacity.

However, past may not be prolog. Backbone traffic in carrier networks is growing at rates of 30–40% each year but revenues are not growing, at least not at anything approaching that rate. The common practice of engineering failover capacity at 100% of carried traffic may be replaced by a more discerning set of design principles to protect higher priority, revenue producing traffic. Most routes in carrier networks are protected on a 1:1 basis, either through SONET, MPLS, IP reroute, or other means. Since the majority of traffic on most carrier networks is Internet traffic, engineering backbones to protect VPN, VoIP, and other sensitive revenue producing traffic on a priority basis should yield significant savings. It does require an investment in systems engineering talent and tools to achieve the necessary sophistication, but the rewards in capital savings can more than offset the costs. Carriers choosing to use QoS in engineering backbone failure scenarios are likely to take a different approach than those engineering for 100% failover capacity. Optical and metallic transport fault detection and reroute solutions are of limited use in a QoS design since those layers have no knowledge of the priority of the traffic. Segregating traffic according to QoS at those layers forfeits some of the value of statistical multiplexing. There are mechanisms at the IP and MPLS layers that better lend themselves to failover designs that discriminate based on QoS.

6.3.4   Traffic Demand

A-Z traffic demand matrices with growth projections are the starting point for demand analysis, so having robust data collection and an archive in place are a prerequisite. Developing and deploying technology for measuring traffic demand accurately and rigorously in modern backbone and aggregation networks is complex; the value of the data can only be realized by applying analytics constructed by engineers well versed in network design and traffic theory [11]. Demand matrices grow in complexity as different layers of demand, transport, link, IP, and services, are considered and affect some layers of the backbone but not others. If QoS engineering is undertaken as described previously, the problem is expanded by an additional dimension. Demand matrices are normally parameterized, from best to worst case, possibly with variants, and used to evaluate the cost and efficiencies of alternative designs.

6.3.5   Control Plane

Some aspects of control plane design have been touched upon in other sections. Regional separation and segregation of private and public domains were described as well as the use of BGP route reflectors for networks of scale. A thorough analysis of control plane design is beyond the scope of this text and is ably treated elsewhere [12]; however, it is important to understand that control plane designs have direct implications for backbone bearer path design and topology.

Within the IP backbone there are hundreds of routers relying on OSPF for backbone routing updates. Unlike BGP which is session based, OSPF updates are performed on a hop-by-hop basis using LSA. Within the backbone are a relatively small number of very large scale P routers, surrounded by hundreds of much smaller edge routers. OSPF requires each router to compute a shortest path first tree (SPF Tree) to every reachable destination in the OSPF domain and to maintain routing and forwarding information in a LSDB. Changes in the network, such as loss of a link at an edge router cause the two affected routers to publish a connectivity update via an LSA to their adjacent routers, updating their LSDBs. That processes, called LSA flooding, continues until all of the routers in the OSPF domain have received the updates and recomputed the SPF Tree. For networks of scale, that means P routers could spend a great deal of time recomputing routes to relatively unimportant edge routers. Recall bearer traffic reaches edge routers via MPLS, not IP routing.

Fortunately OSPF supports segregation within the OSPF domain. OSPF Areas can be created, with Area 0 at the center, surrounded by adjacent Areas. Area identifiers are 32 bit unsigned values, referenced as either an integer (Area 0) or in an IPv4 format (Area 0.0.0.0). When the state of connectivity changes within an OSPF Area, LSA flooding occurs only within the affected Area. Routers serving as bridges between Areas, called OSPF Area Border Routers (ABRs), summarize and share routing information, dramatically reducing the effects of LSA flooding. Bearer traffic passes through ABRs, with some exceptions, and so the use of OSPF Areas enforces a hierarchy on bearer routing design. Non Area 0 routers generally have to send traffic through Area 0 to reach destinations outside of their Area.

6.4 Summary

Carrier networks rely on optical systems to deliver reliable transport on a global scale. These optical networks are organized in a three tier hierarchy, the backbone, regional networks, and metropolitan networks. Metro networks are largely SONET/SDH ring designs; regional and core networks are DWDM mesh networks using optical add drop multiplexers. Above the optical layer the backbone uses MPLS and IP protocols to deliver IP and VLAN services to a wide range of aggregation and access networks. Control plane design for the backbones is complex using a combination of internal and external gateway protocols to achieve the right degree of coupling between the backbone and connecting networks. The backbone itself is divided into different link state areas to achieve scale and ease management. Reroute on failure is accomplished in the backbone by MPLS-TE, FRR, and by optical bypass in the metropolitan networks. Regional networks use a combination of the technologies to achieve sub 1 second reroute. P routers at the center of the backbone provide scale and high speed LSPs; edge routers provide feature functionality including QoS, VPN, multicasting, and dual stack support for IPv4 and IPv6.

References

1. ANSI T1.105 (1996) *Synchronous Optical Network (SONET) – Sub-STS-1 Interface Rates and Formats Specification*, ANSI.
2. ITU Recommendation (1996) G.707 *Network Node Interface for the Synchronous Digital Hierarchy (SDH)*.
3. Telcordia (2007) GR-496. *SONETAdd-Drop Multiplexer (SONET ADM) Generic Criteria, Telcordia*, Issue 2, Aug 2007.

4. Willis, P. (ed.) (2001) *Carrier-Scale IP Networks: Designing and Operating Internet Networks*, Institution of Engineering and Technology, BT EXACT Technologies.

5. Rosen, E. and Rekhter, Y. (2009) IETF RFC 2547, BGP/MPLS VPNs, March 2009.

6. Pepelnjak, I. and Guichard, J. (2001) *MPLS and VPN Architectures*, Cisco Press.

7. Lakshman, U. and Lobo, L. (2006) *MPLS Traffic Engineering*, Cisco Press.

8. Moy, J. (1998) IETF RFC 2328. *OSPF Version 2*, April 1998.

9. Bates, T., Chandra, R., Katz, D., and Rekhter, Y. (1998) IETF RFC 2283. *Multiprotocol Extensions for BGP4*, February 1998.

10. Cisco (2006) MPLS Traffic Engineering (TE) – Fast Reroute (FRR) Link and Node Protection, Cisco IOS 12.0 S Release Notes, December 19, 2006.

11. Crovella, M. and Balachander, K. (2006) *Internet Measurement: Infrastructure, Traffic and Applications*, John Wiley & Sons, Inc., New York.

12. Kalmanek, C.R., Misra, S., and Richard Yang, Y. (eds) (2010) *Guide to Reliable Internet Services and Applications*, Springer-Verlag, London.

7

Cloud Services

The move from Time Division Multiplex (TDM) to Internet Protocol (IP) infrastructure has been well underway for over a decade. In the mid-twentieth century data transport was seen as an ancillary service. As data traffic volumes grew and TDM volumes flattened or even declined near the end of the century, engineers set as their goal the repurposing of the TDM network to better carry data. In the longer view of network evolution those efforts, ISDN (Integrated Services Digital Network), ATM (Asynchronous Transfer Mode), and Frame Relay provided bridges between voice and data networks and set the stage for the emergence of IP networks. Far from an ancillary service, data transport in the form of IP is the de facto base upon which services are now built. With the concept of cloud, IP enabled devices are free to move anywhere, use any access and draw services from data centers connected to global backbones. This notion of cloud services breaks the dependencies of location, provider, and technology. In this chapter several cloud services are examined, along with the specific engineering, and operations challenges they pose.

7.1 Competition

Carriers still battle among each other for services and customers, but the coming war is really between carriers and over-the-top (OTT) service providers, aided by smart Customer Premises Equipment (CPE). The number of truly competitive global carriers with a tier 1 designation and offering a suite of IP services globally is shrinking. Here are some examples of how OTT companies compete with carriers for these services. The examples[1] are by no means exhaustive.

- **Voice over Internet Protocol (VoIP)** – Vonage®, Magic Jack®, Skype™, and Google Talk™ compete for the consumer; Cisco®, Avaya™, and ShoreTel™ among others compete for the enterprise.
- **Video services** – Netflix™, Amazon Prime™, Hulu™, Apple TV®, Roku™, and SlingBox® compete for the consumer; Bloomberg™, CNBC™, and the Wall Street Journal™ compete for the enterprise.

[1] The names in these examples are registered trademarks and their reference is not an endorsement or comment on their service.

Global Networks: Engineering, Operations and Design, First Edition. G. Keith Cambron.
© 2013 John Wiley & Sons, Ltd. Published 2013 by John Wiley & Sons, Ltd.

- **Conferencing and collaboration services** – Google®, Yahoo!®, and Skype™ compete for the consumer; WebEx™, GoToMeeting®, and Adobe Connect™ compete for the enterprise.
- **Content delivery networks** – Akamai™, Limelight Networks™, Rackspace™, and Edge-Cast Networks™ are a few of the providers.

In designing IP services carriers must identify and enable network capabilities that differentiate their service from OTT offerings, or have an economic advantage by virtue of network scale and connectivity. Without a network differentiator or advantage carriers run the risk of competing in an arena where they are at best another player; forays into non-competitive services siphon off resources and talent better applied to their core business.

7.2   Defining the Cloud

From the perspective of global carriers, clouds are defined by the architecture, the infrastructure, and network functionality that distinguish it from competitors and other services.

7.2.1   Architecture

Recall the general three layer model for carrier networks (see Figure 7.1).

Legacy services, like TDM voice and cable TV service, have a high degree of functionality located at customers' premises, in the access networks, or both. Cloud services generally have a lightweight IP client at customers' premises and applications and data repositories located in large data centers interconnected by backbone networks. Public cloud services are enabled by the Internet. In the enterprise private cloud services are enabled by enterprise Virtual Private Networks (VPNs) or federated VPNs that interconnect private enterprises and private cloud service providers. Consumer private cloud services are best defined by

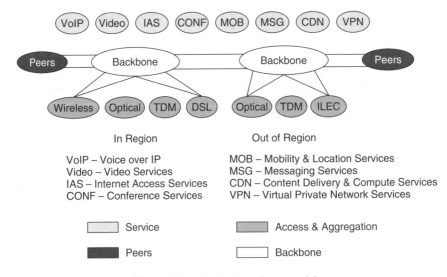

Figure 7.1   Carrier three layer model.

walled garden services, like America On Line (AOL) service that was so popular in the early 1990s. The best examples of successful consumer private cloud (walled garden) services today are the application libraries enabled by Apple and Google on their respective OS's (Operating system), iOS and Android. But even there, consumers have unfettered access to the Internet. Cable broadband, FiOS, and U-Verse Internet Protocol Television (IPTV) all offer a private video and phone service alongside open access to the Internet and its public cloud services. In essence, service providers are enabling their competitors. Netflix, Amazon and Hulu compete with IPTV and cable TV offerings. Skype and Google Talk compete with the carriers' phone services.

Cloud service is not a take all or take nothing proposition. Voice service is moving to the cloud but most enterprises and consumers will have a mix of both for a long time. Cable TV, IPTV, and satellite TV services are being challenged by Internet TV but the legacy services are likely to survive and possibly prosper, although they will have to adapt. Compute and storage as a service have made real inroads as an alternative to premises mainframes and servers, but enterprises retain critical and sensitive services within their own infrastructure. Private clouds are attractive to enterprises seeking to take advantage of the economics and ubiquity of cloud and retain a high level of security and control.

Much has been written about the advantages of cloud services. The list that follows summaries some of those attributes.

- **Ubiquity** – once an application is hosted in the cloud it is available anywhere there is IP connectivity.
- **Device agnostic** – cloud services work through thin clients, either Internet browsers or applications that can be ported to PCs, smart phones, and tablets.
- **Centralized installation and maintenance** – desktop and site administration are dramatically reduced.
- **Security** – data loaded on laptops and mobile devices are exposed to theft and malware introduced outside enterprise firewalls. Clouds protect data through multiple layers of security.
- **Economics** – for many applications, pay-as-you-go is more economic than fixed costs of acquisition and recurring costs of maintenance.
- **Standardization** – sites across an enterprise don't have an option to build or customize applications in ways that reduce commonality and frustrate interworking.
- **Interworking** – applications built in clouds have a far better chance of collaborating than applications built on a site distributed model. Applications work within the confines of a data center, simplifying communication, security, and administration.

7.2.2   Infrastructure

Success in deploying cloud services depends on scale and flexibility. Global carriers achieve both by moving compute and applications into, or as close as practical to the network backbone. Greater compute scale is achieved by concentrating computing into as few locations as possible. That has to be balanced against incurring unacceptable latency and increased risk of site failure or isolation. Effective bandwidth is limited by:

- **Speed of the physical transport facilities** – since backbones have very high bandwidth, end-to-end speed is limited by access networks and round-trip time (RTT).

- **Processing delay of servers and applications** – carriers provide content delivery services that compete with companies like Akamai and Limelight.
- **RTT** – caused by the finite delay of the propagation of light and electrons over the distance between the consumer and the provider of content. This is more important than some engineers recognize. RTT is limited by the design of cloud and the extent of dispersion carriers build into cloud infrastructure.

Carriers reduce RTT by moving cloud data centers closer to demand and by optimizing routing. Collateral benefits are a reduction of backbone traffic and increased protection against site failures or overloads. Carriers place applications and services in the following facilities:

- **IP/MPLS (Multiprotocol Label Switching) backbone points of presence (POPs)** – applications that consume high bandwidth need to be as close to high capacity transport as possible. Building data complexes directly in POPs is ideal but facilities may constrain the size of the center.
- **National Data Centers (NDCs)** – carriers design large data centers as close as practical to their backbone switching centers. High bandwidth diverse access with full site power and High Voltage Alternative Current (HVAC) protection are engineered into the facilities. The centers tend to be multi-purpose, housing wireless, wireline, enterprise and consumer services, and equipment.
- **Regional Data Centers (RDCs) and metro facility hubs** – incumbent wireline carriers often have a few hubs in regions and metropolitan areas that serve as the nexus of optical and IP networks. These are excellent locations for building data complexes that specialize in caching or serving applications on a dynamic basis.
- **Central offices (COs) and Mobile Telephone Switching Offices (MTSOs)** – while carriers have thousands of these locations they are less desirable locations for most cloud services. Many are unmanned during most hours and local craft may lack the skill mix needed for cloud data services. Exceptions are COs in metropolitan centers that are information intensive. Manhattan COs are logical sites for financial applications where volumes are high and RTT matters. There are also specific consumer services, like IPTV and gaming that can benefit from being distributed down to the CO. MTSOs serve some mobility cloud functions because they have a high degree of connectivity to the Radio Access Network (RAN).

7.2.3 Intelligent Networks and Intelligent Clouds

The term intelligent network is well worn but has new meaning when applied to cloud services. Apart from their extensive infrastructure, carriers have networks that can turbo charge cloud applications, improving the experience of cloud services, reducing costs, and tightening security. Some of the advantages described here have not been fully exploited by carriers, largely because of a lack of market focus and integration. However, that should change in the next few years.

The essence of the carrier advantage is that their global IP networks are predictable and adaptable. The Internet is neither. Internet traffic is not differentiated and traffic surges are only limited by the size of the access networks they travel. In the case of Distributed Denial of Service (DDoS) that is no limitation. Hackers use botnets to generate DDoS attacks and recent history has shown how vulnerable public clouds can be to those attacks. To exploit their technological advantages carriers must make networks more aware of applications and enable them to respond to changes in cloud demand and capacity in real time. The complement of that proposition is that carriers must publish in real time the state of networks allowing cloud applications to respond to network conditions. Technologies addressed in the following describe some of the capabilities that are readily available to carriers today.

7.2.3.1   Intelligent DNS

There is no single industry accepted definition of Intelligent Domain Name System (DNS); the term is used for a number of technologies that affect routing, caching, security, traffic analysis, or load balancing. Figure 7.2 illustrates how DNS servers affect routing.

In Figure 7.2 the two lower clouds represent two different Internet Service Providers (ISPs) in different geographic regions. Subscribers served by CE#1 enter domain name or subdomain name Uniform Resource Identifiers (URIs) into their browsers which are

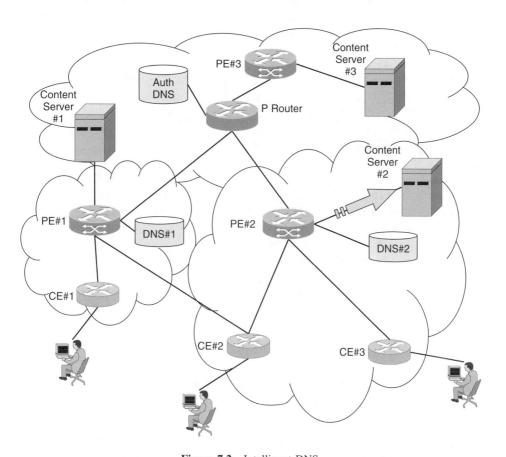

Figure 7.2   Intelligent DNS.

forwarded to DNS#1 where a record is retrieved that is resolved to an IP address by the host. Subscribers served by CE#2 and CE#3 are served by DNS#2 where their URIs are forwarded and return records resolved to an IP address that may be different from records returned from DNS#1. Content services can use DNS to direct subscribers in different geographies to different caching or content servers. In the example content server #1 serves DNS domain #1 and content server #2 serves DNS domain #2 for a particular web site. Content providers often assign IP addresses based on the location of the domain name server. In the case of a subscriber served by CE#2 they may have a more direct route to content server #1, but if they are using DNS#2 for address resolution they will be directed to content server #2. Assuming Border Gateway Protocol (BGP) routing weights the Application Server (AS) trees appropriately, the chosen path would be through PE#2. Specific records are populated by authoritative name servers that map domain names to IP addresses. When a host request is to an ISP DNS for an unknown URI, it contacts the authoritative name server to request a record. The authoritative name server can send different records based on the source address of the ISP DNS. That record is then cached by the ISP DNS for a time specified in a time to live parameter populated by the authoritative name server.

There's an effort in the industry named the Global Internet Speedup Initiative to use the address of the host originally making a DNS request rather than the address of intermediates such as DNS#1 and DNS#2 to determine the location of the subscriber [1]. DNS queries for domain names managed under this regime are forwarded by intermediate servers DNS#1 and DNS#2 to an authoritative DNS, as shown in the diagram. The forwarded query includes the high order bits of the IP address of the requesting host. The authoritative name server provides a response based on the partial host address rather than the addresses of the forwarding DNS servers. Partial addresses are forwarded to maintain a degree of anonymity. Only participating content sites are affected by the service. The great majority of sites will continue to route using single address entries.

While the speedup initiative is an important step forward, carriers are in a position to do much more. Consider the case where the link from PE#2 to content server #2 is congested, represented by an arrow in the diagram, and latency is 10 times the normal delay. If that information were available in near real time the authoritative name server could begin code gapping by sending two out of three requests normally destined for content server #2 to content server #3. The time to live of these records could be reduced so that as soon as congestion is relieved traffic could be restored to content server #2.

Using an authoritative DNS for near real time events is problematical since popular sites are likely to see more traffic coming from hosts with cached addresses rather than ones coming from hosts querying a name server. Using multiple IP addresses for a given URI is also expensive in an IPv4 constrained world. A more effective solution is to manage routing, not address resolution.

7.2.3.2   Intelligent Route Service Control Point (IRSCP)

Routing in the Public Switched Telephone Network (PSTN) is managed centrally in real time using a Service Control Point (SCP) for Toll Free Service and other services as described in Chapter 5. An SCP acts as a real time routing database that instructs switching systems how to route traffic for supported services. SCPs are a central point of control in a network of switches with individual forwarding databases. Intelligent Route Service Control Point

(IRSCP) was invented as an IP corollary of a PSTN SCP [2]. IRSCP is an outgrowth of an approach based on a Route Control Platform (RCP) introduced in 2004 [3] which explored the idea of an intelligent route reflector. IRSCP's elegant design does not require any changes to existing routers in either carrier networks, peer networks, or subtending networks. It enables centralized control through BGP, inserting itself between external and internal routers. As the central exchange of all routing information it not only has a complete view of the network, it is able to use more complete information and exogenous criteria to build more sophisticated forwarding rules. Using the control plane in an opaque way is superior to other approaches using the management plane. Router management interfaces are varied and some are obtuse. BGP is uniform, well understood, and timely. Figure 7.3 is based on the previous example, but it substitutes an IRSCP solution for one using an Intelligent DNS.

The design is cleaner than the DNS solution although in practice multiple IRSCPs are used for scale and they are deployed in pairs for fault tolerance. Not shown on the diagram are the relationships between the IRSCP and the routers. The IRSCP is an exterior border gateway protocol (eBGP) peer for each CE router and an interior border gateway protocol (iBGP) peer for each Provider Edge (PE) router, and so is the only element in the network with a complete view of network routing.

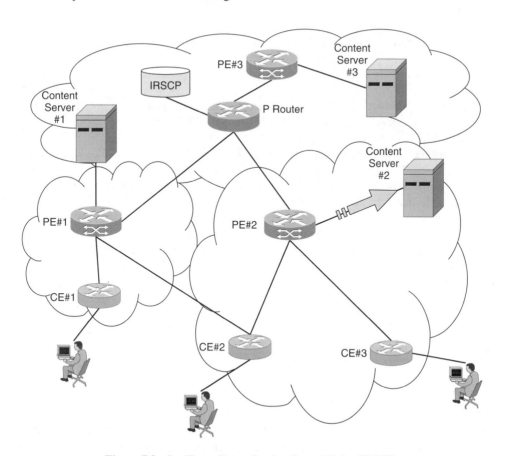

Figure 7.3   Intelligent Route Service Control Point (IRSCP).

The IRSCP has three elements:

- a BGP route reflector that implements eBGP and iBGP,
- a rules engine built on top of BGP rules. The engine conforms to BGP but allows operators to build rule sets that distinguish between traffic flows and can engineer network paths better suited to those flows and to the current conditions in the network,
- a set of external interfaces that can be used to sense changes in the network, changes in services clusters or to communicate changes in network routing.

Advantages of an IRSCP network over other approaches like an intelligent DNS are:

- **Fine grained routing control** – each router receives routing tables designed specifically for it.
- **Routing control is in near-real-time** – updates can be published through BGP and take effect immediately.
- **Flexibility** – the rules engine can respond to a wide range of network and application conditions.
- **Delegation and coupling** – IRSCP doesn't mix routing with address resolution or other non-routing functions.

A few applications that are possible with IRSCP routing control are:

- **Load balancing** – traffic load can be spread evenly and predictably across different points of egress.
- **Site failover protection** – during periods where a site is having a failure of a particular application or when maintenance is needed the IRSCP can be directed to shift traffic to available sites.
- **Network congestion management** – using external data reflecting the congestion on egress links, IRSCP can adjust traffic to more lightly loaded sites.
- **DDoS protection** – if a DDoS attack is detected IRSCP can redirect traffic to scrubbing facilities in real time and meter the amount of traffic that is redirected.

In the congestion example diagramed above the IRSCP has a flexible set of responses. It can direct traffic from CE#2 to content server #1 via PE#1 and traffic from CE#3 to content server #3 via PE#2. In effect the IRSCP can route traffic that have the same IP destination on different paths by publishing availability through BGP. In this solution IPv4 anycast addressing can be used, assigning the same IP address to multiple servers. Anycast is used in other services, such as authoritative name server addressing. However, using it with the IRSCP removes some of the unpredictability of using anycast and relying on individual routers with a limited view of the network to make routing choices. The IRSCP can control route advertisements and so insure that only the desired destination is advertised.

7.2.4   Internet Protocol Multimedia Subsystem (IMS)

The objectives of the Third Generation Partnership Project (3GPP) Internet Protocol Multimedia Subsystem (IMS) initiative are far reaching. Convergence is the central theme of IMS.

- **Converged networks** – migrate the core networks of the PSTN and the Public Land Mobile Network (PLMN) to a single converged IMS core.
- **Converged services** – services work in the same way across all networks and interwork with each other in the same ways.
- **Device agnostic** – users of IMS capable devices find that services work the same way on tablets, phones, and PCs, and they interwork with each other.
- **IP enabled** – not strictly an objective but rather a condition. All communication and services are IP based.
- **Legacy interworking** – IMS should interwork with legacy networks and deliver at least the quality of service (QoS) of those networks.
- **Application abstraction** – applications and services can be added without incurring network or systems design changes.

In a broad sense 3GPP is trying to achieve with IMS what IETF achieved with IP/MPLS standards, except IMS operates at the session, presentation, and application layers of the Open Systems Interconnection (OSI) stack. IMS uses IETF protocols but focuses on architectures, functions, and interfaces. This section describes IMS at a high level, but it is a most complex and ambitious architecture that is treated in depth in other texts [4].

As shown in Figure 7.4 IMS is organized into three layers, the user plane, the control plane, and the application plane.

In the diagram Long Term Evolution (LTE) is used to illustrate the user plane, but IMS is intended to be access agnostic. It is meant to enable multimedia IP services for consumer wireline broadband, such as Fiber to the Home (FTTH) or Packet Cable and enterprise VPNs equally well.

The descriptions here are a great simplification but convey primary IMS elements and their functions. IMS standards break architectural elements down into more atomic functions than previous standards. In taking that approach the number of interfaces and functional elements expands dramatically. As network systems developers set about implementing the standards they have to make choices about which functions to aggregate into network elements. Those decisions are driven by costs, technology, and the application of sound hardware and software engineering. Suppliers may choose to draw those boundaries differently; designing network elements composed of differing functions. If that occurs it is likely to eliminate the possibility of supplier interworking within the IMS core. Databases and media gateways (MGWs) are likely to remain interoperable since backward compatibility is a goal of IMS, but the core of the control plane is likely to be supplied by a single supplier within a regional deployment. The number of interfaces also dramatically increases the difficulty of integration for the supplier and the carrier. Each formalized interface has to be tested for sunny day and rainy day cases. Formalizing internal interfaces takes design latitude away from developers and puts it in the hands of architects who don't have the benefit of understanding all of the practical constraints of design. For example, if an internal interface is formally defined between two functions as a communications protocol, it removes the design choice of a shared database for information exchange rather than a communications protocol.

The majority of IMS elements and functionality are in the control plane but there is usually a bearer plane counterpart that interworks with and executes the functions of its control plane mate. The sections that follow are grouped according to function. Functional placement in the diagram does not imply the elements are physically collocated.

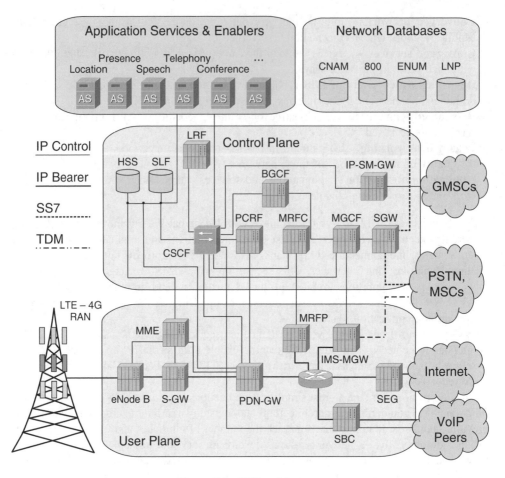

Figure 7.4   IMS architecture.

7.2.4.1   Databases

There are two databases of significance at the architectural level, the (Home Subscriber Server) (HSS) and Subscription Locator Function (SLF).

Home Subscriber Server (HSS)

Recall the HSS subsumed the responsibilities of the Global System for Mobile Communications (GSM) Home Location Register (HLR) with the emergence of LTE and added specific functions to support IMS. The HSS is the repository of all subscriber information for authentication, authorization, and services for IMS, the evolved packet core, the legacy packet core, and circuit switched services. The interfaces and functions provided for legacy 3G and 2G network elements are as described in Chapter 5. For IMS services HSS interworks primarily with the Call Session Control Function (CSCF) to authenticate and authorize Session Initiation Protocol (SIP) based sessions and services.

Subscription Locator Function (SLF)

The SLF acts as a point of redirection, steering the CSCF, and other IMS elements to the correct HSS. A sub-function of the CSCF, the Interrogating Call Session Control Function (I-CSCF) receives incoming requests for connections to a subscriber, based on a public identity. IMS defines public identities as SIP URIs (similar to an email address) and tel Uniform Resource Locators (URLs) (similar to an E.164 telephone number). The SLF returns the identity of the serving HSS.

7.2.4.2   Session Control

The CSCF is responsible for SIP session management. In VoIP softswitch terminology the CSCF performs many of the same functions as a Call Agent or Session Controller. IMS standards break session management into four sub-functions. In the diagram above all of the sub-functions are represented by the CSCF. The PDN – GW (Packet Data Network) is the bearer plane element that enables routing, QoS, and services controlled by the CSCF.

Proxy Call Session Control Function (P-CSCF)

Acting as a proxy between the User Equipment (UE) and the carrier network, the Proxy Call Session Control Function (P-CSCF)

- manages security between the UE and the network,
- compresses SIP messages to conserve RAN bandwidth and decompresses them for the network,
- facilitates QoS and charging control by interworking with the Policy Charging Rules Function (PCRF) and PDN-GW,
- forwards registration requests to the Serving Call Session Control Function (S-CSCF).

Application Level Gateway (ALG) functionality can be included within the P-CSCF control plane functions.

Serving Call Session Control Function (S-CSCF)

At the center of routing control and service management, the S-CSCF performs the following functions.

- accepts registration requests from UEs and authenticates the subscriber with the HSS,
- temporarily stores subscriber and service profiles received from the HSS in a fashion similar to an Serving General Packet Radio Services Node (SGSN),
- coordinates services with authorized application servers based on the user's service profile,
- converts Mobile Subscriber Integrated Digital Subscriber Numbers (MSISDNs) to SIP tel URLs for routing by consulting E.164 Number Mapping (ENUM),
- working with the application servers determines how to route the session. The session may be completed via the packet core or forwarded to the Breakout Gateway Control Function (BGCF) for completion in the 2G/3G circuit switched core,
- forwards charging information to the Charging Data Function (the CDF is not shown in Figure 7.4).

Interrogating Call Session Control Function (I-CSCF)

Just as the P-CSCF acts as an intermediary for UEs accessing the carrier network, the I-CSCF acts as an intermediary between the carrier network and other networks. It performs the following functions:

- acts as the external gateway for incoming SIP session requests (INVITE messages),
- locates the P-CSCF hosting the called UE and the next hop S-CSCF by querying the HSS,
- forwards the incoming request to the S-CSCF,
- selects, via the HSS, the assigned terminating S-CSCF for incoming calls from the PSTN/PLMN signaling gateways (SGWs).

Emergency Call Session Control Function (E-CSCF)

The Emergency Call Session Control Function (E-CSCF) is a Public Service Access Point (PSAP) routing function for emergency calls, acting much like the 911 tandem offices deployed in the PSTN.

7.2.4.3 Media Management

SIP services, like legacy TDM voice services, use a range of audio announcements. However, SIP services go much further in IMS. Video conferencing, audio response services, transcoding services, tone decoding, and collaborative multimedia services can be provided as separate offerings or within the context of a larger service. The Media Resource Function Controller (MRFC) operates under the direction of the CSCF to arrange bearer path connections to the right media at the right time. The MRFC instructs the Media Resource Function Processor (another MRFP) to establish those paths to Application Servers, media servers or other sources with the required quality.

7.2.4.4 Interworking

Any new network technology that replicates legacy services faces the same challenge; it has to replicate and interwork with legacy networks so that existing services work across both networks with the same quality and functionality experienced in the legacy network. 3GPP defined IMS interworking functions to enable Voice Over Long Term Evolution (VoLTE) to interwork with 2G/3G voice and messaging services and with PSTN voice services. Service interworking is managed by four functions, the BGCF, the Media Gateway Control Function (MGCF), the IMS-MGW, and the SGW.

Circuit Switched Calls

Incoming calls arrive at the SGW from the PSTN or Mobile Switching Centers (MSCs). Common channel signaling, SS7, is used by circuit switched networks; SIP signaling is used in the IMS infrastructure. The SGW is the signaling interworking facility, translating from and to SIP and SS7. The first notification of an incoming call to IMS is an SS7 Initial Address Message (IAM) received at the SGW. On receipt of an IAM the SGW notifies the MGCF which instructs the IMS-MGW to establish a circuit switched connection with the

Circuit Identification Code (CIC) in the IAM. The MGCF translates the IAM into a SIP INVITE message and forwards it to the I-CSCF. The I-CSCF queries the HSS to determine the assigned S-CSCF. The S-CSCF then queries ENUM to translate the E.164 address to a URI, which DNS uses to map to a destination IP address. The S-CSCF signals the P-CSCF to contact the UE and sets up an Real-time Transport Protocol (RTP) bearer path between the UE and the MGW.

Outgoing calls begin when a UE dials a called number that terminates in the PSTN or mobility circuit switched core; the request is forwarded to the CSCF serving the UE. ENUM is queried by the CSCF and when no IP address is returned the call is marked for completion as a circuit switched call. To determine where in the circuit switched network to forward the call the CSCF forwards the SIP INVITE request to the BGCF. The initial BGCF may choose to forward the request to another BGCF in a different network for media gateway selection. The responsible BGCF sends a request for completion to the MGCF that is the best choice for completion to the circuit switched network. In turn the MGCF reserves a trunk group member on the appropriate IMS-MGW and instructs the SGW to send an IAM with the called number in the SIP INVITE to the PSTN switching system or MSC. On receipt of an SS7 address complete message (ACM) a call path is reserved between the UE and the called party in the circuit switched network.

Emergency Calls

When the outgoing call is an emergency call an additional step is needed. When the CSCF receives the call, 911 as an example, it has to determine the correct PSAP designated to serve the call for the UE's current location. A location request is sent to the Location Retrieval Function (LRF) which obtains the UE's location from the RAN and sends a response to the CSCF containing the location and the PSAP address. The CSCF completes the call to the PSAP as described above for outgoing calls, marking it as an emergency call. PSAPs may also have connections to the LRF so they can independently query the node for location information and tracking.

VoIP Calls and Session Border Controllers

VoIP calls invoke the CSCF as described in the sections above. At the IMS edge, however, a Session Border Controller (SBC) is used to peer with other VoIP networks and protect the IMS core from untrusted networks. IMS standards define two internal functions in an SBC, they are IBCF and TrGW. The Interconnection Border Control Function (IBCF) is a control plane function and the Transition Gateway (TrGW)[2] is a bearer plane function. In practice, the SBC may perform the following functions:

- Resolve protocol differences such as IPv4 to IPv6 translations and between incompatible implementations of SIP and/or RTP.
- Provide a Network Address Translation (NAT) function that also includes needed VoIP-layer adjustments.
- Provide security from DDoS attacks (via intelligent throttling) and from attempts to hack through to the IMS core (via back-to-back user agents).

---

[2] The TrGW is also called the Border Gateway Function (BGF) in some standards.

- Provide UE registration caching such that all UE registrations need not propagate to the HSS.
- Host the P-CSCF.
- Effect transcoding in the media layer if necessary.

In Figure 7.4 the more general case is shown, an SBC that incorporates the IBCF and TrGW functions. The SBC may interconnect through the Security Gateway (SEG), or it may have SEG functions built in.

Short Messaging Service (SMS)

Messages originated by the UE are sent to the CSCF using SIP for completion. The CSCF forwards the messages to the IP Short Message Gateway (IP-MS-GW) where a check is performed to verify the subscriber is authorized to use the service. The IP-MS-GW formats an SS7 Mobile Application Part (MAP) message and sends it to the Short Messaging Service-Gateway Mobile Switching Center (SMS-GMSC) for completion.

Incoming messages arrive via the SMS-GMSC. The SMS-GMSC sends a MAP message requesting routing information to the IP-MS-GW. The IP-MS-GW queries the HSS/HLR for the location of the UE identified in the SMS address. Based on the HSS/HLR result a routing response is prepared and returned to the SMS-GMSC. At that point the SMS-GMSC forwards the MAP SMS message to the indicated IP-MS-GW. After verifying the subscriber is authorized for SMS completion a SIP SMS message is originated and forwarded to the serving CSCF for delivery to the UE.

7.2.4.5   Policy and Charging

The PCRF, is an IMS architecture component that has been deployed in advance of IMS in some cases to implement a forward looking policy infrastructure in Universal Mobile Telecommunications System (UMTS) suitable for LTE. It is described in some detail in Chapter 5 and has the same responsibilities in LTE as UMTS. The PCRF interworks with the Policy and Charging Enforcement Function (PCEF) which is either interconnected with or embedded in the PDN-GW. The PCRF sets admission and control rules on a per session, per service basis, and is on the control plane. The PCEF takes those session rules and applies them on the bearer plane.

7.2.4.6   Security

The SEG acts as an edge security router protecting the security zone defined by the operator for IMS. Operators may segregate their IMS security areas according to the ASs they operate in or other administrative criteria. All control packets entering or leaving the security zone pass through the SEG. Typically it implements firewalls, filters, and surveillance functions.

7.2.5   Application Servers and Enablers

ASs are not part of the IMS standard in that the services they perform and how their service logic is designed are left open to the carriers and third party developers. IMS does specify how ASs interwork with the IMS core. ASs and the CSCF both implement a half call model. One half of the call is the incoming leg of the call and the other half is the outgoing leg. When the CSCF invokes an AS the incoming leg at the CSCF is matched to the incoming leg at the AS and the two outgoing legs are matched in a similar fashion. In effect call control for both legs passes to the AS, assuming the service uses both call legs. In general ASs are not connected to the user plane. Bearer functions are delegated to the MRFP. When an AS needs to send a tone or announcement or set up a new call leg or bridge, it communicates the request to the CSCF which instructs the Media Resource Function Controller (MRFC) to set up the connections. The MRFC in turn selects the serving MRFP and instructs it to set up the media paths.

Application Servers can take several forms, SIP AS, OSA (Open Services Architecture), and IM-SSF (IP Multimedia Service Switching Function), each of which is described in the following.

7.2.5.1 SIP AS

SIP Application Servers use native SIP protocols to implement services, working with the CSCF. They can be SIP proxies, back to back SIP user agents, redirect servers, terminating user agents, or originating user agents initiating sessions based on application specific logic. Enablers are facilities with open Application Programming Interfaces (APIs) that ASs can invoke to gain functionality without having to build it from scratch. Text to speech is an example of an enabler. Services can add text to speech functionality by passing a text document to the enabler and receiving a wave file in response. The difference between an enabler and an AS is determined by the degree of conformance to the IMS requirement for AS interworking with the CSCF. Text to speech can be implemented as an AS if it conforms to the half-call model and interworks with the CSCF directly.

Of the three types of ASs described here, SIP ASs are meant to be the service workhorses. The prime example of a SIP AS is the Telephony Application Server (TAS). Multimedia telephony service is not enabled by the IMS infrastructure. It is enabled by the UE invoking applications in the TAS, through the IMS infrastructure [5].

7.2.5.2 Open Services Architecture (OSA)

OSA was developed by 3GPP and ETSI with third party developers in mind [6]. APIs are defined as OSA Service Capability Features (SCFs). Applications can be implemented in a language and on a platform of the third party's choosing. The applications integrate with IMS through APIs implemented by the OSA Framework and OSA Service Capability Servers (SCSs). Framework APIs implement common functions required of all applications, such as registration, authentication, and SCF discovery. SCSs APIs expose specific services made available by the carrier or third parties. Examples are location services, notification services, and call control.

7.2.5.3 IP Multimedia Service Switching Function (IM-SSF)

IM-SSF is a service interworking function that bridges services developed for 2G/3G mobile networks using SS7 Customized Applications for Mobile network Enhanced Logic

(CAMEL) into the IMS/LTE network. CAMEL services were developed in the CAMEL Service Creation Environment (SCE). IM-SSF implements SCE compatible APIs for inter-operability across mobility networks.

7.2.6   IMS Design and Engineering

IMS has made strides since it was first introduced into service in 2007, providing consumer VoIP service on AT&T's U-Verse platform. The network elements were largely designed by a single supplier, Alcatel-Lucent; the IMS system was designed by AT&T and Alcatel-Lucent. It does not fully conform to the 3GPP IMS standards that are being used in LTE deployment, but incorporates many of the functions and elements. As a new technology there are five areas, described in the following that need to mature before IMS can be placed in the same class as the TDM voice network. Fortunately IMS elements are designed by suppliers experienced in telecommunications systems and standards, and it is being deployed by seasoned engineering and operations groups in carriers around the globe. While the standards may lack some of the features on initial deployment, the carriers working with the suppliers will bring the systems up to the level of legacy TDM systems.

7.2.6.1   Scalability

Complexity and scale are usually enemies. The large number of network functions and interfaces specified in IMS increase the engineering complexity and operational risks when an in-service IMS system approaches engineered limits and needs to be augmented or upgraded. Every element, link and engineered resource has to be tracked and engineering rules have to be written and applied by network engineers. Suppliers or carriers have to organize IMS into subsystems that can be engineered as modules and deployed as integral units rather than try to augment IMS complexes element by element.

7.2.6.2   Fault Management

3GPP standards for IMS have sparse treatment of fault management. Stateless IP systems, like most Internet applications, tolerate failures well because transactions are atomic; if a system fails the client can simply be routed to the next available server with no loss of continuity. IMS has multiple stateful systems with the CSCF at the center. States have to be consistent and coordinated among the systems, making fault management, and recovery more complex. TDM voice systems are fully integrated with a substantial portion of the code devoted to state synchronization and session verification across processes, network, and periphery. To the extent complex stateful services are deployed using OSA, the need for session audits and fault management extends into those applications as well.

7.2.6.3   Network Management

Most SIP deployments of 1 million subscribers or more are subject to registration storms and other high volume events. OSA or SIP AS applications can also generate focused overloads if the developer fails to recognize a condition that causes sessions to become overly chatty, generating unexpected messaging loads on specific IMS elements. Automatic congestion controls need to be built into core and application procedures to protect systems against a cascade of events triggered by UEs or external events, such as high volume call-ins.

7.2.6.4 Feature Compatibility

When subscribers invoke a feature, such as adding a party to an existing session, the request has to be checked against a compatibility matrix to insure it is authorized and allowed *within the current context*. Understanding context and using it to test feature interaction is a difficult undertaking and it grows exponentially as the feature list increases. Consider the idea of adding an audio stream to an existing session. It makes sense if you're on a session with another person and you want to share a clip of audio. But if you're on a conference call and other audio sessions are running, it doesn't make sense. Context matters. Devices affect context as well. Subscribers have multiple devices with different capabilities and how those devices respond across features will vary. If a session has been established on one device type and moves into a different access network or switches to a different device, features and applications may need to behave differently.

7.2.6.5 Service Management

VoIP service by its nature is far more difficult to manage than TDM voice service. IMS, with its highly distributed design, adds complexity to service management. Consider some of these comparisons between TDM voice and IMS.

- **Session tracing** – TDM voice systems perform control and transport functions within the same system. Service can be traced by starting with a single physical circuit. An operator can find the connecting circuit and the details of the session (call) associated with the circuits in short order. In an IMS system session information exists in the CSCF, but other relevant session information can be in other elements. The session information contains URIs, but not necessarily complete information about the actual bearer path.
- **Independent call paths** – TDM transport is bidirectional and fully associated. The talk path and the listen path ride the same facilities throughout the connection. IP transport is unidirectional. Talk and receive packets often take different routes between two points. An operations technician can run a route trace toward a destination but it's more difficult to trace the path of an incoming stream. Latency or errors on a received stream are difficult to diagnose.
- **Variability** – TDM paths are synchronous. Calls may get blocked, but if they complete latency variations are low on a call by call basis. IP voice paths can be affected by network traffic and so vary call by call. Resource reservation and QoS in IMS should reduce that variability dramatically over Internet VoIP service, but VoIP calls that complete outside the IMS infrastructure can still be subject to IP variability.
- **Transport dependency** – TDM transport with circuit continuity checking has proven to be very reliable. Failures in transport quickly translate into carrier group failures recognized by switching systems and calls are routed around the fault. When a black hole event is experienced in MPLS or another part of IP transport, there are no comparable mechanisms for recognizing those failures and taking automatic recovery actions without

external intervention, either by craft or by a combination of probes and purpose build fault management systems.

Service management continues to improve for VoIP services through the experience of the operators and the application of specialized systems and software. The challenge is to continue the improvements to keep pace with the rapid deployment of IMS.

7.3   Cloud Services

Global carriers were in the cloud services business long before the term cloud became popular. In a broad sense cloud services are those which are moved away from dedicated physical assets and a subscriber ownership model to a shared virtualized model. VPNs began replacing dedicated enterprise networks well over a decade ago. Going back even further, Centrex was offered in the mid-twentieth century as an alternative to a premises Private Branch Exchange (PBX) service. Today carriers have a range of services, focused primarily on enterprise markets that build upon the network infrastructure described in the previous sections.

7.3.1   Network-Based Security

There are three essential security elements global carriers have that distinguish their network security services from third party or premises solutions:

- **Dedicated staff** – security centers are usually manned within or logically adjacent to Global Network Operations Centers (GNOCs). They actively monitor their global network for traffic anomalies and attacks using purpose built systems and tools.
- **Managed firewall service** – carriers actively track sites and port addresses that represent suspected threats and publish firewall updates frequently under managed service agreements. Having firewalls at the ingress edge of the network allows carriers to filter malicious traffic before it even enters the customers' networks; carriers are able to physically locate and block the source at the port ingress using detailed flow data rather than rely on a source IP address.
- **Denial of Service (DoS) scrubbing** – enterprise gateways can use local defenses like firewalls to protect against intrusion but have no real way of dealing with a DoS attack that overwhelms access links. Carriers have scrubbing facilities in the core of the network that can filter traffic to a site so that legitimate traffic is passed through and malicious traffic is discarded.

Global carriers with these capabilities have distinct advantages in the never ending battle to prevent security breaches and attacks. Hackers are software developers that usually perfect their designs by trying early versions on the Internet. Carrier security teams closely monitor global Internet traffic and detect signatures of prototype attacks early; they analyze the attacks and signatures to insure their defenses are adequate, or they need to be quickly enhanced to deal with a new threat.

7.3.2   Voice over IP (VoIP) Services

VoIP may well be the largest and most common cloud service that few think of as cloud. Wireline TDM and mobile voice prior to LTE are based on distributed voice switching systems but VoIP is much like any other cloud service and is often a component of cloud services such as meeting and conferencing services.

Carriers approach VoIP network design and evolution differently, largely because they are starting with different core networks and in some case, different business models. Implementation varies by which standards are adopted, the degree of centralization, and how open the carrier is to interconnection for alternative services and service providers.

7.3.2.1   Cable MSOs

Few Mobile System Operators (MSOs) had a legacy TDM network of scale when the regulatory environment changed for both cable and telephony providers in the mid-1990s. Voice service was an important revenue source used to build a business case to deploy IP as a transport base on Hybrid Fiber Coax (HFC) television networks. To meet that demand CableLabs® wrote an evolving set of standards under their DOCSIS® and PacketCable™ initiatives that have been adopted by the MSO and supplier communities. PacketCable 1.0 [7] is focused on the cable consumer market and interconnection with the legacy PSTN network. The architecture and functionality of that standard is relatively clean and less complex than that of a wireline/wireless network provider, primarily because it is a greenfield design without a set of legacy interfaces to address. As cable providers expand their customer base beyond small business to enterprise customers, they have the opportunity to evolve their networks to PacketCable 2.0 [8]. PacketCable 2.0 is written with an objective of achieving compatibility with 3GPP IMS release 7, enabling mobility and enterprise interworking. By seeking to achieve interoperability with IMS the complexity naturally increases.

7.3.2.2   Wireline Carriers

Legacy wireline voice carriers first built VoIP networks for the business market, in response and in some cases to interwork with the introduction of IP PBXs, and then evolved those networks to serve a range of business and consumer services. Different wireline carriers have taken different approaches, some driven by tactical business needs other by strategic initiatives to speed the introduction of a Next Generation Network (NGN). Two wireline carriers that illustrate different approaches to network evolution are Nippon Telephone and Telegraph (NTT) and British Telecom (BT).

NTT is a pioneer in the deployment of FTTH and fiber to the building (FTTB), bringing data speeds in excess of 50 Mbps with targets now exceeding 200 Mbps. By most measures they, South Korea and Verizon are the leaders in the deployment of FTTH and FTTB. Delivery of high speed IP services to the home in Japan was motivated by competition and championed by regulators as a national imperative. Regulators in Japan moved from one of the world's most vertically integrated franchise models in the late 1980s to a regulated market based model in the late 1990s. By regulated market-based I mean that network demarcations were set by regulators to force facility sharing and resale of NTT's copper distribution. The idea was to create market competition by using NTTs copper infrastructure where possible, but also to allow power companies and other providers to build parallel infrastructure. As a consequence investment was aimed at providing high bandwidth to the home as a competitive response. So while VoIP and the evolution of the TDM network to an IP services network was recognized as directionally correct, it was not the primary goal. The regulatory structure for TDM copper voice service did not encourage NTT to put VoIP conversion high on their priority list. In my personal view, NTT's approach was to develop an NGN plan based on IP services, but to build toward that plan as market forces dictated, making tactical investments to respond to the market place, but insuring they fit within the long term plan.

BT took a different approach from NTT, committing to a complete network makeover, replacing TDM technology with all IP technology. Their program of network modernization is named the twenty-first Century Network, or 21CN. By undertaking a complete network redesign the belief is that operational costs will be reduced and a new generation of services will emerge to fund the significant investment needed for the transformation. While NTT's investment emphasizes fiber distribution in the last mile, BT's investment is aimed at replacing the switching and services infrastructure, going from TDM to IP. In effect BT is moving from a distributed TDM voice architecture to a cloud VoIP architecture. To enable the transition BT elected to reuse existing copper distribution by deploying ADSL2+ to achieve data rates of 12–25 Mbps downstream and 1–3 Mbps upstream.

Both approaches are designed to meet the market and regulatory environments of the two carriers. Time will tell how successful each approach is in their market place. The most significant conclusion to be drawn from the two stories is that the introduction of IP services and VoIP in particular, is being approached by the wireline carrier industry in very different ways than significant network technology transformations of the past. Stored program control switching, time division switching and the evolution to common channel signaling were largely undertaken by carriers in the developed world with common objectives, architectures, and time frames. That is not true of the evolution of wireline carrier networks to IP services. Carriers are going about it differently across the globe; although there are common themes there are significant differences in the pace and direction of investments and technology adoption. Irrespective of the strategic approach, every carrier must meet the move to IP services in the market place. In the next sections carrier VoIP services that build upon the IP infrastructure are described. Irrespective of where the carriers start and how they go about it, in the end the services determine winners and losers.

IP PBXs and IP Centrex

More equivalent lines were shipped with IP PBXs than TDM PBXs by 2005 according to Infonetics Research. Telcos with large installed bases of Centrex (hosted wireline business service) needed to respond to the trend out of wireline voice services into IP voice service. The responses came in two forms. Telcos sourced IP PBXs from companies like Avaya and Cisco and sold them as managed premises solutions. They also developed hosted VoIP solutions, sometimes going by the generic names of either IP Centrex or BVoIP (Business voice over Internet Protocol). Both solutions enabled VoIP services over private IP networks and interconnection with the PSTN. IP PBX interconnections were often enabled with ISDN Primary Rate as their PSTN umbilical.

SIP Trunking

To augment offers of IP PBXs for resale and as a managed service, telcos developed national VoIP networks and offered connectivity to that network with SIP trunking products. With SIP trunking enterprises connect their IP PBXs to a network that is capable of routing on network calls (IP PBX to IP PBX) or off network calls (IP PBX to PSTN) via a common IP connection. SIP trunking over a carrier dedicated voice VPN is superior to Internet completion because it is secure, managed, and enforces QoS.

IP Call Centers

A third significant application for wireline VoIP is the call center market. Call centers host service representatives, technical support and sales agents for enterprises, often using toll free service for inbound calls. Those centers were served by legacy TDM Automatic Call Distributor (ACD) systems that offered queuing controls and measurements to aid in the design and staffing of the centers. As TDM PBXs were displaced by IP PBXs, TDM ACDs were also replaced by IP ACDs. IP ACDs have queuing and call management features comparable to their TDM counterparts, but have two additional advantages. Because they are IP systems, new call centers do not need two sets of wired infrastructure, voice and IP; only IP is needed. The second advantage is the flexibility of moving the work force anywhere IP connectivity exists within the enterprise, so moving call loads to the east coast in the morning and west coast in the evening to take advantage of time zone differences is relatively simple in an IP system.

IP Toll Free Service

IP Toll Free Service is an inbound 800 service that compliments IP Call Center Service, IP PBX, and IP Centrex Service. Inbound calls from the PSTN are processed at an IP media gateway where the 800 number is resolved to an IP address. An advantage of the service is that transfers that are often necessary in an inbound service center can be made using SIP signaling and calls pivot from the media gateway or a SBC rather than hair pinning through the first call center.

7.3.2.3   Wireless Carriers

Carriers with substantial wireless networks who are moving to LTE are implementing the 3GPP standard VoLTE using the IMS infrastructure described earlier in the chapter. Wireless carriers are transitioning from 2G and 3G technologies which are TDM based to VoLTE, which is SIP based. That transition is different for the two dominant wireless technologies, GSM and CDMA2000, but the objectives for both networks are broad and bold.

- Move voice from a TDM circuit switched bearer and infrastructure to an all IP SIP based network.
- Insure smooth interworking with 2G and 3G voice and data (GPRS, General Packet Radio Service and UMTS). This means that when the radio access bearer (RAB) changes from 4G to 2/3G, voice must move from IP to TDM or use the GPRS/UMTS network for handover of an IP call.

- Interworking and compatibility with TISPAN, the European NGN standard [9]. This standard is feature rich, supporting a wide range of media over IMS, security, transport, regulatory requirements, and management functions.
- Support for multimedia session continuity, the ability to maintain multimedia sessions over different bearers.
- Application and services layer support. Third parties or carriers can add applications that work within the IMS framework and use subscriber profiles, with feature interworking brokered by the IMS framework.
- IPv4 and IPv6 compatibility and interworking. This is a non-trivial requirement. In the US the 2G and 3G infrastructure is almost entirely IPv4. With the depletion of IPv4 addresses, IPv6 is mandatory for LTE if not initially, then soon after launch. LTE devices are dual stack; content will largely be IPv4 for some time, but the devices and much of the transport and management infrastructure will be IPv6.

7.3.3 Conferencing

Global carriers offer a wide range of conferencing and collaboration services. Voice conferencing remains one of the most popular services and calls hosting hundreds of participants are not unusual. Legacy conference services are provided via TDM conference bridges but they are being replaced by VoIP bridges at a brisk pace. VoIP conference bridges are more flexible and integrate more naturally with meeting services and other IP collaboration services.

Cisco's TelePresence® video conferencing service has experienced a significant growth in the enterprise market. TelePresence is offered as a managed service by many carriers and it interworks with other lower end systems and services. High definition teleconferencing requires network bridges that can join multiple 5 Mbps H.264 video streams with audio and switch them based on voice response or specific commands issued by conferees. TelePresence sessions are usually carried over dedicated VPNs to assure quality and reserved bandwidth. An operations group is usually assigned to the service to facilitate connections and federations with other service providers.

7.3.4 Compute and Storage

Carriers offer compute and storage with a focus on private cloud service. Enterprise customers select carriers for ease of network integration, their knowledge of the enterprise network and for the network security services the carriers can deliver with service. Public cloud services are susceptible to DoS attacks and are far more likely to experience a higher volume of probing attempts than private clouds in secure carrier switching hubs or data centers.

7.3.5 The Mobile Cloud

Wireless tablets and smart phones accelerated what was already a growing trend, mobilizing the work force. Increasingly carriers are providing comprehensive support for enterprise customers that are enabling their work force with mobile applications hosted in private clouds. The introduction of IMS is intended to improve the quality of mobile media sessions and better enable VPN and cloud integration. As application services and enablers grow on IMS enterprises may be inclined to build applications within the IMS infrastructure rather than use the Internet.

7.4 Summary

Carriers enter the cloud market for two reasons. OTT service providers compete with consumer and enterprise offerings at the core of the carriers' businesses: voice, messaging, and video. The second reason is opportunity; carriers own substantial assets that enable cloud service including their global networks, data centers, and deep understanding of enterprise businesses. Carriers can exploit unique advantages of their global MPLS/IP networks by shaping technologies such as DNS and IRSCP to enable fine grained routing and control of cloud services. The resulting QoS, reliability, and economies cannot be matched by OTT providers.

IMS brings the opportunity to merge IP media with advanced services and features across converged wireless and wireline networks. A wide range of IP devices can take advantage of IMS and third party services can gain access to network features such as location services and be hosted within the IMS constellation. But with the opportunity come challenges of complexity, scale, fault management, and service management. No doubt those challenges will be met, but the pace of adoption will be set by how well service and quality expectations are met.

References

1. Contavalli, C., van der Gaast, W., Leach, S., and Rodden, D. (2011) IETF Draft, Client Subnet in DNS Requests draft-vandergaast-edns-client-subnet-00, January 27 2011.
2. Patrick, V., Dan, P., Aman, S. *et al*. (2007) Wresting control from BGP: scalable fine-grained route control. USENIX Annual Technical Conference.
3. Feamster, N., Balakrishnan, H., Rexford, J. *et al*. (2004) The Case for Separating Routing from Routers, ACM SIIGCOMM FDNA, August 2004.
4. Poikselka, M. and Mayer, G. (2009) *The IMS IP Multimedia Concepts and Services*, 3rd edn, John Wiley & Sons, Ltd, Chichester.
5. 3GPP (2011) TS 22.173. *Version 12.0.0, 3GPP; Technical Specification Group Services and System Aspects; IP Multimedia Core Network Subsystem (IMS), Multimedia Telephony Service and Supplementary Services*, December 2011.
6. 3GPP (2009) TS 23.198 *Version 8.0.0 Release 8, UMTS; LTE; Open Service Access (OSA); Stage 2*.
7. PacketCable™ 1.0 (1999) Architecture Framework. Technical Report PKT-TR-ARCH-V01-991201, Cable Television Laboratories, Inc.
8. PacketCable™ 2.0 (2006–2009) Architecture Framework. Technical Report PKT-TR-ARCH-FRM-V06-090528, Cable Television Laboratories, Inc.
9. ETSI TR 180 001 (2006) *Telecommunications and Internet converged Services and Protocols for Advanced Networking (TISPAN), NGN Release 1*.

8

Network Peering and Interconnection

When the subject of network peering is raised, most engineers think of Internet Protocol (IP) peering. But global networks peer in a variety of ways for the range of services they share. Historically interconnections between network peers are categorized as a network-network interconnection (NNI) and as user-network interconnection (UNI). ITU-T makes a further distinction between interconnection between a carrier's internal domains (e.g., autonomous systems) as I-NNIs and external carriers as E-NNIs. Carriers interconnect with each other on E-NNI and UNI bases. As enterprises continue to expand globally their data services move with them; increasingly that means enterprise Virtual Private Networks (VPNs) span international boundaries. Enterprises often select a single carrier to provide global VPN service, but local access is dominated by the franchise carrier in any particular region. International carriers establish points of presence (POPs) out of region but those hubs need intra-region transport for interconnection among the POPs and access to customer locations. The same is true in territory, but out of region. Carriers in the US purchase a range of services from each other, buying from tariffs or through special services requests.

8.1 Wireline Voice

Wireline voice interconnection dates back to the nineteenth century but was institutionalized in the US in the Telecommunications Act of 1934. In the US AT&T was awarded a national franchise and assumed responsibility for universal service and interconnection with independent telephone companies, which numbered over 1000. The Telecommunications Act of 1996 and the Modified Final Judgment (MFJ) of 1982 dramatically changed rules of interconnection, but the Federal Communications Commission (FCC), and probably more importantly, state Public Utility Commissions (PUCs) still look to franchise carriers, or Incumbent Local Exchange Carriers (ILECs) to meet many regulatory service obligations in return for franchise status. Today, there are over 1200 LECs, interexchange carriers (IXCs), and competitive local exchange carriers (CLECs), using Time Division Multiplex (TDM) and Voice over Internet Protocol (VoIP) technology to interconnect across the US.

Global Networks: Engineering, Operations and Design, First Edition. G. Keith Cambron.
© 2013 John Wiley & Sons, Ltd. Published 2013 by John Wiley & Sons, Ltd.

US federal and state regulatory rules for voice service interconnection are different for TDM and VoIP technology because they are regulated by different branches of the US Government. TDM services are regulated by the FCC, but the US Congress has never granted broad regulatory oversight for Internet service to the FCC. The US Court of Appeals for the District of Columbia reiterated that fact in a ruling in 2010 [1] and to date the US Congress has been reluctant to grant further authority to the FCC. The US Commerce Department has authority of regulation under the Interstate Commerce Acts. To add to the mix, the US State Department regulates some aspects of interconnection with international carriers. The descriptions that follow reflect today's industry practices in the US.

Under the rules of the 1982 MFJ, two competitive markets were created from the previous franchise monopoly of AT&T, local exchanges, and long distance interconnection. Local markets, defined as Local Access Transport Areas; LATAs, are served by exchange carriers. The former AT&T Operating Companies and the other 1200 independents still operating under a franchise tariff model are designated as ILECs. Other local exchange carriers (LECs), competing with ILECs, are known as Competitive Local Exchange Carriers. Prior to 1996 ILECs were allowed to provide local service, within the LATA, but were not allowed to provide service between LATAs. The 1996 Telecommunications Act opened the way for LECs to compete in the long distance market with the provision that interconnection is on a non-discriminatory basis.

8.1.1   Interexchange Carriers (IXCs)

Subscribers can make two choices when they sign up for wireline phone service, the choice of a LEC and the choice of an interexchange carrier. These are the presubscribed carriers. It is possible to select an interexchange carrier on a call by call basis, by dialing a 1010 code plus a Carrier Identification Code (CIC), but that is seldom used. It is not possible to bypass the LEC on a given line.

8.1.1.1   Originating Calls

Intra LATA calls originated by the subscriber are completed by the LEC. Inter LATA calls are routed to the IXC based on the CIC associated with the originating line in the class 5 switch line code data. SS7 signaling is used by the originating switch to send an Initial Address Message (IAM) to the IXC using the SS7 Integrated Services Digital Network User Part (ISUP). The IAM includes the CIC, called and calling numbers, billing number, and optional parameters, depending on the nature of the call. Originating inter LATA call processing is rather straightforward, because the initial routing depends only the CIC code associated with the line. Not so for routing terminating calls.

8.1.1.2   Terminating Calls

IXCs use the NPA−NXX (Numbering Plan Area) of the called number to route the call to their terminating tandem and terminating LATA. But the IXC cannot determine how to complete the call once the call reaches the terminating LATA because the call has to complete to the local service provider (LEC) chosen by the subscriber. In FCC parlance

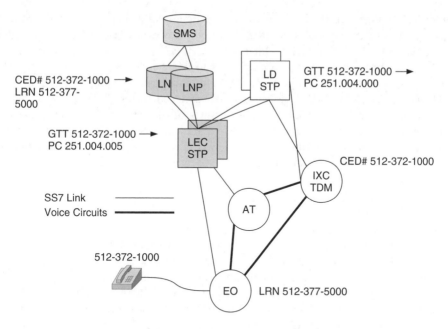

Figure 8.1 IXC call completion via LNP.

the last network to handle the call before it completes to the subscriber's LEC (ILEC or CLEC) is called the N-1 network, N being the last network. The IXC terminating tandem, shown as IXC TDM in Figure 8.1, suspends the call while it initiates an SS7 Transaction Capabilities Part (TCAP) query to the IXC STP (Signaling Transfer Point) pair responsible for Local Number Portability (LNP) service. The IXC STP performs an intermediate SS7 Global Title Translation (GTT) using the called number to determine where to forward the query. The resulting SS7 Point Code (PC) in the example is 250.004.000, the PC of the LEC STP pair. When the query is forwarded to the LEC STP pair a final GTT is performed, with a resulting PC of 251.004.005, which is a LEC LNP SCP (Service Control Point) pair. The TCAP query is forwarded to the SCP and a query of the called number yields a local routing number (LRN), 512-377-5000. The TCAP response is returned directly to the originating IXC TDM which then routes the call based on the LRN. IXCs often have direct trunk groups to high volume end offices, as shown in Figure 8.1. The trunk groups are engineered for higher busy hour blocking rates, as much as 20%. Overflow is directed to the completing group from the IXC tandem to the LEC access tandem.

8.1.1.3 Physical Interconnection

While Figure 8.1 is a logical view of IXC interconnection, in practice, fiber "meet me" points in LEC networks can be found at any end office or tandem office, subject to agreement by the parties [2]. SONET (Synchronous Optical Network) Unidirectional Path Switched Rings (UPSRs) are often used. IXCs with significant traffic in an office may choose to pay for collocation. Advantages of collocation include the ability to place terminal equipment and monitoring equipment in the LEC central office, dramatically improving monitoring, and remote test access. It also allows the IXC to combine voice and data traffic onto a common optical transport system, improving efficiency.

TDM interconnection requires both networks to adopt industry guidelines for synchronization. TDM networks must operate in a synchronous fashion with a fundamental clock operating at multiples of 8 kHz, the North American standard for TDM operation. Without synchronization circuits, interconnecting different systems are subject to buffer overflow and buffer under run since transmit and receive clocks are different [3]. While networks are internally synchronous, boundaries between networks are plesiosynchronous, meaning they are operating with very minor differences in clock rates.

8.1.1.4 Transmission

IXC networks are designed for 0 dB transmission loss. End to end loss is a nominal 6 dB from originating end office to terminating end office. Six decibel of loss are inserted by the LEC in the terminating leg of the call. Echo cancellation is seldom used in a LEC network because local call path delays don't warrant it. Cancellation is the responsibility of the IXC. SS7 contains an echo cancellation parameter to inform the IXC in the event the LEC inserts cancellation. Echo cancellers can work against each other if two are inserted in tandem. One way call path delay within a LATA is 15 ms or less, leaving IXCs about 120 ms of budgeted delay. Call path delay and echo cancellation are effective within these parameters, but problems can develop if customers forward calls over IP networks, dramatically increasing call path delay past the IXC echo canceller. Canceller tails, the amount of delay they can tolerate before breaking down, has moved from 32 ms up to 128 ms, largely because of this problem.

8.1.1.5 Network Management

Network management is a shared responsibility. Different LECs have different offerings for trunking arrangements. A general practice is that larger trunk groups, such as those homing on access tandems, should be one way. Smaller groups, such as those homing on smaller end offices, are more likely to be two way to improve occupancy. One way groups at access tandems significantly improve network management controls. During natural disasters such as earthquakes inbound and outbound traffic volumes spike. Inbound traffic often swamps outbound traffic because of relatives and friends trying to reach loved ones at risk. But outbound traffic can be more urgent and should receive at least a comparable level of service to inbound traffic. One way trunk groups provide a natural choke on inbound traffic and prevent it from overwhelming outbound service, with the advantage of not requiring recognition and invocation by the operators.

Congestion events are of two general types, global and focused. Global events are not geographically bound within a carrier's franchise. Examples of triggering events are financial and political crises. Focused overloads are restricted to a relatively small geography within a carrier franchise. Some overload conditions can be predicted in advance, such as Mother's Day, or a political convention. Predictable events can be mitigated by a range of traffic controls operators can put in place prior to the event. Examples are one way trunk groups, rerouting and code blocking by reducing overflow on certain office codes. Unanticipated events are best handled, at least initially, by automatic congestion controls, such as call gapping. Call gapping is the mechanism usually invoked by switching systems when congestion onsets. Operators can pre-designate codes, or set the controls for all codes to invoke call gapping when traffic demand reaches a set level. When that occurs, the switching system invokes a gap, or timeout of fixed length, between calls to the designated codes. Attempts arriving during the gap are routed to reorder.

8.1.2   Competitive Local Exchange Carriers (CLECs)

CLECs provide local phone service within the franchise area of an ILEC. Common examples of CLECs are Cable Mobile System Operators (MSOs) and VoIP service providers. Interconnection between CLECs and ILECs includes not only the ability to originate and terminate voice calls, it also includes regulatory obligations for 911 service.

8.1.2.1   Emergency Service

Telephone wire centers and exchange boundaries are not congruent with local and state jurisdictional boundaries. Public Service Answering Points (PSAPs) are call centers that respond to 911 emergency calls. They in turn dispatch the responsible fire department, police, or other emergency service to the emergency. When a subscriber places a call to 911 it is critical the call be routed to the PSAP serving the subscriber's jurisdiction. ILECs maintain a routing database of all subscribers and their Plain Old Telephone Service (POTS) routable phone number in a special switching system called a Selective Router (SR). ILECs are required by PUCs to provide interconnection to CLECs for 911 routing service. CLECs only need to route 911 calls in a format compatible with the SR for proper completion, provided the CLEC has entered the correct information in the 911 database. The ILEC must provide reasonable access to the 911 database, where it is managed by the ILEC, but the CLEC is responsible for the accuracy of the subscriber's number and address information. ILECs may also manage an Automatic Line Information (ALI) database, in lieu of or in addition to the SR database. The ALI database provides address and other information, such as medical restrictions to the PSAP when the 911 call arrives.

8.1.2.2   Network Management

CLEC network management is locally focused. High Volume Call-In (HVCI) events are of particular importance. When a popular radio station decides to offer a call-in give away to a popular concert, the radio station may only have four lines to handle incoming calls, but thousands of calls may be originated within a minute. LECs and CLECs build HVCI networks to funnel these calls onto a restricted network. The adoption of SS7 trunking was a significant improvement in the management of these events since busy or reorder is returned at the originating switch rather than tying up trunks to the terminating switch, only to listen to busy. However, without the HVCI networks the SS7 network can overload on ISUP call attempts. Network management controls are also used in the SS7 SCP network. Toll free SCPs can become overloaded, or the A links serving the SCPs can be congested, usually because of an unplanned event, such as the radio station in the example above using a toll free call in number. Automatic code gapping can also be preset for SCP services. When queries trip on preset demand levels a set of parameters is returned to the originating

Service Switching Point (SSP) to initiate gapping. It operates in a similar fashion to call gapping, but SS7 TCAP queries are gapped, not calls. Because CLECs often use the ILEC for SCP services, their SSPs must honor the gap notifications from the SCPs as a condition of service.

8.2 SS7 Interconnection

SS7 interconnection is used by legacy wireline voice providers, Mobile Network Operators (MNOs), VoIP providers, and can be used by providers of other services. First introduced in the US in 1986, SS7 is likely to remain as the core signaling network for voice and some circuit switched data services for at least another decade.

Interconnection with LECs and IXCs is done in one of two ways. Larger carriers offer direct interconnection via D links and sometimes A Links, or over IP using SIGTRAN. Wholesale network providers act as aggregation networks. They provide SS7 networking for smaller independent carriers and interconnection to all of the LECs and IXCs. Wholesalers provide a valuable service because without them interconnecting all of the smaller carriers, SS7 port costs would soar in ILEC networks. When connecting directly with an ILEC the two networks meet at a Signaling Point of Interface (SPOI). There is a testing and certification process in the US governed by Telcordia requirements [4]. When smaller carriers are located within a larger ILEC's serving area, an A link connection works well, eliminating the need for the smaller carrier to invest in any SS7 infrastructure.

8.2.1 Services

SS7 interconnection is used for a range of services. Companies that are pure VoIP companies or mobility companies, or even content providers use SS7 services. There are two general categories of SS7 services, call routing services and value added services.

8.2.1.1 ISUP

The Integrated Services Digital Network (ISDN) User Part is used for circuit switched call completion. ISUP's IAM carries a good deal of information beyond the minimum needed for call setup. The following is a list of some of the information carried in an IAM.

- **Calling party's number and charge number** – those can be different when calls are made from a Centrex group.
- **Calling party's indicator** – can be used to identify operator calls, voice band data, test calls, and emergency services.
- **Redirecting numbers** – indicates that the call has been forwarded from another number.
- **Originating line information parameter** – when properly populated this is a valuable indicator. It can be used to identify operator calls, pay phones, test calls, emergency services, calls originated from prisons, calls originated from hotels, and wireless originated calls.
- **Echo control indicator** – indicates whether echo cancellation has been invoked on the connection.

The Session Initiation Protocol (SIP) supports mapping the ISUP IAM into the SIP Invite message.

8.2.1.2 Local Number Portability (LNP)

LNP, described in Chapter 5, is another call routing service. Voice networks attempting to complete calls to a LEC or CLEC query their LNP databases to determine the routing number of the switching system hosting the subscriber line.

8.2.1.3 Toll Free Service

Sometimes called 800 Service, this call routing service is used to derive a POTS routable number from a toll free number. The 800 number portability predates LNP, but like LNP it allows customers to change carriers and take their number with them. The serving LEC or IXC maintains an SCP database containing the toll free number mapping to the POTS routable. Calls originating in other networks use SS7 GTT to determine the SS7 address of the serving SCP so that a database query can be launched across interconnecting SS7 links. Carriers have added vertical features to toll free calling, so the POTS routable can be changed by customers directly or preprogrammed for different days and different times.

8.2.1.4 Calling Name (CNAM) Service

Calling Name Service (CNAM) is a value added service. When a call terminates to a subscriber that has CNAM service the terminating network queries the originating network of the calling party using a TCAP query to the originating network's CNAM SCP. The TCAP response contains a text representation of the name if the name is found and presentation is allowed. The IAM may mark the number as restricted or presentation now allowed, which means a query should not be sent.

8.2.1.5 Custom Local Area Signaling Services (CLASS)

These value added services are part of the Advanced Intelligent Network (AIN) initiative launched by Telcordia in the 1980s and 1990s. They are SS7 based and use the class 5 switching system to implement call processing primitives to enable advanced services using controls in a remote signaling point. Examples of these services are automatic callback and automatic redial. Some service providers, like VeriSign®, sell services based on Custom Local Area Signaling Services (CLASS). They interconnect with LECs, typically using D links, to send TCAP queries to the Signaling Points providing CLASS services.

8.2.1.6 Mobility

SS7 MAP (Mobile Application Part) messages are exchanged between MNOs to obtain subscriber information from Home Location Registers (HLRs) and mobile station roaming numbers (MSRNs) from Visiting Local Registers (VLRs). Interconnection is needed for voice and short messaging services (SMS). Location updates are sent from Serving General

Packet Radio Services Nodes (SGSNs) and Mobile Switching Centers (MSCs) to HLRs as well. ISUP connectivity to the Gateway Mobile Switching Centers (GMSCs) is provided as well.

8.3   IP Interconnection

IP network to network interconnection includes five general categories: VPN peering, private peering, public peering, transit services, and Internet Access Services (IAS).

8.3.1   VPN Peering

VPN interconnections are restricted to carriers that have strong business and technical relationships because of the inherent operational and engineering challenges. For VPN interconnection to work, carriers must map their service classes across the boundary and exchange information on traffic and policing policies, traffic engineering rules, and control plane design and practices. Because of the risks and complexity, the accepted approach is for the dominant carrier to use layer 2 and transport services of the supporting carrier, minimizing control plane, network management, and traffic engineering issues. Requests for VPN peering can arise when enterprises merge or there is a large acquisition of a multinational company. The two companies may each have large VPNs that are served by different providers and so their first reaction is to try to directly interconnect the two VPNs to speed integration. Classes of service and policing policies may not map directly. A phased integration with an end state of a single VPN and the interim use of bridging technologies, such as tunneling, selected use of Network Address Translations (NATs) and federation of applications is often the best path.

8.3.2   Internet Peering

Connectivity, bandwidth, and settlements are key points of discussion in any dialog about Internet peering. In the regulated voice TDM network arrangements between carriers are governed by regulatory agencies and the standards they adopt. Obligations of voice interconnection come with the franchise granted by the sovereign. The Internet is quite different. Carriers serve as Internet backbones providing connectivity to other players, and as Internet Service Providers (ISPs) for their consumer and business subscribers, connecting them to other carriers offering a similar service to their customer base. For Internet carriers the ability to connect and exchange traffic without charge is governed by size, bandwidth, and connectivity, and is negotiated, not regulated (see Figure 8.2).

8.3.2.1   Tier 1 Networks

Internet networks belong to one of three tiers. Designation of a tier is informal and is subject to some interpretation. The accepted definition of tier 1 networks is that they are directly connected to every other tier 1 network and are able to reach all hosts on the Internet, without paying transit fees. Transit fees, in this context, are usage charges for the bandwidth consumed, not facility connection charges. In Figure 8.2 AS 7777 is a fictitious

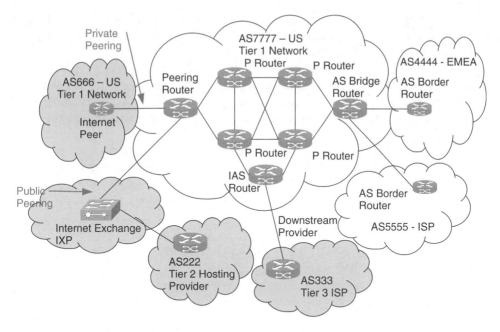

Figure 8.2 Internet peering arrangements.

tier 1 network. In the example the light-shaded clouds belong to a single tier 1 carrier. The darker shaded clouds are other carriers or entities. Each tier 1 network has its own guidelines for settlement-free interconnection and providers can apply for interconnection, but the tier 1 provider is the final arbiter. Here is a summary of the current guidelines AT&T publishes for settlement-free interconnection [5].

- Peers must operate a US-wide IP backbone whose links are primarily OC192 (10 Gbps) or greater.
- Peers must meet AT&T at a minimum of three mutually agreeable geographically diverse points in the US. The US interconnection points must include at least one city on the US east coast, one in the central region, and one on the US west coast, and must currently be chosen from AT&T peering points in the following list of metropolitan areas: New York City/Newark NJ, Washington DC/Ashburn VA, Atlanta, Miami, Chicago, Dallas, Seattle, San Francisco/San Jose, and Los Angeles.
- Peers must interconnect in two mutual non-US peering locations on distinct continents where the peer has a non-trivial backbone network. Non-US peerings are with AT&T's regional AS only.
- Peer's traffic to/from AT&T's AS must be on-net only and must amount to an average of at least 7 Gbps in the dominant direction to/from AT&T in the US during the busiest hour of the month.
- Interconnection bandwidth must be at least 10 Gbps at each US interconnection point.
- A network (AS) that is a customer of AT&T's AS for any dedicated IP services may not simultaneously be a settlement-free peer.

- Peers must maintain a balanced traffic ratio between its network and AT&T. A balance is defined as no more than a 2 : 1 ratio of traffic in : out of AT&T on average each month with a reasonably low peak-to-average ratio.
- Peers must abide by the following routing policies.
    - They must use the same peering AS at each US interconnection point and must announce a consistent set of routes at each point, unless otherwise mutually agreed.
    - No transit or third party routes are to be announced; all routes exchanged must be the peer's and peer's customers' routes.
    - They must filter route announcements from its customers by prefix.
    - Neither party shall abuse the peering relationship by engaging in activities such as but not limited to: pointing a default route at the other or otherwise forwarding traffic for destinations not explicitly advertised, resetting next-hop, selling, or giving next-hop to others.

The rules are meant to avoid inefficiencies in routing and route management and to minimize transit traffic which increases costs for no revenue.

8.3.2.2 Tier 2 Networks

Tier 2 networks lack full backbone networks within a region and rely on other providers for some portion of connectivity. Size and traffic volume does not directly correlate to the tier designation of a network. A local cable MSO, for example, may have a significant customer base and traffic volumes, but be restricted to a franchise area that is does not encompass the entire region. The MSO could not connect as a tier 1 network and would pay transit fees out of their serving area to a backbone provider.

Tier 1 providers in one region are often tier 2 or tier 3 providers in other parts of the world. British Telecom (BT), for example, is a tier 1 provider in the UK, but uses CenturyLink™ as a partner in the US. Without knowing the details of the BT-CenturyLink business arrangement it's not certain that settlement fees are passed; this and other arrangements may be a strategic agreement in which a no-fee swap is arranged.

8.3.2.3 Tier 3 Networks

Tier 3 networks have no Internet backbone and purchase Internet access from a service provider. An example is a smaller independent telephone company that has a Digital Subscriber Line (DSL) subscriber base but their network only spans what is necessary to serve their in-franchise customer base. While their network has "eyeballs," it has no access to content unless it purchases Internet access from a provider.

8.3.2.4 Private Peering

The term private peering has nothing to do with private networks; the exchanged traffic is public Internet traffic. When used in the context of IP network interconnection, private peering refers to a bilateral agreement between two carriers for the exchange of Internet traffic on a settlement-free basis. Tier 1 networks generally peer on a private basis and negotiate the terms of peering, but it is on a settlement-free basis by definition.

The three key elements of peering, described earlier, are addressed in the bilateral peering agreement.

- **Settlements** – by definition private peering is settlement free.
- **Bandwidth** – subject to the agreement of the parties.
- **Connectivity** – if we define connectivity as the ability to reach a destination in a peer network, then it can be divided into three elements.
  - **Physical connectivity** – destinations must be served with sufficient bandwidth and a highly available network connection.
  - **Route adve**rtisement – each network operator sets policies for which routes they will advertise with each peer using exterior border gateway protocol (eBGP), at each peering location.
  - **Route imports** – each network sets policies for advertisements received via eBGP and filters out routes according to those policies before distributing them internally via interior border gateway protocol (iBGP). Unless the routes are distributed internally, no traffic will be routed to the destination.

An example of private peering is shown in the diagram above, between AS666 and AS7777. AS7777 is the US backbone, which also serves a US ISP AS5555 and a European–Middle East–Asia (EMEA) tier 2 backbone AS4444. Private peering in the example is restricted to US locations, and so routes to customers that purchased US IAS from AS7777 or AS5555 would be advertised via eBGP to AS666. No AS666 routes would be advertised to AS4444 or any subtending networks served by AS4444.

Routes advertised by AS666 via eBGP would be distributed to peering routers within AS7777 and these would be distributed via iBGP within AS7777 and AS5555, policies permitting. Recall Border Gateway Protocol (BGP) advertisements include the IP address prefix and the list of ASes used to reach the destination. Routers receiving those updates verify their AS is not in the list, and in turn advertise AS666 reachability via eBGP to their eBGP peers.

8.3.3 Public Peering

Public peering occurs in more than 300 Internet Exchange Points (IXPs) worldwide [6] connecting over 30 000 ASes on more than 44 000 links [7]. IXPs, originally called Network Access Points (NAPs), are usually layer 2 interconnection points with peers choosing interconnection on a pair wise basis. Interconnection may be settlement free, or for fee, as negotiated by the parties. IXPs obtain subnets from their Regional Internet Registry (RIR). IP addresses are in turn assigned to each partner's router to enable eBGP sessions. IXP interconnection is cost effective. IXP providers are often not for profit organizations whose fees only cover the cost of maintaining and administering the site. They are particularly attractive for content delivery and web hosting organizations since they can deliver content directly to ISPs without having to negotiate transit fees from backbone and transit providers. In Figure 8.2 a fictitious AS222 is a web hosting provider operating as a tier 2 network at the IXP. As a Tier 2, AS222 also purchases transit or IAS service from other providers, but uses IXPs where economics are favorable. By expanding the number of IXPs a hosting provider uses it not only lowers settlement fees with backbone providers, it moves content closer to the end user, reducing latency which also improves delivery bandwidth since the two are closely related by Transmission Control Protocol (TCP) round trip times. Of course it requires the content provider to provide sufficient storage and processing near IXP locations.

8.3.3.1   Transit and Internet Access Services

In the final example of this section an ISP, AS333 purchases IAS from the tier 1 provider AS7777. Service need not be restricted to the tier 1 region, in this case the US. IAS service can be provided to EMEA, AS4444 as well. eBGP policies are set up much as they were for Tier 1 private peering, but in this case routes advertised in the EMEA AS are forwarded to the customer, AS333. Destinations advertised by AS333 are also forwarded to the EMEA AS, assuming the customer wants them advertised. Since the route between the US and EMEA ASs is via a subsea cable, we can assume the service fees between AS333 and AS7777 are higher to compensate for the expense of transit to EMEA.

8.3.3.2   Interdomain Routing

Strictly speaking BGP is not a routing protocol. Rather it communicates reachability, the ability to connect to a destination. Routing protocols, like Open Shortest Path First (OSPF), are generally designed to take a shortest path or otherwise minimize the "cost" of reaching a destination. But OSPF is an Interior Gateway Protocol (IGP) and so it typically optimizes within an AS, not globally. If we think about two tier 1 networks that are interconnected we can begin to see the potential for arbitrage in the design of routing, as shown in Figure 8.3.

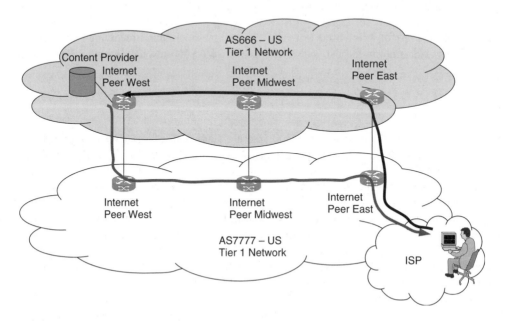

Figure 8.3   Interdomain routing.

In the diagram a content provider, represented by the database symbol, is physically hosted in the western part of a tier 1 network AS666. The routes of the content provider are advertised across peering points in the West, Midwest, and East to a tier 1 peer network AS7777. An Internet access customer of AS7777 in the East initiates a request to stream a video from the content provider. When BGP routes are weighted equally, as they are in this case with one AS on the path list, the IGP least cost route is used. Assuming the request is routed to the lowest cost peering point, it will leave AS7777 via the eastern connection and use AS666 to transit the country to reach the content server. If the ISP prefix is advertised across the three peering points, AS666 may choose the western peering point as the lowest cost route for the video stream, meaning AS7777 will carry the stream across the country. Since the size of the stream for the video is likely to be 1000 times the size of the request stream, AS7777 will carry the bulk of the traffic but AS666 may realize much more revenue. Unless there are mirror image cases to balance the flows, AS7777 may start to see one sided ratios of traffic in to traffic out, possibly violating their peering guidelines.

Possible solutions to this problem include the use of MEDs or AS prepending. Multi-Exit Discriminators or MEDs, are a mechanism for a network to advertise the preferred route for incoming traffic to a specific destination within the AS [8]. There is no guarantee the other network will honor the MEDS, so prepending multiple copies of the network AS in BGP advertisement is another approach. In our example, one copy of AS7777 would be placed in the eBGP route advertisement from the east coast router, two copies for the Midwest router and three copies for the West router. AS666 routers receiving the BGP updates would discard the West and Midwest routes in favor of the shorter eastern path.

8.3.4   Mobility Peering

Mobility networks interconnect with other Public Land Mobile Network (PLMN) and Public Switched Telephone Network (PSTN) networks using the voice and SS7 peering arrangements described above. For Internet services they use the Internet peering arrangement described directly above or purchase IAS, but for intranet mobility interconnection they use a special interconnection arrangement called the GPRS Roaming eXchange (GRX) [9]. GRX offers interconnection for GPRS roaming, Universal Mobile Telecommunications System (UMTS) roaming, Multimedia Message Service (MMS) interworking, IP Multimedia Subsystem (IMS) interworking, and Wireless Local Area Network (WLAN) data roaming. The GRX concept has been extended to include wireline network operators, ISPs, and Application Service Providers (ASPs) in an offering called Internet Protocol Packet eXchange (IPX). GRX service is restricted to MNOs. IPX service is available to all network operators, ISPs and ASPs. Both GRX and IPX use backbone IP providers for transport service. GRX providers must be fully interconnected to all other GRX providers. Transit routing from one GRX to a terminating GRX through an intermediate is discouraged. In effect a GRX service provider can use their own backbone or purchase VPN service from a global backbone provider. There are several hundred MNOs connected to about 20 GRX service providers today. Route advertisements between GRX and IPX providers are performed using eBGP at border routers. A GRX specific Domain Name System (DNS) hierarchy is used to resolve wireless IP addresses, such as APN URIs (Access Point Name Uniform Resource Identifiers). The GRX and IPX service providers appear as a federated VPN except public routable IPv4 addresses are used instead of private addresses. The specification for GRX and IPX does not support IPv6 at this time.

8.3.4.1 GPRS, UMTS, and LTE Roaming

As described in Chapter 5, IP sessions for mobile devices are anchored to the Radio Access Network (RAN) by the SGSN and Signaling Gateway (S-GW) for 3G and 4G networks, respectively. The sessions are anchored to the Internet or other service networks by the Gateway General Packet Radio Services Support Node (GGSN) and Packet Data Network Gateway (PDN-GW). When a mobile subscriber roams in a foreign network the local SGSN or S-GW establishes a GPRS Tunneling Protocol (GTP) tunnel to the home network GGSN or PDN-GW. When the User Equipment (UE) launches a service request to the visited SGSN, a query is made to the home HLR via SS7. If roaming and the service is authorized the APN URI is resolved to an IP address by issuing a query to the GRX DNS hierarchy. Once the IPv4 address is returned the SGSN can request a Packet Data Protocol (PDP) context from the home GGSN and set up a GTP through the GRX to continue with session establishment.

8.3.4.2 WLAN Roaming

MNOs can also use GRX for transporting Remote Authentication Dial In User Service (RADIUS) and TAP (Transferred Account Procedure) messages for authentication, authorization, and for accounting. Fraud monitoring and remote provisioning can also be accomplished over this service. Login and message procedures are specified by Global System for Mobile Communications (GSM) [10].

8.3.4.3 SMS Interworking

MNOs interconnect with other operators and with External Short Messaging Entities (ESMEs). The architecture used for interconnection is shown in Figure 8.4.

Terminating messages destined for a MNO's subscribers can originate in other mobile networks or be forwarded by ESMEs. MNOs designate SMS gateways where other MNOs, ESMEs, or SMS hub providers that follow GSM standards [11] can interconnect. Short Messaging Peer to Peer Protocol (SMPP) sessions are established with Short Message Service Centers (SMSCs) via a validation manager that authenticates and authorizes sessions to the regional SMSCs.

Mobile originated messages are collected by the SMSCs and delivered to ESMEs and other MNOs via the same SMS gateway. In mobile networks of scale SMSCs are distributed regionally and the SMS validation managers are located in data centers. The message delivery infrastructure behind the SMSCs is described in Chapter 5.

8.3.4.4 MMS Interworking

MMS traffic is exchanged via GRX using Simple Mail Transfer Protocol (SMTP) and Multipurpose Internet Mail Extension (MIME). Procedures are specified by GSM [12].

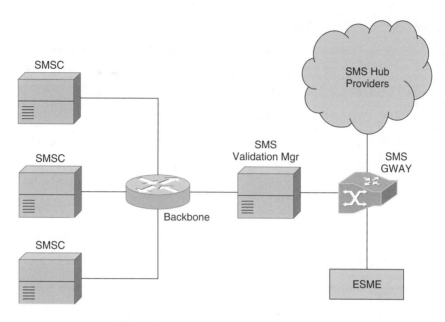

Figure 8.4   SMS interworking architecture.

8.3.4.5   IMS Interworking

GSM specifies IPX interconnection for IMS, presumably since wireline operators, ISPs, and ASPs will want to interconnect with each other and MNOs. The DNS infrastructure used for GRX is enhanced to include E.164 Number Mapping (ENUM) capability. While the concept is relatively new it will be interesting to see if ENUM and SIP interconnection occur on more of a bilateral basis along voice TDM historical interconnection agreements or whether mobility will lead the way with Long Term Evolution (LTE) and IPX becoming the leading global interconnection point for IMS and SIP services.

8.4   Summary

Carriers interconnect with each over a range of technology to provide voice, Internet, and messaging services worldwide. Unlike the passage of people and goods, electronic communication in its many forms crosses carrier and sovereign boundaries without interruption or delay. Regulatory changes that began in the 1980s introduced competition and redefined network boundaries. The industry responded with network changes that insured interworking among the newly created competitors. Standards bodies, national and international, worked to reconcile differences and achieve global consensus. Where consensus was not entirely achieved international gateways were established to intermediate between differences in protocols and signaling standards, hiding differences from subscribers. Voice, Internet, and mobility networks interconnect under different rules and with different technologies, but each has achieved a high level of connectivity and uniformity of service across the developed world.

References

1. Comcast Corp. vs. Federal Communications Commission (2010) US Court of Appeals for the District of Columbia, No. 08–1291, April 6, 2010.
2. AT&T (2010) 22 State Network Interconnection, Version: 4Q10 – CLEC ICA – 19 November, 2010.
3. Bellcore (1986) Digital Network Synchronization Plan. Technical Advisory TA-NPL-000436 (Issue 1), November 1, 1986.
4. Telcordia (1991) Common Channel Signaling (CCS) Network Interface Specification, Technical Reference TR-TSV-000905, Suppl. 1, Issued July 1991.
5. AT&T (2011) Global IP Settlement-Free Interconnection Policy, http://www.corp.att.com/peering (accessed September 28, 2011).
6. PeeringDB Project (n.d.) http://www.peeringdb.com/ (accessed June 1, 2012).
7. Augustin, B., Krishnamurthy, B., and Willinger, W. (2009) *IXPs:Mapped?* Université Pierre et Marie Curie, Paris and AT&T Labs–Research, Florham Park, NJ, IMC'09 November 4–6, 2009.
8. McPherson, D. and Gill, V. (2006) IETF RFC 4451. *BGP MULTI_EXIT_DISC (MED) Considerations*, March 2006.
9. GSM Association (2008) Inter-Service Provider IP Backbone Guidelines, Version 4.4, GSM Association Document IR.34, 19 June 2008.
10. GSM Association (2003) WLAN Roaming Guidelines, Version 3.0.0, GSM Association Document IR.61, April 2003.
11. GSM Association (2010) SMS Hubbing Handbook, Version 4.2, GSM Association Document IN.08, November 25, 2010.
12. GSM Association (2003) MMS Roaming Guidelines, GSM Association Document PRD.52.

Part Two

Teams and Systems

9

Engineering and Operations

To compete, a network must have:

- **Connectivity** – it must reach you and those with whom you want to communicate.
- **Capacity** – it must have sufficient bandwidth to send and receive the information you want.
- **Superior performance** – network errors should be corrected or be below the threshold of annoyance; network delays should be imperceptible or at least immaterial.
- **High availability** – services should work when called upon.

As engineers strive to achieve these goals, many forces are working to deny them:

- **Variable demand** – traffic demand is far more variable than it was when networks only carried voice. Cellular service not only adds the dimension of mobility, it dramatically increases personal connectivity so more time is spent communicating.
- **Multiple modalities** – as we saw in mobile networks, devices can have broadcast video, voice, and data sessions all active at the same time, using different sessions and gateways. In many cases the sessions can be competing for the same limited computing and bandwidth.
- **Applications** – new applications can become viral quickly, causing dramatic changes in demand patterns with little or no notice.

Engineering and Operations teams work hand in hand to design and deploy network technology and to solve systemic technical problems that degrade network capacity or performance. In this section the makeup of teams is treated first, followed by a listing of the key functions of the joint teams. Each carrier will have variations on these teams, but the core functions and roles described here are common.

Global Networks: Engineering, Operations and Design, First Edition. G. Keith Cambron.
© 2013 John Wiley & Sons, Ltd. Published 2013 by John Wiley & Sons, Ltd.

9.1   Engineering

The main divisions of labor in Engineering are among teams that:

- introduce new technology into the network;
- design the networks;
- build the networks; and
- augment the networks.

Engineering teams undertaking these tasks are organized by function and by skill set.

9.1.1   Systems Engineers

University trained electrical engineers, communications engineers, mechanical engineers, software engineers, physicists, mathematicians, statisticians, operations research, industrial engineers, and human factors specialists make up this group. A bachelor's degree is mandatory, most have Masters or PhDs. If you want to find these engineers in numbers, look for the laboratories. They are either testing and certifying new technology or reconstructing a field problem with Operations to find the root cause of some service failure. A key strength of Systems Engineering is that it is multi-disciplined. On a given project you find an electrical engineer with a specialty in communications protocols working with an operations research analyst and a software developer solving a network design problem.

Unlike other groups in Engineering, the demands of mastering the science and technology usually lead these engineers to specialize in a particular network area or design role. Engineers develop and enhance their skills by working with new technology every day, working with suppliers and colleagues and actively participating in industry forums and professional organizations such as IEEE, IETF, ITU, and ACM.

Prior to the breakup of Bell Labs in 1982 there was one title for all engineers and scientists, Members of Technical Staff (MTS). The most recent hire and distinguished scientists held the same title, but were compensated quite differently. After the breakup, Human Resource (HR) departments in the various telephone companies and Lucent introduced a structured tier of titles that was more in line with the rest of the job titles in the companies and more familiar to the HR organization. The titles that follow are typical for Systems Engineers in carriers.

- **Distinguished Member of Technical Staff (DMTS)** – engineers that are often recognized as preeminent throughout industry or at least across the company for their accomplishments and contributions in their fields. They are technological thought leaders that are out in front of new technologies and alternative network designs. Twenty years or more experience is typical and they often hold multiple patents or have published extensively in technical journals.
- **Lead Member of Technical Staff (LMTS)** – LMTS are assigned responsibility for major projects or areas of design. As their title suggests, they lead teams of engineers, usually between three and eight members of staff. They also head joint efforts with suppliers for equipment specification, certification, and trouble shooting. Network design depends on LMTS, usually working with teams of engineers and mathematicians who specialize in traffic theory, algorithms and optimization, or communications protocols. LMTS are not

supervisor, but rather technical team leaders. The composition of their team changes with the task. In their leadership role they assign technical tasks to each of the team members and are responsible for overall design or problem resolution. LMTS usually have at least 10 years of relevant experience.

- **Principal Member of Technical Staff (PMTS)** – in some sense these engineers are the heart of the organization. They are highly skilled individual contributors that take full responsibility for solving technical problems within their specialties. LMTS or supervisors can give them a problem without having to describe it in detail or formulate lines of investigation. PMTS often work across teams, bringing their expertise to bear on multiple projects. They are excellent mentors for junior engineers and represent Engineering in cross functional teams with Operations, Marketing, and others. PMTS usually have at least four years of relevant experience.
- **MTS** – this entry level position is populated with new graduates or employees from outside Engineering that recently attained their degree, or ones that have the necessary skills and are seeking a different career path. They work under the technical supervision of PMTS and LMTS who usually frame the problem the MTS needs to solve, with the benefit of guidance and consultation.

Managers in Systems Engineering are not generally considered MTS, as HR categorizes employees. However, in effective organizations these managers are all former MTS. Other departments in a carrier value diverse experience, and so a manager may have spent time in Marketing, Finance, and Operations. Success in managing MTS, and solving the challenging technical problems faced each day are not compatible with a general management background. Trained professional engineers that have acquired management skills are the most effective managers in Systems Engineering teams.

The list below captures the most common Systems Engineering groups, but is not exhaustive. Engineers working in these groups are not only responsible for the network elements and network design, they are also responsible for the Element Management Systems (EMSs) and for supporting IT and Operations in introducing the technology.

9.1.1.1 Transport Engineers

Transport engineers are responsible for specifying, evaluating, and certifying optical and metallic layer 0 and 1 technologies, and for the design of those networks. Optical engineers working on advanced optical transport are often physicists or electrical engineers with strong mathematical foundations. The scope of their work spans from optical components to Dense Wave Division Multiplexing (DWDM) network design. Transport systems are evaluated for their functional, performance, and capacity characteristics, as well as their conformance to operational design criteria. Long haul systems in particular require complex design rules and transport engineers often end up working with software engineers to build specialized systems to plan and design the networks. Metallic transport engineers are still needed to specify and evaluate the metallic interfaces to optical systems, and to verify software and hardware changes that are still made to legacy systems. Point to point radio, or microwave systems are also engineered by this group. Microwave is used increasingly out of region for mobile cell site backhaul. Layer 1 services, such as Fiber Channel and dedicated Synchronous Optical Network (SONET) rings are also vetted and designed by this group.

9.1.1.2 Layer 2 Engineers

While this area is dominated by Ethernet today, there are still large deployments of Asynchronous Transfer Mode (ATM) and smaller deployments of Frame Relay technology. Ethernet is both a technology and a service. It is the technology of choice for high speed access for Virtual Private Network (VPN) and Internet service, and it operates over different metallic, optical, and radio transport technologies. Switched Ethernet service has grown beyond the metropolitan area and is now a wide area service, riding on Multiprotocol Label Switching (MPLS) backbones. The network design of switched Ethernet is complex, rivaling VPN service design. While ATM is not growing, there are still substantial deployments, particularly as the Digital Subscriber Line (DSL) aggregation network. As traffic grows in the DSL network, route augmentation, and the conversion to Ethernet where the technology permits are steady work demands on ATM engineers. Ethernet engineers are often called in to support complex data center designs as well. Load balancing technology and some specialized layer 2 appliances are usually assigned to these engineers.

9.1.1.3 IP/MPLS Backbone Engineers

Backbone routers, aggregation routers, peering routers, border routers, route reflectors, and their associated support systems are specified, certified, and integrated by these engineers. Testing and certification of these systems is crucial and complex, particularly analyzing how these systems perform under various types of bearer path and control plane traffic loads and failure modes. All of the protocols described in Chapters 6–8 have to be verified and pairwise testing must be performed with all network elements that interface physically or logically with these systems. Network design of MPLS and Internet Protocol (IP) layers for Internet, VPN, and VPLS (Virtual Private LAN Service) services are also the responsibility of this team. Supporting systems, such as Domain Name System (DNS), and IPv4 and IPv6 systems and administration are managed by the team. The team either has members or access to traffic engineers and reliability engineers that assist in simulations or other special studies needed to understand and quantify the effects of conditions that cannot be readily created in the physical labs. Special software systems are often developed in this group to design and monitor these networks, with special attention to their scale and global reach. Equipment suppliers usually lack the network expertise to develop the cross platform tools and systems that are needed.

9.1.1.4 IP Services Engineers

Driven by enterprise customers, these engineers work primarily on edge routers and managed service routers. They specify new IP services, such as multicast video or security services such as Directed Denial of Service (DDoS) protection, and design, certify, and test the solutions. As part of their edge responsibilities they also work across teams, supporting global Voice over Internet Protocol (VoIP) designs, teleconferencing, and mobility IP Multimedia Subsystem (IMS). Their workloads peak as new services or new edge routers are introduced. Those new technologies must be vetted against all current configurations and systems. The range of managed premises routers makes this task time consuming and complex. When a new edge router is introduced, or an existing edge router gets a new software release all combinations of supported managed premises routers and the services they support must be verified if the change is judged consequential. A lot of this team's time is spent working with IT teams to design and verify the proper provisioning of these services. Not only are they responsible for insuring the services can be provisioned, they are the only ones with the expertise to write or review the detailed provisioning requirements.

9.1.1.5 Voice and Collaboration Services Engineers

Voice networks began moving to IP in 2000. That move is accelerated by the increasing adoption of VoIP Private Branch Exchanges (PBXs) in the enterprise market, and by the introduction of Long Term Evolution (LTE) in mobility networks. All of those technologies must be integrated with and interwork with legacy voice systems and the Public Switched Telephone Network (PSTN). Because VoIP is supported under different protocols and different versions within each protocol, the tasks of design and certification of these services continues to grow. In addition to plain old voice, the voice engineering group is responsible for voice mail, teleconferencing, and web collaboration services. Hosted voice service, sometimes called IP Centrex, is their responsibility as well. Scaling the voice network globally and the associated IP and Time Division Multiplex (TDM) designs fall under this set of engineers. They are responsible for the migration strategy away from TDM voice, and the integration of all voice platforms into a unified architecture.

9.1.1.6 Mobility Radio Engineers

The three generations of mobile wireless platforms described in Chapter 5 each have different radio technology and they operate on different frequencies, meaning they have different propagation characteristics and are subject to somewhat different impairment sets. LTE with Multiple In Multiple Out (MIMO), for example, benefits from a multipath environment such as downtown canyons, where Global System for Mobile Communications (GSM) and Universal Mobile Telecommunications System (UMTS) are degraded by the same multipath environment. Radio Access Network (RAN) engineers design and manage the deployment areas for all three technologies. They select the controller serving areas, design the sectors and set the antenna direction and azimuth. Radio base station and controller specification and certification are also their responsibilities. A separate group of radio engineers tests and certifies all mobile devices for their respective radio technologies. Each device must be tested under a range of Radio Frequency (RF) conditions in Faraday rooms, and Inter-Radio Access Technology (IRAT) handover and handback must be measured as well.

9.1.1.7 Consumer Access Engineers

Consumer and small business access technologies including Passive Optical Network (PON), Asymmetric Digital Subscriber Line (ADSL1), ADSL2+, G.SHDSL (Symmetric High-speed Digital Subscriber Line), and Very High Bit Rate Digital Subscriber Line (VDSL2) are the responsibility of this group. These technologies generally have three sets of components which have to be specified, certified, and incorporated into network designs. They are the multiplexer, a customer premises termination or gateway, and a central office router,

Broadband Remote Access Server (BRAS) or other aggregator. Customer Management Systems (CMS) and EMSs must be certified as well. Deployment studies and recommendations are usually undertaken by this group and special support systems are often defined and developed to support operations forces in the installation and maintenance of outside plant systems and customer premises systems.

9.1.1.8 Network Systems Engineers

Heavily populated with operations research analysts, mathematicians, and other specialists, this engineering group brings reliability, traffic engineering, and network design expertise to other systems engineering teams. They are adept at cross network service and reliability analysis, which is valuable in making tradeoffs for the placement of functions, data centers, and facility hubs. Their expertise in service and network analysis also comes into play in designing performance measurement plans. They can recommend data sets and measurement periods that are best suited to measure network and service quality.

9.1.2 Network Planning

Where the primary role of Systems Engineering is technical, the role of Network Planning Engineers is economic and functional. They are experienced in their respective fields in the placement and operation of network systems and facilities. The plans they develop are for network expansion, or for technology refresh or introduction. They take into account capital, supplier, and operational constraints in developing network wide plans for network element upgrades, new data centers, new fiber and undersea cable routes, and deployment of new technology and services such as LTE. Network planning is a centralized, not a regional market function in most companies. The teams are integrated by project or network, but have specialists in transport, switching, voice, and mobility. They produce equipment specifications and schedules for procurement teams, templates, and schedules for network design teams, and timelines for installation and turn up for operations teams.

9.1.3 Network and Central Office Engineers

This engineering group is generally organized first by market area, and then by technology in a fashion similar to Systems Engineers. It is the largest group in Engineering and they are responsible for the detailed designs for equipment upgrades, new equipment placement, power, grounding, heating, air conditioning, facility placement and scheduling all suppliers, and installation teams. Capacity management is also their responsibility. Voice trunk groups must be augmented; Ethernet, ATM, and Frame Relay links are monitored and augmented as well. Switch capacity, port and memory exhaust, and all other resources that can limit a system are monitored by these teams. They have solid technical understanding of the equipment and network, work closely with suppliers and procurement, and manage a myriad of scheduling and logistical details. To support the individual teams, a program management office may be set up to assist in coordinating particularly large projects, like the introduction of cloud computing or LTE.

9.1.4   Outside Plant Engineers

Network and Central Office Engineers have responsibility for the core network and all its elements, up to the cable vault in the Central Office. That is where the responsibility of Outside Plant Engineers begins. While they work with Consumer Access Systems Engineers, their day to day interactions are with outside plant construction and engineering, and installation and maintenance teams. Whether they are placing new fiber routes or planning a rehabilitation job for copper feeder, civil permitting, placement, and construction are their primary considerations. The technology they deploy is standardized and so time and energy is spent in planning for, and obtaining permission for the actual placement. In addition to the placement of fiber and copper, their responsibilities extend to outside plant electronics; remote terminals for DSL, pair gain systems, fiber to the curb, and sometimes point to point radio systems fall within their responsibility.

9.1.5   Common Systems Engineers

Power, Heating Ventilation and AC (HVAC), and specialized systems such as fire suppression and security systems are designed and placed by this engineering group. They are called upon to service central offices, data centers, and outside plant enclosures. A good deal of their time is devoted to power, battery and generator selection, design, and placement. They usually monitor and inspect these systems as well.

9.2   Operations

Operations has responsibility for monitoring all networks and services, for managing network routing and configurations, and for upgrading the network and turning up new technology and systems. Organization of operations groups varies, but the following sections describe general functional areas and responsibilities that are found in carriers of scale. Smaller carriers, and even large enterprise networks perform these functions, but are likely to be organized differently. Individuals perform a wider range of tasks in smaller network organizations. Frank or Sue may be the IP routing guru, so they not only make all of the changes to routing configurations, they monitor the IP network and even assign IP addresses to individual ports. Networks of scale have to break those responsibilities down and build repeatable processes and systems, as well as train operators in more depth on specific technologies.

In networks of scale forces are organized into Operations and Care teams. Operations is focused on the health and capacity of the networks. Care is focused on solving customer problems. If a customer loses DSL service, that is a Care problem. If a neighborhood loses DSL service, that is an Operations problem. Care responds to customer inquiries and complaints, and to automated alerts of individual service problems. Operations responds to alarms and notifications generated by network equipment, network probes, and performance monitors.

Within Operations different groups have different responsibility in support of the Virtuous Cycle (see Figure 9.1).

When a failure of sufficient scope occurs, the NOC (Network Operations Center) receives an alarm notification. Critical alarms and some major alarms bring the attention of NOC

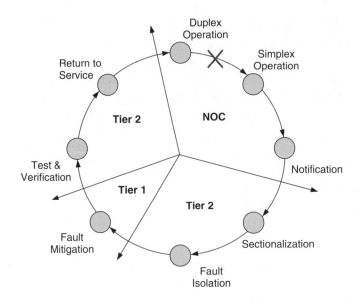

Figure 9.1   Tiered operations support.

staff. Because significant outages may generate notification storms, the NOCs first job is to identify which technology and network elements are causing the outage. Once the trouble is traced to a particular technology, such as a fiber route, the trouble can be referred to the transport work center where a tier 2 technician can remotely access the affected systems to determine if the problem is in a network element, or an outside plant physical facility. Once the fault is isolated to the point where a specific action, such as a circuit pack replacement, can be identified, a tier 1 technician can be dispatched using a trouble ticketing system. When the technician completes the work and updates the trouble ticket, the tier 2 technician is notified. The tier 2 technician can then test the circuit and verify the alarm has cleared.

9.2.1   Network Operations Center (NOCs)

When people think of Operations they often picture a NOC like the one in Figure 9.2.

9.2.1.1   Origins

Today's NOCs have a lineage back to Switching Control Centers (SCCs) designed in the 1960s and deployed in 1974 [1] based on a Bell Labs designed management system for stored program control switching systems, the Switching Control Center System (SCCS). By 1982 there were 300 SCCs in the AT&T companies. The duplex design of stored program control switching systems like the No. 1 ESS (Electronic Switching System) enabled offices to be unmanned, or at least unmanned most of the day. Remote maintenance coupled with the greater line capacity of the No. 1 ESS dramatically reduced field maintenance forces. The concept at that time was that maintenance, traffic engineering, and administration could be centralized around the SCC. The SCCS logged all alarms and informational messages

Figure 9.2   AT&T Global Network Operations Center (NOC).

from the switching systems. Technicians accessing any switching system logged into the SCCS first and then opened a session with the switching system through a telnet session. Different users had different privileges, such as maintenance, traffic, or provisioning. Traffic engineers could look at traffic reports, but they could not execute commands affecting the processor. All sessions were logged and visible to the SCCS root administrator. Experience with the SCCs laid much of the groundwork for what we find in NOCs today, although a great deal has changed.

Today instead of 300 SCCs AT&T has one Global Network Operations Center (GNOC) that is site redundant. There are many differences between the operations design of the SCCs and the GNOC. Although the AT&T of 1982 had more subscriber lines, the SCCs only managed the US voice network and not all of that; the SCCs did not manage a substantial number of lines that were still served by electromechanical systems like No. 5 Crossbar and Step by Step. The GNOC monitors a worldwide network with voice, Internet, VPN, teleconferencing, video conferencing, and an array of other services. The important difference is that the NOC is more narrowly focused than the SCCs. It is responsible for monitoring the health of the network, taking immediate service restoral action in the event of outages, and bringing Operations forces to bear when and where needed. Other Operations groups are responsible for network upgrades, network configuration, and traffic management. Node, link, and port failures that do not isolate or threaten to isolate critical network segments and elements are worked by other groups.

9.2.1.2   Responsibilities

In addition to fault management NOCs also monitor traffic levels and network congestion. In the 1980s that function was separate from the SCCs, in an work center called the Network Management Center (NMC). Today the NOC usually has that responsibility. TDM voice networks have a range of controls that can be exercised when an unexpected event, such as an earthquake, causes traffic loads to soar beyond network capacity in a particular region. IP networks have fewer controls, and have the added threat of Denial of Service (DoS) attacks. While some controls are activated automatically as congestion is detected, IP overloads generally have to be diagnosed before controls are put into play.

As shown in Figure 9.2, NOCs are generally organized into sections or tiers, usually by technology. Typical sections in a NOC include,

- **Transport** – monitoring fiber backbones, Ultra Long Haul (ULH) optical systems, DWDM systems, Reconfigurable Optical Add Drop Multiplexers (ROADMs), Synchronous Optical Network (SONET) systems, and undersea cable routes.
- **Voice** – monitors TDM and IP voice systems, SS7 network, voice mail systems, and SCP/NCPs (Service Control Point/Network Control Points).
- **MPLS/IP** – monitors core IP backbone, national, regional, and metro points of presence (POPs).
- **Mobility** – monitors RAN and mobility packet and support cores.
- **Public emergency services** – monitors E911 switches, systems, and network. May also monitor Communications Assistance for Law Enforcement Act (CALEA) and government emergency communications systems.

Specialized NOCs or operational centers of excellence are often established for networks such as Internet Protocol Television (IPTV) video networks, video conferencing, or teleconferencing networks. A carrier's primary NOC will generally not monitor premises systems or systems below a certain size. D4 channel banks, as an example, don't warrant monitoring by the primary NOC. Monitoring smaller systems can distract the staff from the larger risks they need to manage.

9.2.1.3 Automation

Automation is covered more completely in the chapter on support systems, but NOCs depend on it to keep their force size manageable and to avoid trying to solve problems that can be handled directly by others. Over 90% of the alarm notifications and troubles reported in the network should never reach the NOC. These alarms have recurring patterns enabling automation of the Virtuous Cycle. Operations support systems should have rules which enable sectionalization and notification of the responsible tier 2 or 1 team. Automatic trouble tickets and work orders should be prepared enabling direct dispatch. An example is a port alarm on an ATM aggregation switch where an ADSL remote terminal is homed. The failure can affect 500 customers or more, so action needs to be taken quickly. When an alarm is received via an EMS it should be correlated with other related alarms to look for a pattern. For example, a T carrier alarm, taken with an ATM alarm, may point to a root cause. Alarm correlation and suppression should be augmented by a rules based support system that invokes a port test at the ATM switch and at the remote terminal if it can be reached. In the majority of cases the results of the tests point to a problem in an ATM switch, the outside plant facility, or the remote terminal. Rules-based triage patterns are developed by the operators from their experience and codified in the support system.

9.2.1.4 Outage Reporting

Local exchange companies (LECs), interexchange companies (IXCs), wireless Mobile Network Operators (MNOs), satellite companies and cable companies (MSOs), and all have Federal Communications Commission (FCC) reporting requirements for outages meeting specified thresholds [2].

- **MSOs** – any outage of 30 minutes or more that,
    - Potentially affects at least 900 000 user minutes of telephony service,
    - Affects at least 1350 DS3 minutes,
    - Potentially affects any special offices, or
    - Potentially affects a 911 special facility.
- **Satellite providers** – any outage of 30 minutes or more that,
    - Results in a loss of complete accessibility to at least one satellite or transponder,
    - Results in a loss of a satellite communications link that potentially affects at least 900 000 user-minutes of either telephony service or paging service,
    - Potentially affects any special offices and facilities other than airports, or
    - Potentially affects a 911 special facility.
- **LECs and IXCs** – any outage of 30 minutes or more that,
    - Potentially blocks 90 000 calls or more,
    - Affects at least 1350 DS3 minutes,
    - Potentially cuts off 30 000 calls or more,
    - Potentially affects any special offices, or
    - Potentially affects a 911 special facility.
- **MNOs** – any outage of 30 minutes or more that,
    - Causes the loss of an Mobile Switching Center (MSC),
    - Affects at least 1350 DS3 minutes,
    - Potentially affects at least 900 000 user minutes of telephony,
    - Potentially affects at least 900 000 user minutes of data or paging service,
    - Potentially affects any special offices, or
    - Potentially affects a 911 special facility.

Operators usually have 72 h from the onset of the outage to prepare the report. In some cases, such as satellite failures, notification is required in 2 h. Reporting is automated through the Network Outage Reporting System (NORS) [3]. Carriers can file reports, but not review reports. Only registered members of the Department of Homeland Security can access them. FCC reportable outages are taken very seriously by carriers. There are usually well-defined escalation processes within the carrier's operations team and notifications are sent to the leadership team, affected enterprise customers, and consumer care teams.

The FCC lags technology and the market place in defining essential communications using the categories above. All of the providers above are facility based providers; that is, they own and manage the interconnecting facilities over which the service is provided.

Recently the FCC issued a Notice of Proposed Rule Making (NPRM) asking for comments on the inclusion of VoIP and Internet services [4]. The Telecommunications Acts of 1934 and 1996 are designed to regulate facility based providers. Cloud providers, companies that provide service via the Internet, have not been included within the FCC's regulatory jurisdiction. Microsoft's acquisition of Skype should serve as an interesting case to watch relative to this NPRM. Potentially Microsoft and others will need to begin reporting outages. Other services provided by facility based providers, such as Short Message Service (SMS) are mentioned in the NPRM as well. If added to the service list, outages like the one experience by RIM (the company, Research In Motion) recently could come under the reporting regime. Then the question arises of whether Instant Messaging (IM) provided by Yahoo or others falls within the definition. Although IM is not mentioned in the NPRM, it certainly behaves much like SMS.

9.2.2   Tiered Maintenance

Operations is usually organized into tiers, with skills and responsibility increasing in the higher tiers. The definition of each tier is not standardized by the industry; the ones described here are meant to be typical. Talent is hard to come by and so the objective of the carrier is to push as much work as possible down to the lower tiers, without risking service or network quality. Automation, tools, and process design can turn high tier jobs into lower tier jobs; Operations staff should be working continuously to design those processes and implement tools that accomplish that goal. Assignment of tiers should bring with it a disciplined escalation process. An escalation process is meant to insure that difficult service affecting troubles do not linger at lower tiers too long, and that problems that don't require high tier attention get solved at lower tiers, although it may take longer and the path to resolution may not be a straight one. A typical escalation policy is shown in the following.

- All failures that affect customer service and have the potential to become an FCC reportable, and all failures that affect 10 000 or more subscribers should be worked by tier 2 technicians for no more than 2 hours; after that they must be escalated to tier 3 and the NOC Watch Officer notified.
- All failures escalated to tier 3 should be worked by tier 3 technicians for no more than 4 hours. If the trouble is not resolved or bypassed, tier 3 will escalate to the equipment or software vendor's technical assistance center (TAC), or to tier 4. On escalation the NOC Watch Officer will be notified.

9.2.2.1   Operations Tier 1

Tier 1 technicians perform directed work that has little chance of causing network harm. Their work is either repetitive, or they work under close supervision of their manager and possibly a higher tier technician. Access to systems is often on a read only basis so that critical configurations or elements cannot be accidentally changed. Some examples of tasks a tier 1 technician might be asked to do are:

- verifying circuit connectivity and configurations,
- reviewing and archiving log files,

- verifying port configurations and performance (e.g., discards, carried packets),
- assisting IT in turning up a maintenance link to a remote system,
- reviewing documentation,
- assisting enterprise customer care in trouble shooting a customer port, and
- changing out-circuit packs.

9.2.2.2   Operations Tier 2

Tier 2 technicians resolve the great majority of problems in the network and also handle most of the administration that is not fully automated. Unlike tier 1 technicians they have the ability to go into network systems and change port and routing configurations. They may be restricted from changing critical configuration information and may be prohibited from changing configurations on critical systems like core MPLS routers and Signaling Transfer Points (STPs). Some examples of tasks a tier 2 technician might be asked to do are:

- using an existing procedure and design, configure an ATM Private Virtual Circuit (PVC) end to end,
- verify and change an Ethernet access circuit on an edge router to match a customer's gateway configuration,
- sectionalize and isolate a fault on an MPLS route, and
- verify a DNS server caching and access to the correct top level domain servers.

9.2.2.3   Operations Tier 3

Tier 3 technicians work at a design level when they are not engaged in network restoration or working on chronic or systemic problems. They often work closely with Systems Engineers in designing routing plans, complex network procedures for new system turn up or upgrades, or system interconnection plans. After an outage they are the ones that generally pore over log and alarm files to figure out what failure was at the root of the outage. They are proactive, always looking for ways to improve reliability, operational integrity, and automate trouble resolution. Some examples of tasks a tier 3 technician might be asked to do are:

- lead the design of a new operations work center or NOC,
- develop, test, and certify a complex procedure[1] for replacing a legacy edge router with a new generation,
- find the root cause of chronic call drops in a mobile network in a given geography,
- stabilize the network during a cascading failure, and
- develop automation to implement wide-scale network changes or upgrades.

9.2.2.4   Operations Tier 4

Most carriers have a tier 4 organization. In AT&T two organizations had responsibility for tier 4 support, the Electronic Systems Assurance Center (ESAC), and AT&T Labs. ESAC stands for Electronic Systems Assurance Center. They were formed in regional telephone

---

[1] Procedures are usually called Method of Procedures (MOPs), maintenance operations procedures in carriers.

companies when stored program control switching systems, like the No. 1 ESS and No. 4 ESS were introduced. Those systems had a level of complexity far above their predecessors and because they were computer controlled, required a completely new skill set. Membership in ESAC was restricted to the most talented and experienced engineers in Operations.

Systems Engineers often act as tier 4 teams. These are the same engineers that specify and certify the network elements, and so they have a depth of knowledge not necessarily acquired by other engineering or operations personnel with broader responsibilities. Systems Engineers also have access to captive systems and sophisticated test equipment. If a failure occurs in the network that seems congestion related, it is extremely difficult for operations technicians to recreate that problem in a live network. Captive, or lab systems, can be instrumented and load applied with test equipment specifically designed to simulate various types of live traffic.

9.2.2.5 Technical Assistance Centers (TACs)

Equipment suppliers in the carrier market understand they have to man TACs 24 hours a day. Escalation to TACs is controlled; only Operations tier 3 or tier 4 personnel, usually identified on a select list, are authorized to refer a trouble to TACs. Design engineers in the supplier organizations do not man the TACs; systems engineers, preferably ones with carrier experience are better choices for the TAC. TAC systems engineers, working with carrier engineers share alarm and log files, core dumps if they are available, and any other information that helps identify or eliminate potential causes of the failure. Supplier labs are drawn into the process when possible. Carrier labs tend to be better prepared to replicate network problems however, simply because they are more likely to replicate the configuration and interconnection arrangements used in the carrier's network. Equipment supplier's labs do not have their competitor's equipment interconnected with their equipment, with the same software, firmware, and configurations found in the live carrier network. TACs do have several tools and capabilities that can speed problem resolution. They have debug tools and can set software traps and additional logs in live systems that carrier engineers, with the exception of ESAC, are generally not qualified to use. A unique asset of suppliers are the design engineers. In the end, their in-depth knowledge of the design of the equipment is not replicated anywhere else.

9.3 Summary

Carriers manage their networks differently than most enterprises. The scale, technical complexity, and wide range of services drive carriers to develop technical tiers, processes and systems that specialize along technology and service lines. Engineering teams are organized into systems engineering teams that are responsible for technology approval and integration; network planning teams that develop economic guidelines for equipment selection, reuse, and placement; network and central office engineers that order and follow each system and facility installation in the field; outside plant engineers who plan and care for the extensive distribution networks; and common systems engineers that assess and add power, HVAC, and protection systems in the thousands of facilities that house network systems.

Operations is also organized into tiers with lower tiers responsible for individual trouble resolution and higher tiers responsible for network monitoring, restoration, and operational methods standardization. Operations automation is essential for consistency and repeatability of operational processes, and to contain operational costs and reduce down time. Engineering and Operations work together to define network and operational models that can be standardized and repeated around the globe. Standardized designs and operations enable automation and limit exposure to unproven software and systems. Suppliers are critical partners in the standardized models and the relationships with Operations and Engineering are formed early in the adoption and introduction stages so that when escalations to the supplier's TAC occur, each organization is familiar with the designs, network applications, and technical expertise of the other.

References

1. Bell Labs (1982) *Engineering and Operations in the Bell System*, 2nd edn, AT&T Bell Laboratories, p. 637.
2. FCC Rules (2009) Outage Reporting Requirements – Threshold Criteria. 47CFR4.9 – Sec. 4.9, Part 4.
3. FCC (2009) FCC Network Outage Reporting System User Manual, Version 6, April 9, 2009.
4. FCC PS Docket No. 11–82. (2009) The Proposed Extension of Part 4 of the Commission's Rules Regarding Outage Reporting to Interconnected Voice Over Internet Protocol Service Providers and Broadband Internet Service Providers.

10

Customer Marketing, Sales, and Care

Marketing, Sales, and Care are important partners of Engineering and Operations. Sales and Care are usually organized along three lines in global carriers, Industry, Consumer, and Enterprise.

10.1 Industry Markets

In Chapter 8 network peering and interconnection were described along with many of the services enabled by those arrangements. Industry Marketing and Sales are the organizations that serve industry competitors that purchase services from the carrier. Industry competitors are sometimes called "compartners" because they are simultaneously competitors, customers, and partners. There are two general categories of compartners, Competitive Local Exchange Carriers (CLECs) and other carriers (Incumbent Local Exchange Carriers: ILECs, Interexchange Carriers: IXCs, and Internet service providers: ISPs).

10.1.1 Competitive Local Exchange Carriers (CLECs)

The US Telecommunications Act of 1996 mandated broad unbundling requirements for the ILECs [1]. Facility unbundling meant that ILECs were required to provide non-discriminatory access to ILEC facilities, equipment, and services. The Federal Communications Commission's (FCC's) implementation of the Act identified seven areas for unbundling under the regulatory terms UNE-L and UNE-P. UNE-L is an acronym for unbundled network element – loop; UNE-P means unbundled network element – platform. The UNE-P concept was that logical functions, such as switching and transmission, or even more granular functions under Advanced Intelligent Network (AIN) could be unbundled into a platform for resale under regulated pricing. The seven areas identified by the FCC for unbundling are:

- local loops;
- network interfaces;

Global Networks: Engineering, Operations and Design, First Edition. G. Keith Cambron.
© 2013 John Wiley & Sons, Ltd. Published 2013 by John Wiley & Sons, Ltd.

- local and tandem switching;
- interoffice transmission facilities;
- signaling networks and call related databases;
- operations support systems; and
- operator services and directory assistance.

Under the UNE-P concept it was possible for a CLEC to enter the market and never put in place a single facility or piece of equipment. Since pricing was based on regulatory rules using a long term incremental model, business cases could be constructed for doing just that. The order as originally released, and as modified and interpreted by the FCC and state Public Utility Commissions (PUCs) was consider unduly burdensome and non-compensatory by the ILECs. After eight years of regulatory review and several court cases, ILECs were granted some relief by the US D.C. District Court in 2004 [2], and the rules in place today are relatively unchanged from that ruling [3].

10.1.1.1 Local Loops

Today CLECs can lease local loops and transmission facilities (DS1/DS3), but at prices much closer to market. They can no longer purchase unbundled switching.

CLECs such as Covad® and MegaPath® provide voice, broadband, and transport services by combining their switching, typically Voice over Internet Protocol (VoIP), with leased loops and facilities from ILECs. While each state in the US has their own rules regarding loop unbundling, the following rules taken from AT&T's broadband services handbook for CLECs in the Missouri franchise are typical [4].

- The loop must be qualified for broadband services. Loops that have load coils, bridge taps, or excessive attenuation are disqualified. A simple test is that if the ILEC can offer Asymmetric Digital Subscriber Line (ADSL) service, then the CLEC should be able to as well.
- The high frequency portion of the loop is unbundled. Time Division Multiplex (TDM) voice if provided, is served by the ILEC. When provided by the ILEC, only Plain Old Telephone Service (POTS) service is supported. Centrex, ground start, and other voice services are specifically prohibited.
- The handbook specifically mentions ADSL but does not offer unbundling for other transmission technologies, like Very High Bit Rate Digital Subscriber Line (VDSL2).
- The CLEC must have established collocation in the office, either virtual or physical.
- The CLEC must have a cross connect point in the serving central office (CO).

When working with a CLEC to serve ADSL on unbundled loops, spectral compatibility needs to be managed to minimize interference. T1 and High Bitrate Digital Subscriber Line (HDSL) are dominant disturbers in the downstream (CO to premises) direction on customer loops; HDSL and Symmetric Digital Subscriber Line (SDSL) are dominant disturbers in the upstream direction. If circuits are provisioned with these services where existing CLEC loops are in service there is a potential for service degradation. One way of mitigating interference is by binder management. Provisioning CLEC loops in one binder and T1 and HDSL in another. While that is a reasonable approach, loop assignments sometimes end up being changed in the outside plant because of availability or because of impairments such as common mode noise. Usually the dominant impairment is self far end crosstalk (FEXT). As more ADSL service is added, self-crosstalk increases, potentially affecting the highest data rates customers receive.

CLEC unbundled loops can also limit or complicate engineering and service options. For example, if a distribution area is a good candidate for deployment of VDSL2, planners and engineers have to consider the effects of CLEC broadband loops on the service. The ILEC can choose to swing their entire ADSL customer population to VDSL2, but they cannot mandate CLEC moves. VDSL2 and ADSL are often incompatible in the same cable, so binder management or some form of segregation may be needed before deployment of VDSL2 can begin. The same kind of issue arises when considering remote terminal deployment. If a distribution area at an extreme loop length is a good candidate for a remote terminal but some loops are unbundled, the ILEC needs to work with the CLEC to figure out if and how a remote terminal can be placed while preserving the CLEC's customer service.

10.1.1.2 Dark Fiber

With the modifications of the 1996 Act coming after the D.C. Court's decision, unbundling dark fiber is no longer a regulatory obligation of ILECs. Where ILECs offer it, they do it under restrictions, limiting the conditions so that only spare capacity is leased.

10.1.1.3 Voice Interconnection

Interconnection between a CLEC and ILEC looks much like interconnection to an IXC from a technical viewpoint. Here interconnection applies to calls that originate on a CLEC switch and terminate on an ILEC switch, or originate on an ILEC switch and terminate on a CLEC switch in a given Local Access Transport Area (LATA). Interconnection can occur at tandems or local switching systems and Signaling System No. 7 (SS7) is used for out of band signaling. When completing calls to each other, Local Number Portability (LNP) queries must be made since they are the N-1 network. The companies work together on transmission designs to conform to industry standard end to end loss and delay objectives. Network management, particularly planning and designing for high call in volume events is recommended and required by some ILECs.

10.1.1.4 Voice Transit Service

ILECs have extensive completion fields in a LATA. CLECs can use the ILEC for transit service. Calls originating on a CLEC switching system and destined for a mobile carrier, IXC, or another CLEC switch can be completed by forwarding the call to an ILEC tandem because the tandem will have the necessary connectivity. ILECs sell this service under the name of transit service or unbundled common transport.

10.1.1.5 Operator Services and Directory Services

ILECs often offer operator services and directory services to CLECs. CLECs generally need to order one way trunk groups from their switching systems to the ILEC directory assistance or toll operator systems and conform to the technical specifications for calling party identification and signaling. CLEC calls are identified to the operator system and can be branded with the name of the CLEC so that the calling party knows the CLEC is providing the service, albeit indirectly. White and Yellow Pages® listings and services can also be ordered from the ILEC in many cases.

10.1.1.6 E-911 Service

Enhanced 911 (E-911) service is mandated in virtually every US jurisdictions. There are two essential elements to E-911 service, emergency call routing and Automatic Line Identification (ALI), as described in Chapter 8. For smaller CLECs purchasing this service from an ILEC makes economic sense because of the cost of the infrastructure and the operational challenges of maintaining service to all of the Public Service Answering Points (PSAPs).

10.1.1.7 SS7 Services

CLECs can usually purchase SS7 connectivity from the ILEC. 64 kbps A links or SIGTRAN links are provisioned between ILEC STPs (Signaling Transfer Points) and CLEC switching systems. The ILEC works with the CLEC to make the necessary notifications to other carriers via the Local Exchange Routing Guide (LERG).

10.1.1.8 Database Services

Line Information Database (LIDB) and (LNP) database services are available to CLECs. CLECs can populate ILEC databases with their data, saving them from having to build the infrastructure themselves. If the CLEC is purchasing SS7 services from the ILEC, then using these databases is a logical choice. If the CLEC is using a third party SS7 aggregator, they may choose to purchase LIDB and LNP services from them.

10.1.1.9 Engineering and Operations

Cooperation between ILECs and CLECs helps both operators and it begins with planning and forecasting. As competitors there is a reluctance to share market information, but sharing forecasts helps each provider to better plan capacity and not be caught short with orders and due dates that can't be met. The same is true for engineering and operations. Having periodic sessions to discuss plans and best practices as well as well documented procedures and contact points serves both companies and their customers well.

10.1.2   Interexchange Carriers (IXCs)

Interconnection arrangements for IXCs are covered in Chapter 8. IXCs can purchase the services described above under CLECs and can also purchase switched access service as a registered IXC, or purchase VoIP interconnection, which is not regulated like switched access.

10.1.2.1 Switched Access Service

Switched access service for ILEC to IXC interconnection was defined as Feature Group D under the terms of the Modified Final Judgment in 1982. Feature Group D was a Multi-Frequency signaling protocol[1]; that same functionality was later translated into SS7 [5]. Switched access service allows ILEC subscribers to select an IXC as their Primary Inter Local Access Transport Area Carrier (PIC). When ILECs received authority to compete with IXCs, they formed their own IXC organizations that operated at arm's length from the ILEC, but carried the brand of the ILEC. With the mergers and acquisitions of the last decade, most subscribers choose the same carrier for local and long distance service, although the option of choosing another carrier for local or long distance service is still there. Subscribers can also select an IXC other than their PIC on a call by call basis by dialing 10-10XXX before dialing the called number. The XXX code identifies the carrier of their choice.

10.1.2.2 Voice over IP Wholesale Service

ILECs also offer VoIP interconnection to the Public Switched Telephone Network (PSTN), but unlike switched access it is not regulated by the FCC because it is an Internet Protocol (IP) service. Some carriers offer originating and terminating services and some offer only terminating service. Terminating service is popular with providers like Vonage® because they can peer at a few points and have access to the PSTN which assures completion to any subscriber.

10.1.2.3 Billing Services

ILECs generally offer billing services to IXCs and other services providers. It simplifies the operation of the IXC and subscribers generally prefer one bill with all telecom charges.

10.2 Consumer Markets

Once limited to phone service and broadcast TV, the mix of consumer products offered by carriers today is a rich one. Products include:

- **Wireless voice service** – global roaming with Global System for Mobile Communications (GSM) and selected roaming with CDMA2000. Prepaid and postpaid products, family plans, and a wide range of features, depending on the device.
- **Wireline voice service** – TDM voice has long had a rich feature set, conference calling, call waiting, automatic call back, calling name and number, call screening, call blocking, and so on.
- **VoIP** – available to broadband customers, including cable MSO customers; the primary attraction is the ability of customers to use Internet access to manage features and their account.
- **Voicemail** – visual voicemail, voicemail to e-mail.

---

[1] The author wrote the original specification in 1979 at Bell Labs with advice from Bob Keevers and others.

- **e-mail** – still a product provided by ILECs for some consumers, but more important in the enterprise space.
- **Telephones, wired and cordless, and inside wiring**.
- **TV** – Internet Protocol Television (IPTV), Switched Digital Video (SDV), Video On Demand (VOD), Pay per View, and a wide range of features such as Digital Video Recorders (DVRs) are offered by MSOs and some telcos. There are also partnerships between telcos and satellite TV providers.
- **Home networking and computer support** – a service similar to the Geek Squad™ in the US. Internet based automatic backup services are also provided by some carriers.
- **Home security**.
- **Wireline broadband** – a range of speeds, products, and services for Internet access, including home Wi-Fi.
- **Wireless broadband** – air cards and service for laptops, tablets, e-readers, feature phones, and smart phones; access to the Internet and applications.

10.2.1 Product Marketing

Product Marketing has two significant forums in common with Engineering and Operations, long range planning and product development.

10.2.1.1 Long Range Planning

Product Marketing plays the lead role in forecasting and planning consumer products and services. Access networks, for both wireline carriers and cable MSOs, are the most costly part of the network and the least tractable. Significant upgrades of access networks take three to five years and large capital investments, so it is crucial that Product Marketing layout a five-year projection of the direction services will take, how much bandwidth will be needed, and what quality of service will be needed. A list of products and services isn't essential but a clear statement of trends is foundational for network design and engineering. A topic that always needs to be addressed in a five year plan is bandwidth to the home. Consider the historical trends shown in Table 10.1 and chart in Figure 10.1.

Historically, as high-end users pursue technology, they have come to expect an increase in downstream bandwidth at an annual rate of 40–50%. This phenomenon was first noted by Jakob Nielsen and has been named Nielsen's Law.[2] Like Moore's Law, it is more correctly an observation or even a conjecture, but historical data certainly support the observation. While these high end users don't affect the market directly, they do set trends and the mass market eventually follows. Engineering teams meet the demand in two ways, technology extension and technology replacement.

Copper loops continue to benefit from technology extension. Integrated Services Digital Network (ISDN), ADSL1, ADSL2+, and VDSL2 are all examples of extending the capability and investment of copper pairs well beyond the 3 kHz bandwidth originally intended. Current technologies to further extend copper fall into two categories, pair bonding and advanced processing. Pair bonding is in early deployment and essentially doubles data rates by using two pair rather than one. Advanced processing includes technologies like dynamic

---

[2] Postulated by Dr Jakob Nielsen as a corollary to Moore's Law.

Table 10.1 Consumer broadband access speeds

Technology	Rate (kbps)	Introduced
V.22	1.2	1980
V.22bis	10	1984
V.32bis	36	1991
ISDN BRI	144	1992
ADSL 1.5	1500	1999
ADSL2+	8000	2003
BPON	50 000	2003
VDSL2	23 000	2005
DOC 3.0	80 000	2007
GPON	100 000	2008

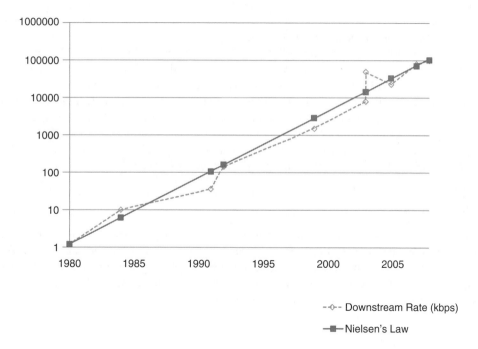

Figure 10.1 Historical chart of consumer data rates.

spectrum management (DSM) and expanded bands of up to 30 MHz. Advanced processing yields substantial benefits when VDSL is used in fiber to the building (FTTB) and fiber to the curb (FTTC), but is far less effective when used in fiber to the node (FTTN) and CO based applications.

Data Over Cable Service Interface Specification (DOCSIS) is another example of technology extension. DOCSIS 3.0 has demonstrated achievable consumer data rates approaching 100 Mbps by improving electronics at the terminal and home, and without significantly affecting the design of the coaxial distribution plant. Like FTTN deployments, DOCSIS pushes fiber deeper into the network, in effect replacing coaxial feeder networks with fiber.

Technology replacement in the consumer access network means deep fiber deployment, either to the curb, the building, or the home. Consumer Marketing has to develop a compelling business plan to warrant that commitment. Engineering can help by aggregating demand from other sectors, such as mobile cell site bandwidth growth and enterprise access growth.

10.2.1.2 Product Development

As used here, Product Development is the process of designing and deploying a new product or service. Engineering and Operations usually have members that are part of new product teams. They can bring significant value to the design of the product, particularly when the product has a strong network connection. Consumer products that have a network connection and particularly ones that are part of a bundle or have a link to existing products tend to be the most successful. Broadly speaking, carriers are networking companies, not home entertainment, content providers or device companies.

Systems Engineering, working with Operations, Information Technology (IT), and Customer Care can design complete solutions for new products that serve the market well functionally and make it easy to install, provision, maintain, and troubleshoot. Service analysis, described in Chapter 4, is an important prerequisite to setting bandwidth and quality objectives of new products. Human factors engineers working with systems engineers can improve usability and dramatically reduce customer installation errors. They can also address critical environmental, home networking, and feature interaction issues that are easy to solve in the design phase but very difficult once the product is launched.

10.2.2 Consumer Care

Customer Care is one of the largest organizations staffed by carriers. They are responsible for responding to customers inquires about products and services. Care operates call centers, self-help web sites, e-mail, and instant messaging (IM) care groups and tiered technician organizations. These teams serve the entire consumer product set described earlier. High volume customer contacts are in the following areas:

- **Service inquiry** – service availability, price, terms, and features,
- **Service changes** – new service and service discontinuation, change, or modify existing service,
- **Billing** – often the biggest single source of contacts; in many jurisdictions wireline carriers are obligated to allow third parties to bill via the carrier. Inquiries include disputes, requests for an extension of payment, explaining the bill, and fees, refunds, combining bills,
- **Installation and operation** – some products like ADSL and voice mail are usually self-installs and can generate a high volume of contacts, particularly for customers that are not web users or technically savvy. Some services, like IPTV and mobile smart phones have complex feature sets,

- **Service complaints** – anything from a broken device to outages. Triaging these and getting the right technician or resolution is one of the biggest challenges of the carriers.

Customer Care can be organized into three tiers.

1. Tier 0 is the lowest skill level usually serving as an entry point for callers who do not use directed numbers, web, e-mail, IM, or interactive voice responses (IVRs) to direct the inquiry to a specific class of agents. Tier 0 agents refer customers dialing a general toll free number. They transfer the customers to service specific tier 1 agents. Sales agents in retail stores also act as tier 0 agents for in store customer complaints that cannot be resolved immediately.
2. Tier 1 agents are service specific agents skilled at resolving the most common questions and complaints. They service customers via phone, e-mail, and IM. In general they can access customer information, service histories, and some limited support systems; they can make changes to customers' service packages within the bounds of the defined services. They cannot make changes to network parameters or trouble shoot customer premises equipment. Their authority may extend to "bouncing" a line. That is, forcing a Digital Subscriber Line (DSL) line into a restart and realignment to see if the trouble clears.
3. Tier 2 agents are the highest skill level assigned to resolving individual customer complaints or troubles. Tier 2 is normally called in by tier 1 and does not respond directly to customers unless a referral is made. These agents can make network changes that affect particular subscribers but cannot change general network equipment parameters that affect more than an individual customer. They are often skilled at resolving conflicts with services and customer premises equipment and can access network elements remotely and run tests on subscriber lines and equipment such as a residential gateway.

Engineering, working with IT and Operations improves customer satisfaction and helps manage costs in four distinct ways.

10.2.2.1   Network Systems and Services Design

Customer care features, like network operations features, should be designed into network elements and service offerings from the beginning. For wireline services faults need to be isolated to the network, outside plant, or customer premises. The three areas have different work forces so any fault that cannot be isolated is likely to result in a trouble ticket, or worse, a dispatch of the wrong technician. Not only does that generate avoidable costs, it frustrates customers. When premises technicians are dispatched customers have to arrange their schedules to meet them. If the technician cannot resolve the problem, the customer has taken time off from work or otherwise changed their schedule with no resolution. To better diagnose problems in the home, residential gateways should implement a TR-069 [6] or an Simple Network Management Protocol (SNMP) agent with Wide Area Network (WAN) and Local Area Network (LAN) remote configuration, notifications, performance monitoring, and testing functions.

An effective way to codify and add rigor to the design process is to prepare detailed care use cases. There may be multiple residential configurations to consider, but by documenting each expected fault condition and stepping through expected notifications, logs and troubleshooting procedures a set of use cases can be prepared that guides service design requirements, operations support systems requirements, and serves as a starting point for the preparation of care procedures. Use cases are valuable for assessing the necessary skill levels of care technicians as well.

10.2.2.2 Business and Operations Support System Design

Systems engineers can simplify and speed IT design of provisioning systems and improve transactional reliability by building provisioning templates and models. IT developers may lack the domain knowledge needed for setting transmission options on subscriber lines or selecting the right Customer Premises Equipment (CPE) options for different residential network configurations. Systems that qualify subscribers for service are another area where the network and systems knowledge of engineers can contribute. Operations research engineers can also analyze call centers, voice response systems, and other service arrangements for effectiveness. Call centers and voice response systems in particular are queue and routing oriented and can often be dramatically improved by analysis and redesign.

10.2.2.3 Special Tools and Systems

Special tools, like the DSL tools described in Chapter 5, can be prototyped relatively quickly when installation or maintenance issues arise with new technologies. Systems Engineering has the three skills necessary to act quickly to develop solutions, network domain knowledge, operations knowledge, and software development experience. In addition to the contributions to DSL operations sited earlier, the approach of employing systems engineers to develop specialized tools has worked well in consumer mobile care. AT&T engineers developed an application called Mark the Spot. The application is used by mobile subscribers when they have poor or no reception. When the user submits a report with the application the location and time of the report is saved on the device. When the device reestablishes connectivity with the network, a report is forwarded noting the fault. The application accomplishes two objectives. It gives customers an easy way to notify the carrier of inadequate service. It also gives the carrier information it cannot get another way. The application solves a basic problem; network elements do not record failures they cannot detect.

10.2.2.4 Operational Analysis and Information Mining

There are strong patterns and tendencies in customer troubles and customer contacts. Surprisingly, those closest to the problems are often the least likely to see them, simply because they are consumed with the operational pressures of daily demands. Operations research analysts and industrial engineers, paired with domain engineers and software engineers, working together in very small teams are adept at working with Care and making significant strides to improve customer responsiveness and lower costs. This touches on an issue addressed in Chapter 12, information management. Carriers have more data than they realize. Data sits in repositories that are often isolated from each other and so correlation and understanding is marginal since each work group focuses on the data they have and understand.

Measurement and reward structures can also drive poor behavior. In one case our teams investigated we found a customer care contractor was being paid based on two measurements, the average time to clear a customer call and the number of calls handled. Given the incentives, the contractor had done what made sense; they pushed the measurement and reward regime down to the agents handling customer calls. The more calls and the shorter the time to clear the call, the better the rating, and presumably the higher the agent's compensation. An examination of transactions revealed that the agents found they could tell the customer to call another center or otherwise dispose of the customer as quickly as possible, without responding to the complaint. That behavior increased call volumes and shortened holding times. Their metrics improved, but of course the customer suffered.

There is also a common misperception about high level customer satisfaction goals. Customer satisfaction as measured by surveys, are the industry norm for assessing service. That idea is intuitive: if you want to know if someone is satisfied with your service, ask them. There is a long history of significant investment and focus on surveys in the industry. The Telephone Service Attitude Measurement (TELSAM) Program was launched by AT&T in 1971 [7]. Through a series of questions asked to customers by an independent TELSAM group, measurements were taken and an index, a single or small set of numbers created for each work group measured by the program. There were measurements and indices for installation, repair, operator services, voice quality, and business offices. Employees were rated, and eventually compensated in some part, based on those indices. In 1971 TELSAM and customer surveys in general made more sense than they do in today's competitive market place. In 1971 AT&T was a regulated monopoly, so customers had little choice when choosing their provider. If they wanted phone service, they had to use the franchise carrier. Today customers have choices. What matters is whether they choose your service over others. So a customer rating of excellent may give a false sense of progress if the same customer, asked the same question rates your competitor exceptional, or even best in class. Likewise a rating of poor may be fine if your competition is judged to be unacceptable. Customer satisfaction, as measured by surveys, on individual carrier services and products is less important than market share or churn. Collectively product satisfaction matters a great deal if it affects the public's perception of the brand. If a brand is tarnished, customers are less likely to consider the carrier's products and services.

Data mining and analyses, when undertaken by an experienced team, can produce results that can dramatically improve the effectiveness of Care organizations. Typical benefits include:

- **Solving 80% of inquires with 20% of the responses** – The majority of customer inquiries are repetitive. This is particularly true when new products or services are introduced and the installation or operation is non-intuitive, or installation instructions are wanting. Crafting Q&A scripts for call centers, or text instructions for e-mail and IM contacts can solve these problems in a consistent and repeatable way. This is an iterative process to identify the keys that indicate which pattern an inquiry represents, if any, and then to craft responses that are successful.
- **Improved IVR trees** – Analysts generally have sophisticated tools for IVR tree analysis. They can improve the effectiveness of the trees, eliminate dead ends, and put in place permanent metrics and tools. IVR effectiveness changes almost daily because new products and services are introduced, network performance changes, or other exogenous effects impact customer service.

- **Correlate customer churn with other metrics** – When customers abandon a product, associating cause and effect quickly is essential to reversing the trend. Unfortunately managers often react instinctively, and work on the wrong problem. Analysts can mine through data and find correlations to prepare a list of suspects. Correlation is not causation, but it is rare that the cause does not correlate to the effect.
- **Data aggregation and presentation** – Service representatives need the best possible support. When a customer contacts a representative they should have as much information about the customer as possible, but presented in a way that is customized to the context and circumstances. If the customer has just traversed an IVR tree and indicated they have a billing question, the representative's screen should be different than one with a service question. This seems obvious, but analysts will find many ways to improve the information presented to agents. Agents particularly need customer history. If a customer has called before about this exact problem, the agent should have that history as well as trouble ticket histories to be able to respond to the customer without having them verbally repeat the history each time.
- **Agent alerts on network status changes** – Agents are too busy to watch a wall of network alarms and status. When a customer calls and they are being affected by a network outage, that information should come to the front of the agent's screen. Critical and relevant information should go to the agent. They should not have to look for it.
- **Customer service status** – Systems Engineering can build tools that enable the agent to perform real time checks of a customer's service. Ideally those checks are performed automatically when a contact is routed to an agent, and the information presented automatically. Additional information, such as the customer's current bit rate and service performance should also be displayed.
- **Manager and agent training** – When done with care, analysts can find comparable agents, managers, and work groups, and compare their results. If there are statistically relevant differences, then the analysts can search for operational, knowledge, or behavioral differences that account for the differences in results.

10.3   Enterprise Markets

Consumer markets and Consumer Care are transaction oriented. Each sales opportunity or care call is unique and so the goal is to make the most of that opportunity. Enterprise markets are life cycle accounts that succeed or fail based on the relationship the carrier builds with the customer over time. Customers are segregated horizontally and vertically. The horizontal dimension is delineated by the size of the account or the size of the customer. The vertical dimension is industry segment. The financial industry has a different view of reliability and security than the agricultural industry, and so solutions need to be crafted to fit the needs and demands of each industry.

Expertise is a differentiator in this market. Customer contacts in this market are often CIOs, senior IT managers, or telecommunications managers. The range of expertise varies, but in general they are very knowledgeable and current with technology and trends in networks and market places. Account teams in Consumer Marketing study their products. Account teams in Enterprise Marketing study their customer, and their customer's business. A good grounding in enterprise products is important, but Enterprise Marketing should have dedicated product specialist available to the account teams and their customers.

Senior Engineering and Operations managers spend time with enterprise customers, large and small. They engage with customers in several ways:

- **Technical and information forums** – usually held once or twice a year, large and middle size customers spend a day or two with engineering, operations, sales, and marketing teams to gain exposure to new technology, network and product plans, and to give direct feedback on what is working in the relationship and what needs improvement. Customers come to these forums and put a good deal of their own effort into directing the topics and discussion.
- **Web-based seminars and white papers** – web conferences with Q&A sessions are a good way to introduce specific technology topics like IPv6 or cloud computing. An hour seems to be about the right length of time to share key topics.
- **Network Operations Center (NOC) or lab visits** – customers like visiting technology centers, so visits to NOCs, laboratories, or other technology oriented sites generate a good deal of interest and are usually very interactive.
- **Industry analysts** – companies like Forrester, Gartner, and Nemertes Research analyze and rate carriers for product, scope, performance, and strategy. As a group they are knowledgeable and engaging with them enables carriers to communicate their view of the market and technology. Analysts will also ask timely questions about technology direction and offer insights into market and technology trends.
- **Customer site visits** – customers with large account or complex networks rightfully expect individual attention. Engineers making these visits need to set aside a morning or afternoon prior to the visit to study the customer's network, recent operating history, and any conversion or expansion plans.
- **Outages** – if the customer experiences a significant outage or performance degradation, find a way to communicate exactly what happened when the outage is over and root cause analysis is complete. A phone call may be enough or a site visit may be warranted.

Engineering teams can make contributions during the entire customer cycle. Customers and sales teams benefit, but so do the engineers. Participating in the entire cycle is one sure way to know what is working in the network and what is not.

10.3.1 Pre-Sales Support

Data mining and analytical analysis and tools can be applied effectively to identify sales opportunities for enterprise account teams. By examining traffic volumes and netflow data analysts can compare locations within an enterprise customer network and identify locations where sales opportunities are the greatest. A more in depth analysis of a customer network can identify opportunities for upgrades to more current technology. Comparisons of customers within an industry, like health care, may give insight into which companies are leading in leveraging their networks and which are falling behind in adopting technologies that can improve efficiency. If the size of the opportunity warrants, engineers can work with the account teams on network redesigns and submit unsolicited proposals backed by data to show improved performance and efficiency.

10.3.2 Sales Support

Sales engineers are devoted specifically to providing hands on technical advice and information to the account team and to customers. Many customer sales inquiries and requests for proposals (RFPs) directed to the account teams are similar in nature, but the teams have to respond, often by researching network configurations, preparing a preliminary design, pricing it, and putting the information in a response. If they win the bid, they begin the process of translating their design and proposal into an order so that it can be provisioned and prepared for service. Network systems engineers have access to databases that have complete inventories and topologies of each customer and the network in general. The engineers, coupled with a small software development team, can automate the design, estimate, and order processes into a single unified flow for the most common customer product inquiries. Once a process is automated, sales engineers can be trained on how to use the tools, thereby improving response times and consistency in design and ordering.

10.3.3 Engineering and Implementation

The brunt of engineering and implementation falls directly on the account teams and line engineering teams. For the great majority of sales it is simply a matter of project management, timing and executing the ordering, provisioning and turn up of ports, routes and services.

Large customer network sales are a different matter, particularly where managed services are part of the sale. In complex sales Systems Engineering should be engaged early in the process to identify critical design issues and plan the design process. Line engineers tend to work from the bottom of the Open Systems Interconnection (OSI) stack upward, ports to nodes to networks to services. Systems engineers work down the OSI stack, exactly the opposite sequence. Identifying services and communities of interest, current, and planned, is a starting point for laying out designs for connectivity, traffic types, traffic management, and network domains. From those analyses come alternatives for control planes, route topology, ingress, and egress points. When working through this sequence compatibility between the customer's systems and the carrier's systems has to be vetted at each point. If the sale is for a managed network offering, operational planning, and systems planning have to move with the network design. Here again, use cases are a great help. Each potential network failure has to be examined to determine how alarms will be generated, correlated, and managed. Reliability analysis of the network is needed so that all fatal single faults are eliminated and traffic capacity under single faults is sufficient, assuming the customer has elected to pay for the protection. Performance management data collection, reduction, analysis, and reporting are required as well; customer performance Service Level Agreements (SLAs) need a mechanism for conformance.

10.4 Summary

Carriers build their marketing and sales organizations along customer lines. Industry Markets serves CLECs, ILECs, and other carriers that interconnect with the carrier's network. Carriers and their industry customers work together, within the regulatory frameworks set by sovereigns to interconnect over a wide range of services. CLECs and ILECs are competitors, but they are also customers and partners. Within the carrier Industry Markets supports and serves their customer base, working with Engineering and Operations to plan, engineer, and manage interconnecting and resale services. Consumer markets are served by standardizing processes and automating sales and care. Tiered work groups are structured to solve problems at the lowest practical level, while achieving low churn rates and managing the cost of customer acquisition. Enterprise markets are segmented by industry. These customers require specialized talents within and outside Sales and Marketing. Systems engineers can make a real difference in the degree of customization, information, and analyses developed in support of the sales and account teams.

References

1. Telecommunications Act of 1996, Pub. L. No. 104–104, 110 Stat. 56 (1946).
2. D.C. Circuit Court United States Telecom Association v. FCC, 359 F.3d 554 (D.C. Cir. 2004) (USTA II) (2004).
3. Unbundling Policy in the United States, Players, Outcomes and Effects (2005) Quello Center for Telecommunications Management and Law.
4. AT&T (2005) Advanced Broadband Service Carrier Guide, Release 1.0, November 31, 2005.
5. GR-394, LSSGR (2007) *Switching System Generic Requirements for Interexchange Carrier Interconnection (ICI) Using the Integrated Services Digital Network User Part(ISDNUP)*.
6. TR-069 Amendment 3 (2010) *CPE WAN Management Protocol, Broadband Forum*, Issue 1, November 2010.
7. Bell Labs (1977) *Engineering and Operations in the Bell System*, 1st edn, AT&T Bell Laboratories, p. 619.

11

Fault Management

This chapter explores how carriers of scale manage networks, the systems that support fault management and how work groups and systems follow the Virtuous Cycle.

11.1   Network Management Work Groups

Carriers of scale need management processes and systems that are quite different than small carriers or enterprises. Those needs are based on how the work groups are organized, what information they need, what their responsibilities are and what freedom of action they have. Chapter 9 described the roles of Operations and Engineering which depend most directly on Network Management Systems (NMSs) and Operations Support Systems (OSSs). Care and Sales organizations, described in Chapter 10 depend on Element Management Systems (EMSs) and OSSs for subscriber trouble management and configuration. Summarizing the roles of the relevant organizations for the discussion of EMS and NMSs:

1. **Network Operations Centers (NOCs)** – their primary concerns are network fault management and performance monitoring. NOCs are not directly responsible for applications or subscriber management. Their charter is to maintain connectivity and capacity under link failures, node outages, core systems (e.g., DNS: Domain Name System) failures, network congestion, and security attacks. Their primary tools are NMSs.
2. **Operations Tiers 2 and 3** – their principle tools are EMSs and direct access to Network Elements (NEs) but they also use NMSs and OSSs. These organizations are dedicated to specific technologies and services.
3. **Engineering** – data collection for capacity management is performed through EMSs and in some cases directly through network elements. In the next chapter engineering systems are described in more detail.

NMSs and EMSs are used by Operations to monitor and restore networks and network systems. Network management is distinctly different than either service management or subscriber management; it ensures:

- **Connectivity** – the physical pathways that touch all of the network end points,
- **Capacity** – delivery of sufficient bandwidth to carry subscriber services,

Global Networks: Engineering, Operations and Design, First Edition. G. Keith Cambron.
© 2013 John Wiley & Sons, Ltd. Published 2013 by John Wiley & Sons, Ltd.

- **Routing** – choosing the path through the network for a specific session or service, and
- **Connection control** – monitoring signaling and connection management functions that enable end to end session establishment and release on demand, and enforce admission control.

NMSs work in cooperation with NEs and EMSs to achieve these goals. The role of an NMS is a function of the sophistication and effectiveness of underlying elements, their EMSs and how the networks are designed and organized. ITU developed an extensive set of recommendations for network elements and systems under the banner of the Telecommunications Management Network (TMN) [1]. However, the comparison between standards in networks and standards in systems is stark. The telecommunications industry has been effective at standardizing and advancing network systems and network communication protocols. Network elements from different suppliers interwork well, have clearly defined roles and have design rules and constraints that are readily translated into functional networks. The same cannot be said of management systems. ITU, ATIS,[1] and the Technology Management Forum (TMF) have labored for two decades in attempts to standardize support systems but no single set of standards, functional, or technical have taken hold as a practical model. Some of those concepts and nomenclature are used in this section and the next chapter to convey functionality in the abstract, but the majority of this text reflects the author's practical experience and point of view.

11.2   Systems Planes

Understanding network management begins with an explanation of how networks separate management functions from user services and transport. Networks implement and rely on three communication planes, a bearer plane, a control plane, and a management plane. The bearer plane carries user data. Control planes carry signaling that determines the routing of the bearer plane and network management information affecting the real time topology and reachability of network addresses. The management plane, unlike the other two planes is not involved directly in the delivery of user data; it facilitates management of the network elements including all of the Fault, Configuration, Accounting, Performance, Security (FCAPS) functions. The three planes can be separated physically or logically. Figure 11.1 illustrates the role of the classes of network systems and the planes in which they participate.

11.2.1   Bearer Planes

Bearer planes are the largest of the three planes by far carrying several orders of magnitude more traffic than the other two planes. Bearer planes are restricted to NEs. The statement is axiomatic in that a system carrying bearer plane traffic by definition becomes a network element. Systems designated as NEs must conform to network standards described in Chapters 2 and 3, otherwise end to end availability can suffer dramatically. As new services and next generation networks are introduced architects sometimes place an Information Technology (IT) system in the bearer or control plane. It often becomes a weak link in the service path and either has to be removed or adopt network standards for hardware and

---

[1] Alliance for Telecommunications Industry Solutions.

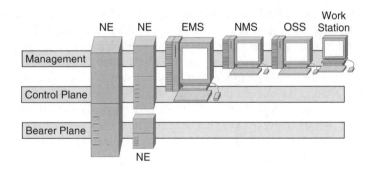

Figure 11.1 Network and support system roles.

software design; it is thereby transformed into an NE. All systems that sit on the bearer plane are NEs, but not all NEs sit on the bearer plane. Some NEs are responsible for control plane traffic without carrying any user data. Examples are route reflectors (RRs), signaling transfer points (STPs), and Mobility Management Entities (MMEs). None carries bearer traffic but a loss of any one affects bearer traffic directly and in real time.

11.2.2 Control Planes

Real time traffic that affects the delivery of bearer traffic but is not user generated travels on control planes. There are two general categories of control plane traffic, network routing and control traffic, and connection management traffic. Network routing and control traffic establish forwarding rules for network elements. Open Shortest Path First (OSPF) and Label Distribution Protocol (LDP) are examples of protocols that establish forwarding rules for bearer plane traffic. Border Gateway Protocol (BGP) communicates reachability, indirectly affecting routing. MMEs manage connectivity between the Radio Access Network (RAN) and the Internet without carrying user data. STPs carry both routing and connection management traffic. Bearer connectivity is enabled using the Integrated Services Digital Network User Part (ISUP) messages carried by STPs but bearer traffic is carried by Service Signaling Points (SSPs) not STPs. STPs do carry route management messages indicating reachability and preferred routes for SS7 messages.

Almost all NEs have control plane connections. Small Time Division Multiplex (TDM) multiplexers and some transport equipment are exceptions. EMSs seldom have control plane connections; Asynchronous Transfer Mode (ATM) EMSs were an exception in the sense that they managed network routing in some cases. ATM was deployed in many networks, like Digital Subscriber Line (DSL) aggregation networks, with permanent virtual circuits (PVCs) with a plan to introduce switched virtual circuits (SVCs) in later releases. PVCs are configured by the network operator and if a link or node failure occurs in the network connectivity is lost unless the operator logs into serviceable nodes and reconfigures the connection manually. Soft PVCs were introduced to provide a degree of protection to PVC networks. EMSs or other network hosts were configured with alternate routes or algorithms to discover alternate routes in the event of a failure; when links failed the EMS or host would reconfigure ATM PVCs just as a network operator might. In practice the implementation of soft PVCs was problematic. ATM PVCs rode optical and metallic TDM facilities and so loss of a facility link often meant that more than one PVC was affected. The algorithms implemented in EMSs to discover and reroute PVCs worked reasonably well when a single PVC failed, but were overwhelmed when multiple PVCs failed. Sometimes the EMS would take an hour or more to converge to a new routing solution and would even reroute PVCs that were not affected in the original failure; customers that were not affected in the original failure had their service disrupted when the EMS tried to reconfigure the whole network to reach optimal routing under the failed condition. Moreover when the failed facility was restored connections had to be torn down again to return to the original routing assignments. The lesson everyone should take away from the ATM experience is that EMSs should not participate in control plane rerouting. EMSs are not network elements.

11.2.3 Management Planes

Bearer and control planes directly affect user services and failures in either of those planes can have immediate effects. FCAPS functions are performed over the management plane by EMSs, OSSs, or work stations directly accessing the EMS or NE. Management plane failures may blind Operations to the state of the equipment and prohibit configuration and subscriber moves, adds, and changes, but management plane failures should not affect transaction services.

We can categorize the degree of sophistication of NEs, EMSs, and NMSs by examining the scope of their responsibilities and their place in the hierarchy of systems (see Figure 11.2).

OSSs include engineering systems, provisioning systems, and performance management (PM) systems. Billing systems are technically not in the group but are sometimes included. OSSs have responsibilities for all of the FCAPS functions with the exception of fault management, which is the province of the NMS. Customer Management Systems (CMS) are also OSSs; they are sometimes called Subscriber Management Systems (SMS) and

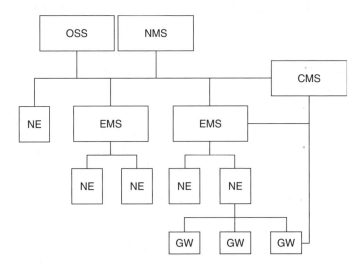

Figure 11.2   Hierarchy of management systems.

may work through an EMS, directly with a customer gateway (GW), or both. Provisioning systems may work with all three when provisioning access services.

11.3   Management Systems

NMSs are developed either by broad network systems suppliers such as Cisco and Alcatel-Lucent, or by third parties such as HP and IBM. Systems developed by third parties are cross domain systems, often with support for plug-ins from equipment suppliers. Network systems supplier NMSs are generally only effective in managing the supplier's equipment.

Suppliers choose whether to develop an EMS and what functions and protocols to support. There are at least four different EMS models in the industry today:

- **Telco EMS** – some network element suppliers build comprehensive EMSs that follow a telco model and the characteristics described earlier.
- **Functional EMSs** – other suppliers develop multiple EMS/OSSs to support different FCAPS functions. Specific systems are built for fault management, PM, and provisioning, for example. The systems are vendor specific so they don't really qualify as a cross domain NMS or OSS. They generally are expensive and redundant since functions like auto-discovery and configuration management have to be built into each system.
- **NE to NMS** – still others rely on FCAPS functionality in their network elements assuming they will interwork directly with NMSs and OSSs. These solutions tend not be scalable because NMSs often cannot scale to support tens or hundreds of thousands of systems in such an arrangement. The entire burden of alarm correlation and suppression falls on the NMS rather than pushing responsibility down into an EMS with deeper domain knowledge. The other risk is to the network element. Carriers tend to perform highly redundant polling. Each organization wants its own data and so elements can be burdened with heavy processing loads just to satisfy the individual needs for data. An EMS polls network elements once and is better suited to handle heavy demand from OSSs.
- **Third party EMSs** – third party suppliers build EMS "platforms"; suppliers and carriers are left with the responsibility of building rules, management objects, visualization, and data structures that represent the domain knowledge to implement a meaningful system. Invariably the effort to keep these systems up to date weighs them down and they become out of date and ineffective.

11.3.1   Network Management Systems

NMSs have the following characteristics:

- **Cross domain capability** – an NMS must be able to manage a range of network technology from different suppliers using a variety of standard interfaces for collection of alarms and performance data.
- **Event correlation** – an NMS must be capable of analyzing a set of notifications arriving from different systems and be able to use rules to determine which notifications are associated with a root event and which are secondary events triggered by the root. Correlation is dependent at the NMS level upon subtending EMSs and NEs performing correlation and suppression and forwarding root events from each of those systems.

- **Event suppression** – an NMS must be capable of suppressing non-root events and craft designated events.
- **Event presentation and management** – an NMS must present a crisp and intuitive visual display of the most critical active alarms and associated information on resolution progress.
- **Network performance monitoring** – an NMS must monitor key performance indicators as signs of normal network operation. Performance indicators exceeding normal bounds may be signs of an unalarmed network failure or unusual spikes in traffic demand.
- **Topographical presentation** – an NMS must create multiple views of networks, by domain, by region, and by hierarchy. Faults and points of congestion need to be indicated clearly in the presentation.
- **Auto-discovery** – an NMS must be capable of discovering all nodes of a designated type or profile within a domain or address range and preferably links between nodes. For most network elements an NMS can invoke standard IETF MIBs Management Information Bases to learn about the element and its interfaces.
- **Technology independence** – an NMS should not depend on knowledge and functions for specific suppliers' equipment; that is the role of the EMS. NMS plug-ins are a good way to incorporate equipment specific knowledge while keeping the NMS independent. Ideally the equipment supplier provides and updates the plug-ins.

Figure 11.3 is an illustration of a general model of an NMS.

An NMS must support multiple southbound interfaces toward EMSs and NEs as described in detail in the following. Data are retrieved from NEs in five ways:

- **Autonomous messages** – alarms, traps, and notifications generated by the NEs autonomously are processed by the alarm manager and stored in an alarm log. A mediation process groups related alarms and suppresses redundant alarms. An NMS should not be burdened by syslog messages. In some cases an NMS, like Cisco's CNote retrieves syslogs and generates alarms using a supplier or carrier built rule set.
- **Scheduled PMs and reports** – some NEs publish performance and traffic reports according to their internal schedule. Legacy equipment using TL1 are more likely to use this option. The performance manager function is responsible for parsing and storing the reports in the metrics database.
- **Polled PMs and reports** – tasks are scheduled in an NMS, including polls of NEs and EMSs. Inventory and communications databases are consulted to retrieve relevant system and interface configurations.
- **Directed requests** – craft can poll active alarms or PMs on demand. When an NE has been off line or is moved from one NMS to another a directed poll is a good way to refresh the relevant databases.
- **Import** – data can be imported from other NMSs, EMSs, or OSSs. Inventory and topology information are examples.

Mediation and analysis functions vary from simple to complex, depending on the NMS. NMSs that are specific to a particular supplier's network elements can be very sophisticated because the company that develops the network elements incorporates deep domain knowledge into the NMS. Third party NMSs are less likely to have complex functions and more likely to rely on third party plug-ins or carrier defined rules. Performance analysis

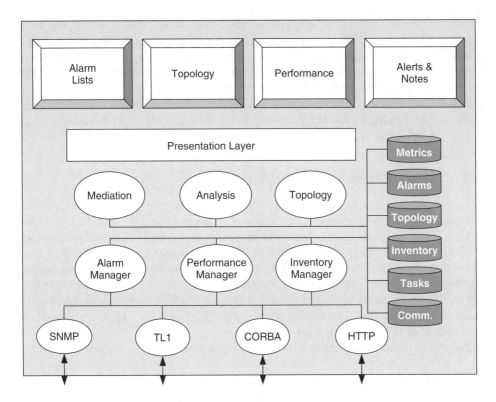

Figure 11.3   Model NMS.

is most useful in an NMS when it is kept at a high level. An NMS is not an engineering system and is unlikely to have rich engineering rule sets for the wide range of network elements monitored. Alarm thresholds based on performance metrics should be instituted down at the NE or EMS, not the NMS. If threshold related alarms are generated by the NMS for a specific NE they will not show up on the NE, breaking the Virtuous Cycle. Craft checking the NE alarm list and managed objects will find no off normal states. Performance monitoring at the NOC or high in the network is useful as an indicator of major shifts in traffic patterns, such as when natural disasters or massive Denial Of Service (DoS) attacks strike. They can also serve as a failsafe indication of node outages or loss of routing to large parts of the network.

Network connectivity is automatically discovered for most Internet Protocol (IP) network elements. Legacy TDM systems are less likely to enable automatic discovery. IP routes can be discovered in a variety of ways depending on the technology. Address Resolution Protocol (ARP) table retrieval, OSPF tables, BGP tables, and Multiprotocol Label Switching (MPLS) forwarding databases are some examples of information that may be polled and incorporated into an NMS topology database, depending on the sophistication of the design. Geographic information, Common Language Location Identification (CLLI) codes and a wide range of other information either has to be entered and maintained manually or it can be pulled from an OSS, provided the OSS and NMS are flexible enough to reconcile schemas and semantics.

The presentation layer(s) of the NMS include the views shown at the top of the Figure 11.3. In a NOC global views are presented on the common billboards viewed by all team members. Different topologies and alarm lists may be posted for optical and IP networks or for voice and mobility, depending on the responsibilities of the NOC. The makeup of the billboards doesn't tend to change; operations technicians observe the billboards but use their local terminals to change NMS presentations to view local networks specific to their technology. They are likely to log into an EMS or even an NE to verify the state of individual systems.

11.3.2 Element Management Systems

EMSs are common among consumer distribution networks where hundreds or even tens of thousands of network elements are deployed. Examples are DSL and Passive Optical Network (PON) systems. In those networks functional concentration and scalability depend on an effective EMS design. EMSs, unlike NMSs are technology specific with a high degree of domain knowledge incorporated into the design. Suppliers often develop a base platform and then adapt it to their different network technologies. Alcatel-Lucent's 5520, 5523, and 5526 are good examples of such a system. The base platform serves as an EMS for the Litespan®, Asymmetric Digital Subscriber Line (ADSL) and Very High Bit Rate Digital Subscriber Line (VDSL) multiplexers.

EMSs need the following characteristics as a minimum. Many are common to NMSs:

- **Event correlation** – an EMS must be capable of analyzing a set of notifications arriving from associated systems and use rules to determine which notifications are associated with a root event and which are secondary events triggered by the root. Root events should be forwarded to an NMS.
- **Event suppression** – an EMS must be capable of suppressing non-root events and craft-designated events.
- **Event presentation and management** – an EMS must present a crisp and intuitive visual display of the most critical active alarms and associated information on resolution progress.
- **Auto-discovery** – an EMS must be capable of discovering all associated network elements within an address range and preferably their connectivity. An associated network element is one the supplier has designated as supported by the EMS.
- **Inventory management** – an EMS must be capable of retrieving all relevant hardware, firmware, and software inventories including versions and physical location for all associated network elements.
- **Bulk configuration** – an EMS must be capable of performing bulk updates to associated network elements from templates or other internal data structures selected by the operator.
- **Audits** – an EMS must be capable of performing audits of network elements to verify conformance to system wide parameters, firmware, hardware, and software versions set by the operator. Where possible an EMS should also verify network consistency by comparing parameters at both ends of a link.
- **System backup and restore** – an EMS must be capable of storing active configuration parameters for network elements and restoring them. This requirement excludes subscriber information which should be restored from an OSS.

- **Engineering resource monitoring** – an EMS must be capable of monitoring all resources in associated network elements that are subject to engineering or exhaust. Summary reports of out of bounds elements and elements projected for exhaust should be available.
- **Network element upgrade support** – an EMS must facilitate network upgrades and migrations. This is one of the greatest shortcomings in the industry. When a supplier discontinues a network element and replaces it with another, the EMS should facilitate the migration by converting the legacy configurations and mappings into the new system, automating the change. Failure to provide this capability is tantamount to asking the carrier to choose another supplier to replace the legacy system.

11.3.3 Network Elements

NEs, as described in Chapters 2 and 3 are responsible for the following as a minimum:

- detecting and reporting all internal hardware faults,
- detecting and reporting all internal software faults,
- invoking internal defensive actions to switch to redundant hardware when needed,
- restarting failed processes using escalating levels of severity,
- auditing internal data stores, reporting errors, and recovering configurations to safe states,
- implementation of the managed object model, and
- correlation and filtering of internal faults.

These fault management capabilities reflect the minimum self-management set for a network element. A very few NE designs relied on EMSs to detect and enforce procedures remotely to meet these minimums. EMSs should not assume any of the responsibilities described above; self-awareness and self-recovery are a primary responsibility of an NE. Some classes of network elements also have network management capabilities. Functionality ranges from network-aware NEs to network-managing NEs. Network-aware NEs change their own behavior as a function of changes in other NEs and the associated links that constitute a network. Network-managing NEs change their own behavior but also inform other NEs of those changes and in some instances order reconfiguration of other NEs in the process of orchestrating changes in network routing or topology. Synchronous Optical Network (SONET) add-drop multiplexers (ADM) are an example of a network-managing NE.

11.3.4 Management Interfaces

Network management interfaces are based on a manager-agent model. The following sections introduce different interface protocols according to when they were developed, ordered by oldest to most recent. The section is only an introduction; full texts are available that describe each of the protocols in depth. Because the industry has been unable to coalesce around standards, none of the protocols has disappeared entirely. The two oldest ones, TL1 and Simple Network Management Protocol (SNMP) remain the most common for fault management. Equipment suppliers usually build three or four interfaces into their network elements and virtually all of them into their management systems.

Management interfaces follow a general pattern composed of the following:

- **Managed elements** – network elements or hosts that are managed using the interface.
- **Managed objects** – ports, cards, databases, or other real resources in the managed element accessible by a manager through an agent.
- **Agents** – software resident on the managed element that provides the standard interface to managers.
- **Managers** – software resident on a host system used by Operations or other work groups to affect changes or retrieve information about managed elements.

For readers coming from an IT perspective the manager-agent relationship is akin to a client–server relationship. Agents listen on well-known Transmission Control Protocol (TCP) or User Datagram Protocol (UDP) ports for managers. Managers establish sessions with agents by connecting and authenticating, retrieve information or make changes to a management element, and then drop the session.

11.3.4.1 Transaction Language 1 (TL1)

TL1 is an extension from an effort by CCITT, now ITU-T to standardize a Man–machine Language (MML) [2]. The idea was to develop a protocol that can be read directly by human beings, but whose syntax and semantics are sufficiently rigorous to enable machine to machine automation. TL1 was developed in the 1970s and set the standard for telecommunications management. TL1 and SNMP are still the dominant interfaces used for fault management. TL1 dominates telco systems and SNMP dominates packet systems. Figure 11.4 illustrates a typical TL1 implementation between an NMS, EMS, and NEs.

In the example three management systems access network elements, such as DSL multiplexers. Engineering retrieves performance and engineering data for capacity management; Operations relies on an NMS for fault management and performance monitoring; Care and Sales use a CMS to check for available ports and to make moves, adds and changes to subscriber services. Craft can access TL1 network elements directly using Telnet from any work station if they have authorization. Each group has privileges that allow them to accomplish their tasks but they should not be allowed to invoke commands that can cause harm unnecessarily. The three management systems' normal access is via the EMS. Each management system has a TL1 manager software module that accesses the EMS by establishing a Telnet session to a well-known port (TCP 3083) that is served by the EMS TL1 agent.

TL1 Commands

TL1 syntax defined by Telcordia [3] has the following structure.

    VERB[-MOD1[-MOD2] :[TID]:AID:[CTAG]::[PARAMETERS];

- **VERB** – is an action word such as RTRV (retrieve), INH (inhibit), SET, RLS (release), or REPT (report).
- **MOD1** and **MOD2** – are dash separated modifiers such as RTRV-ALM-T1 (retrieve alarms for T1 ports). The modifiers are not needed for all commands but if they are part of the command, they are required; the command will not execute without them.

- **TID** – Target Identifiers that specify the network element to which the command is applied. CLLI codes are usually used for the TIDs. TIDs are often optional when a session is established directly with a network element. If the TID is omitted, the NE is used as the default. If the session is with the EMS an empty TID will cause the command to fail.
- **AID** – Access IDentifier that modifies the command to apply to a specific shelf, card, port, or other physical resource.
- **CTAG** – Command TAGs are numbers assigned by the user for each command. Responses to a particular command have the original CTAG embedded so the command and response can be correlated. If the user omits a CTAG the system will generally assign one on a serial basis.
- **PARAMETERS** – a comma separated list of arguments.

The Table 11.1 lists commands pertinent to the example.

Typical TL1 command sets have 30–50 distinct commands. Telcordia publishes recommended sets for standard network elements such as SONET/SDH (Synchronous Digital Hierarchy) add drop multiplexers. Suppliers are free to supplement those recommendations to fit their particular design. Parameter lists in particular are system specific.

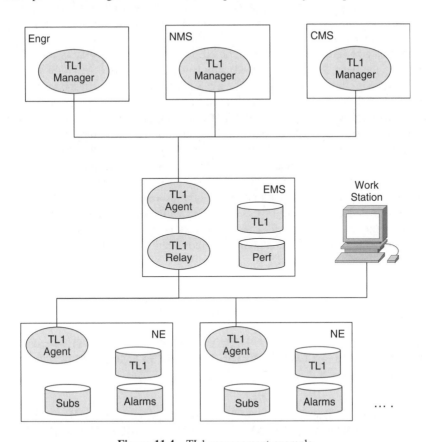

Figure 11.4   TL1 management example.

Table 11.1 TL1 example command set

Command	Description	System	Authorization
ENT-USR-SEC	Add a new user to a security group	EMS, NE	Superuser
RMV-EQUIP	Remove a piece of equipment from service	NE	Maint.
REPT-PM	Publish a performance report	EMS	Maint., Engr.
SET-ATTR-SUB	Set a subscriber attribute	NE	Care
RTRV-ALM	Retrieve active alarms	EMS, NE	Maint.

A database of users and command privileges is kept at the EMS and each NE. Some systems allow operators to declare user groups, much like a UNIX or Linux file system, and a command table is populated with commands allowed for the declared groups in a fashion similar to Table 11.1. Other implementations pre-designate groups and hard code command authorizations to those groups. Restricting commands to authorized groups is a security principle to be applied across all network systems; recall that 40% of all outages are caused by human error.

An EMS implementation often has a TL1 relay process. When TL1 commands are received at the EMS they may be processed locally or relayed to an NE specified by the TID. In the example the REPT-PM command is executed at the EMS by pulling data from a performance database locally; the EMS is the logical system to collect and aggregate performance data periodically and relieve the NE from storage and possibly providing complex views of the data. Subscriber related commands, such as SET-ATTR-SUB are routed via the TL1 relay to the NE. In this case the EMS does not maintain subscriber information; it relies on the CMS to back up customer information. Synchronization becomes overly complex if data are in too many places. CTAGs are used by the EMS when messages are relayed to match responses with commands. If the user at the OSS doesn't add a CTAG the EMS normally assigns one serially before relaying the command to the NE.

TL1 agents respond to commands in two stages. An acknowledgment message is sent within 2 seconds of receipt of a command indicating either an error in the command, or that the agent received the command and it is being processed, or that no response will be sent. The acknowledgment is sent by the TL1 agent without the benefit of knowing the final disposition of the command which is usually executed by a separate process. The second message is a response message containing the requested information or disposition of the response to the original command.

TL1 Autonomous Messages

Autonomous messages are generated by the EMS or the NE and have the following syntax.

> **<SID> <YEAR>** - **<MONTH>** - **<DAY> <HOUR>**:**<MINUTE>**:**<SECOND>**
> **<ALARM CODE> <ATAG> <VERB>**[**<MOD1>**[**<MOD2>**]]
> **<TEXT BLOCK>**;

1. <SID> – a Site ID that is usually a CLLI code and identical to a TID.
2. **<YEAR>** – **<MONTH>** – **<DAY>**   **<HOUR>**:**<MINUTE>**:**<SECOND>** – a date and time stamp for the message.

3. **<ALARM CODE>** – one of the following
   (a) *C – critical alarm
   (b) ** – major alarm
   (c) * – minor alarm
   (d) A – informational non-alarmed message
4. **<ATAG>** – Alarm TAG similar to a CTAG, usually assigned sequentially by the NE.
5. **<VERB>**[**<MOD1>**[**<MOD2>**]] – the subject of the autonomous message. An example is REPT-ALM-T1.
6. **<TEXT BLOCK>** – none, one, or more lines of text. Scheduled reports, like PMs, have lines of data that generally correspond with a database entry. Text blocks can be delayed or extended in which case the message ends with a ">" symbol instead of a terminating ";". Each additional segment will have two header lines identical with the original message.

There are three general classes of autonomous messages: alarms, informational messages, and scheduled reports. Complete implementations of TL1 agents include the command INH-MSG which turns off autonomous messages, including alarms. TL1 sessions are full duplex and it is disconcerting if you are trying to type a command and a performance report begins listing in the middle of typing. It also plays havoc with provisioning and engineering systems using TL1 as the interface. The OSS can't distinguish an autonomous message from a response message.

11.3.4.2   Simple Network Management Protocol (SNMP)

TL1 originated in ITU and SNMP originated in IETF. They are both effective but there are significant differences.

- TL1 uses a text format that can easily be read by machines or humans. A Telnet session can be opened from a simple terminal and commands can be typed directly to an agent.
- Both have formalized syntax rules, but SNMP also formalizes the semantics and grammar through the Structure of Management Information (SMI) and MIBs.
- SNMP Managers import managed object definitions, table structures, and data types by importing MIBs.
- Network elements often use TL1 as the native command line interface (CLI). SNMP managed network elements need a proprietary language for their CLI. Optionally they can use TL1.
- SNMP has well defined data types, INTEGER, COUNTERS, TIME TICKS, and so on. TL1 has no defined data types and uses text messages. Numbers appearing within the text must be parsed and interpreted in the context of that particular message.

Figure 11.5 is an example of an SNMP implementation.

The origin and name of SNMP say a good deal about the functionality and application space of the protocol. SNMP originated in the IETF as a network management tool for IP systems. SNMP is frequently the network management interface of choice for network elements, but is seldom the only interface implemented by the element. Cisco's IOS CLI is the prototypical example. Cisco has broad support of SNMP but has an equally rich CLI for

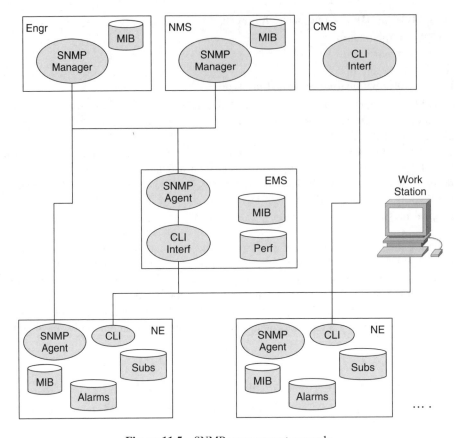

Figure 11.5   SNMP management example.

managing their network systems. SNMP excels at fault management, network management, and performance management. It can be used for configuration management but is generally not well-suited to subscriber management.

Versions
SNMP has evolved through three distinct versions.

1. IETF RFC (Request for Comment) 1067 introduced SNMP Version 1 in 1988 [4]. It defines management stations (managers), network elements, and agents. The protocol itself has only five message types:
   - **GetRequest** – returns a scalar value,
   - **GetNextRequest** – returns a scalar value from the next index in a table,
   - **SetRequest** – sets a scalar value,
   - **Response** – sends a response to the three previous messages,
   - **Traps** – autonomous messages.

2. IETF RFCs 1441 and 1910 [5] introduced SNMP Version 2 and 2c in 1993 and 1996. They added security features using community strings and introduced a new command, GetBulkRequest, significantly improving navigating tables by allowing retrieval of multiple rows. A proxy was also introduced since SNMP Version 2 is not backward compatible with Version 1. Inform protocol data units (PDUs) were introduced which are autonomous messages like traps, but require a notification of receipt issued by the manager.
3. IETF RFC 3411 [6] introduced SNMP Version 3 in 2002 which significantly improved security by adding encryption and authorization. Deployment of SNMP today should be based on SNMP Version 2c or Version 3. Security is non-existent in Version 1.

Management Information Bases (MIBs)

The power and versatility of SNMP is derived from the simplicity of the protocol and from the way structured information is incorporated into network management. A set of IETF RFCs define the SMI [7] and its use in MIBs [8]. A MIB defines a hierarchical data structure of SNMP addressable objects that conform to the SMI. Every object in a MIB has a unique object identifier, or OID. Objects can be read-only, such as a count of the number of received packets, or read-write for those variables which can be set via SNMP. New objects can be created in tables with a read-create attribute by appending rows. Table 11.2 illustrates how OIDs are ordered.

Industry standard MIBs have been written by IETF and other forums. Suppliers of network equipment decide on which standard MIBs they will support and incorporate into their designs. Once they commit to support a MIB they must fully implement it; partial implementations are not allowed. Suppliers also write their own MIBs, called enterprise MIBs, to support their proprietary features and attributes. The list of standard and enterprise MIBs supported by the equipment is made available to prospective operators.

Traps and Informs

Traps were introduced in SNMP Version 1 as autonomous messages issued by the agent to notify managers of state changes or other events. There are seven general categories of events defined by traps:

Table 11.2 SNMP object identifier example

OID	Text Definition
1	OSI
1.3	DOD
1.3.6	Internet
1.3.6.1	Private
1.3.6.1.4	Enterprise
1.3.6.1.4.9	Cisco
1.3.6.1.4.9.1	Cisco Products
1.3.6.1.4.9.1.8	Cisco 7000

- **coldStart** – signifies that the sending element is reinitializing itself in a way that may alter SNMP settings.
- **warmStart** – signifies that the sending element is reinitializing itself in a way that should not alter SNMP settings.
- **linkDown** – signifies a failure in one of the links supervised by the agent.
- **linkUp** – signifies a restoral of one of the links supervised by the agent.
- **authenticationFailure** – signifies that the sending protocol entity (typically the agent) is the addressee of a protocol message that is not properly authenticated.
- **egpNeighborLoss** – signifies that an external gateway protocol (EGP) neighbor for whom the sending protocol entity (typically a border or edge router) was an EGP peer has been marked down and the peer relationship no longer applies.
- **enterpriseSpecific** – signifies that the sending protocol entity recognizes that some enterprise-specific event has occurred. The specific-trap field identifies the particular trap which occurred.

Trap PDUs have the following fields:

- **Enterprise** – identifies the type of agent or element that originated the PDU.
- **Agent address** – the IP address of the element originating the PDU.
- **Generic trap type** – an integer specifying one of the trap types listed directly above.
- **Specific trap type** – an integer assigned by the supplier giving more detail to the event.
- **Time stamp** – time between the last element re-initialization and this event.
- **Variable bindings** – a set of optional SNMP objects that convey additional details.

Informs were introduced in SNMP Version 2. Informs have the same data structure as traps but require an acknowledgment from the receiving host.

SNMP Managers
There are three requirements of all SNMP managers.

- They must be capable of constructing and sending the SNMP PDUs.
- They must be capable of receiving and decomposing SNMP PDUs.
- They must be capable of compiling MIBs that conform to SNMP SMI rules and mapping numerical OIDs to their MIB text equivalent for craft interpretation.

The first two requirements are similar to what is found in a TL1 manager. The third requirement, the ability to compile MIBs and use them to interpret and build SNMP messages and traps sets SNMP apart and accounts for its dominance in network management of IP-based systems.

11.3.4.3 Common Object Request Broker Architecture (CORBA)

Developed by the Object Management Group (OMG) Common Object Request Broker Architecture (CORBA) emerged in the mid-1990s. The principles of object oriented design (OOD) and distributed computing are at the heart of the technology [9]. In CORBA applications are viewed as collections of standard services that are bound together by

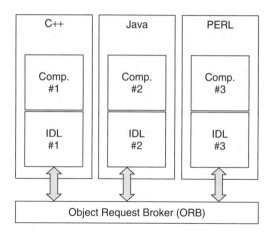

Figure 11.6 CORBA example.

CORBA components interacting through an Object Request Broker (ORB). Services and their associated CORBA components can be written in any language that has a toolkit that translates CORBA's Interface Definition Language (IDL) into a native executable. Figure 11.6 illustrates the concept.

In the example three components written in three different languages make up an application. Each expresses an interface in CORBA's IDL which is then compiled in the native language. Client side components generate a client stub when their code is compiled. Server side components generate a server skeleton when compiled. Clients contact servers through the ORB and invoke a remote procedure call to obtain the desired service. CORBA has many of the same architectural attributes of J2EE: components, distributed design, a messaging buss, directory services, naming services, and authentication services.

In practice, CORBA has been more successful in application management than in network management. Virtually all network elements support SNMP or TL1 and many support both. Adding a new feature in CORBA can require changes to IDLs, often forcing a recompile of the client and server. NMSs and OSSs using CORBA either need software upgrades or new plug-ins each time the underlying IDL changes. That is acceptable for provisioning, inventory and even performance systems but not for NMSs. SNMP only requires that new or revised MIBs be compiled, a low risk and low overhead procedure. EMSs may use CORBA in a northbound direction to interface to provisioning and inventory systems, but are likely to use SNMP and TL1 when interworking with an NMS. CORBA is not often used on network elements. Many network elements are highly tuned for real time performance. CORBA enabled remote procedure calls pose a risk that is more difficult to manage than simpler SNMP or TL1 command invocations.

11.3.4.4 HTTP

Embedded Web Servers

The HyperText Transfer Protocol (HTTP) began showing up on network elements and EMSs in the late 1990s after the arrival of Apache 1.0. In part the introduction of web-based element management was in reaction to the difficulty and performance of the X Window System. Carriers place EMSs and NMSs in either operations centers or data centers and users log into those systems remotely. Hyper Text Markup Language (HTML) has fewer functions than X Windows, such as drop and drag under mouse control, but has adequate graphics and controls for most purposes. X Windows is more difficult to maintain, hard to modify locally and performs poorly over low bandwidth connections. Craft working from remote sites or home over dial up connections found X Windows unusable where HTML works reasonably well. Embedded servers on EMSs and NMSs are effective and promise to be even richer with the introduction of HTML 5. Web servers deployed on network elements work well for small enterprise networks but aren't as useful for carriers of scale. They can help Care agents when deployed on Customer Premises Equipment (CPE) simply because the agent sees the exact same thing the customer sees and so they can better communicate when trouble shooting.

Web Services

Web services adoption into network technology came primarily from the IT move toward a service-oriented architecture (SOA) [10]. SOA is more of a framework than a protocol or specific technology. The technologies SOA depends upon are managed primarily by the World Wide Web Consortium (W3C): HTTP, XML, Web Services Description Language (WSDL), and Simple Object Access Protocol (SOAP). Within the framework:

- HTTP is the transport protocol,
- XML, the eXtensible Markup Language, is the document standard,
- WSDL is a metadata standard for describing web services, and
- SOAP provides a messaging layer for transmitting XML documents through a request-response exchange.

Applications designed using the SOA framework are more loosely coupled than a CORBA or J2EE implementation. Those technologies require recompilation, often on the server and client side when Application Programming Interfaces (APIs) change. SOAP messaging and process invocation of XML files is more flexible and is like TL1 in one respect; XML documents, when properly designed, can be read and interpreted by humans or machines.

SOA is used on northbound interfaces from EMSs and is the basis for TR-069, the standard adopted by the DSL Forum for CPE management. Other home CPE standards bodies are considering TR-069 as well.

11.3.5 Specialized Management Systems

Third party management systems and equipment vendors' EMSs are a necessary starting point for carriers. But global carriers have networks that are too heterogeneous and of a scale that requires specialized systems. The most effective systems are built by the carriers or by third parties working closely with carriers in a directed development. Specialized systems fill roles not filled by commercially available solutions.

- they take advantage of large carrier data stores, many of which are specific to each carrier,
- systems engineers recognize fault and maintenance patterns unique to the carrier and automate responses to them, reducing downtime, and operational costs,

- these systems fill gaps created by new protocol and network elements while waiting for commercial systems to recognize the need and respond,
- carrier built systems undergo continual improvement because of the close relationship between systems engineers and operations analysts. Commercial systems try to respond to the market as they see it, so the ability to evolve quickly and with a narrow focus on important problems is missing.

The next sections describe carrier designed systems that filled crucial needs that were not met by the commercial offerings.

11.3.5.1 Darkstar

In Chapter 14 a major network outage is analyzed with a good deal of discussion of root cause analysis. In an outage one of the first activities is to gather all available data (logs, alarm reports, performance data, usage, etc.) and begin to look for patterns of activity leading up to the failures. Engineers and analysts who have experienced outages have revisited the scene many times. Researchers at AT&T, working with Operations decided the time to gather all the data, filter, and correlate it was before the outage, not after it. That idea was the genesis of Darkstar [11] (see Figure 11.7).

Darkstar collects data from the backbone network and provider edge. Data include the sources shown previously. Collection differs from other NMSs in that it is more diverse, more consistent and stored in a far more scalable and flexible repository. In addition to the usual logs and alarms, monitoring tools like Keynote are used. Workflow data, logs of commands issued by craft to the network elements are also included. All of the data are normalized, timestamped with a consistent Universal Coordinated Time (UTC) format and

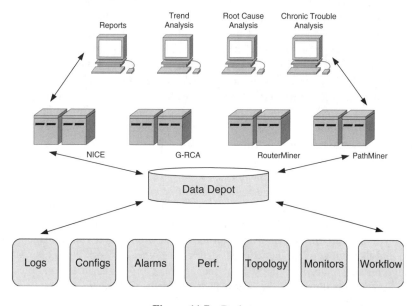

Figure 11.7 Darkstar.

stored in Data Depot, a proprietary data warehouse built on Daytona™ technology. With over 170 data feeds Data Depot stores over 340 million records every day.

Applications are built upon Darkstar to serve particular needs of Operations work groups.

- **RouterMiner** – an application that examines correlated events in a user specified set of routers during a specific time frame.
- **PathMiner** – users designate two or more locations in the network and the application constructs the likely path at the time and looks for correlated events along the path during a period specified by the user.
- **Generic Route Cause Analysis (G-RCA)** – a knowledge-based application that uses heuristic algorithms to identify the most likely root cause of a set of events based on repeated patterns and domain knowledge.
- **Network-Wide Information Correlation and Exploration (NICE)** – a correlation application like G-RCA, but instead of relying on a knowledge base it relies more heavily on statistics. Time series are created from symptoms (detected errors, failures, outages) and diagnostics (logs, alarms, command files). Using transformational analysis NICE learns the relationship between symptoms and diagnostics through iteration over a large corpus of data.

Applications results are available to operations work groups in reports, trend analyses, and interactive web tools. The applications are used in daily trouble shooting, and NICE is used to resolve chronic problems that don't result in hard failures, but rather cause a slow and steady degradation in service performance.

11.3.5.2   Watch7

Systems engineers and network software teams can quickly fill voids created when new technology is introduced without sufficient fault or service management by developing specialized management systems. In Chapter 14 an SS7 outage is chronicled; that outage became extensive in large part because the fault management system that was delivered with the new technology was simply not up to the task. The system, Engineering Admin Data Acquisition System (EADAS), was immature at that point in the deployment and could not process the numbers of alarms and informational messages generated when alarm storms occurred. In the aftermath of the outages it was clear that an interim solution had to be found to give Operations the necessary visibility to manage the SS7 network, particularly in the face of network instability when alarm storms are the rule, not the exception. To fill the gap two Pacific Bell systems engineers[2] developed an SS7 fault management system using a design pattern I call the impostor.

Impostors are specialized management systems that sit on the control plane and pretend to be a member of the community they are monitoring. They have a tremendous advantage over other forms of fault management by their direct participation in the signaling exchanges that occur within the community. They can exploit test procedures in management plane protocols and their view of the network is identical to other network elements, untarnished by an interpretation placed on it by fault management processes in the various network elements. Figure 11.8 illustrates the physical deployment of Watch7.

---

[2] David Fannin, now a Senior Director at Cisco and the author developed Watch7.

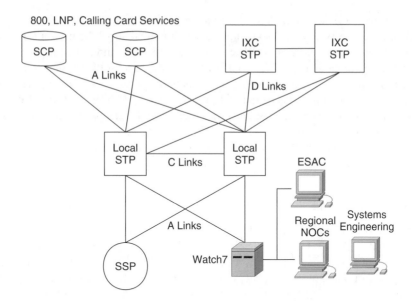

Figure 11.8 Watch7 SS7 network monitor.

Watch7 posed as an SS7 Signaling Point (SP) with links to mated STPs. By posing as a member of the STPs' community it received all routing updates for all SS7 network elements directly connected to the STP pair, typically a community of 40–100 systems. For those directly connected elements Watch7 operated in a passive mode, simply listening for updates. In an active mode Watch7 sent test messages for various services, such as 800 service and calling card service, to the regional STPs (not shown) for global title translation and database dips at the Service Control Points (SCPs). All of the routing in the local network, and all the services in the regional network were verified in a test cycle that took less than 30 seconds.

The scope of coverage was extended by sending SS7 test messages to interconnecting STP pairs, such as the Interexchange Carrier (IXC) STPs shown in the diagram. Routing to all of the SPs served by those interconnecting networks could be tested using SS7 test procedures. This active approach was only possible with the complete support and partnership between Systems Engineering and Operations. Actively probing an in-service network, including adjacent networks, has to be done with great care, otherwise the fault management system ends up creating the faults.

The status of all SPs were distributed to client side applications at regional NOCs, Electronic Systems Assurance Center (ESAC), and Systems Engineering. Multiple views of the networks were presented, exception reports, status, and topographical maps. Important advantages of Watch7 were that changes in the network were viewed directly from control plane messages; a change in the network was reported in less than 10 seconds. Technicians did not have to rely on the network elements to correctly interpret the state of routing. Each network element had a different way of reporting network changes and the messages could be buried among hundreds of other alarms and notifications. Different work groups did not have different views of the network, skewed by the particular system they monitored. All work groups had access to Watch7 and so could see the current status of the network in real time.

The need was immediate and so Watch7 was developed in three months and went into beta deployment immediately. The scope was expanded and software improved over the next three months. SS7 peering partners, other network operators, sometimes called Pacific Bell Systems Engineering to learn the state of their SS7 SPs because Watch7 maintained the status of connectivity across the network boundaries. Developed as an interim system, Watch7 stayed in service for 10 years. It cost less than $200 thousand to develop and deploy. The commercial replacement system cost $50 million, but it relied entirely on a passive design, monitoring every link in the network.

11.3.5.3   Other Systems

Darkstar and Watch7 are not isolated examples. Systems Engineers and Researchers, such as those found in AT&T Labs and Pacific Bell's Network Systems Engineering organizations have created dozens of systems in skunk works that fill crucial roles in managing emerging network technology and services. Equipment vendors, third-party suppliers, and IT organizations often lack the necessary domain and operational knowledge to design these systems and the waterfall requirements, design and delivery processes they use follow network technology by several years.

11.4   Management Domains

NMSs operate within the constraints of the domains defined by the networks they support. The job of network management is complicated by the differences in the way the layers of the network define and manage domains, as shown in Figure 11.9.

Beginning at the bottom of the diagram, optical network domains are defined by the technology and suppliers used. SONET/SDH rings, for example, tend to be populated by a single vendor's equipment with an intelligent EMS that only supports that vendor. Ultra long haul optical systems and regional Reconfigurable Optical Add Drop Multiplexers (ROADMs) are also likely to be vendor specific because of the manner in which different systems provide protection and how individual systems manage wavelengths, optical amplifiers, and transducers.

Network management responsibilities, systems, and work groups are segregated by domains and their boundaries. Vertical separation is according to technology, skills, systems, and work groups. Horizontal segregation is by region and network. Sizing domains is the primary responsibility of Engineering but it must be done in cooperation with Operations. Large domains may reduce capital investment but complicate administration and increase operational risks. Failures are often compartmentalized within a domain. In our example below an OSPF control plane failure in Area 1 is likely to be limited to traffic to and from that area. Larger domains increase the time for network restoral. OSPF convergence is directly affected by the number of nodes and the size of the routing tables, both of which increase as the domain expands. Network and EMSs should align with network domain boundaries. That can be accomplished by using separate systems or common systems with domain views.

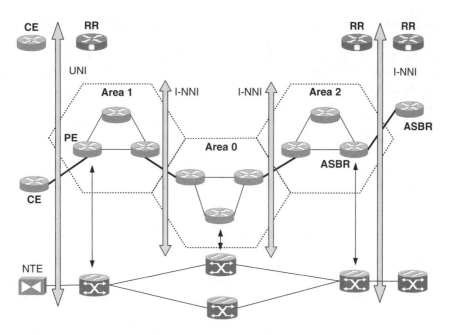

Figure 11.9   Network management domains.

11.4.1   Optical Networks

Optical networks deliver connectivity and capacity upon which all else rests. At present that bandwidth is managed as a provisioned resource, not a dynamic demand driven one. Optical networks do not generally respond in real time to changes in demand and are not affected by individual flows in the bearer plane. Optical networks do reconfigure themselves in real time, but only in response to faults in an attempt to restore the provisioned bandwidth. The introduction of switching of both optical fibers and wavelengths over the last decade opens the possibility for higher control plane functions to take advantage of switched capacity. Efficiencies of transport are presently being driven by other technologies, such as the Optical Transport Network (OTN). OTN is a wrapper technology enabling more efficient bundling of customer data onto optical transport [12]. In a broad sense OTN is the MPLS of transport. It is agnostic carriage, supporting Ethernet and SONET/SDH payloads with equal ease. OTNs gaining acceptance also advances the integration of legacy TDM payload structures with Ethernet frame structures, yielding more efficient and more granular encapsulation and control. At present however there are no wide deployments of call control driven optical networks. Enterprise customers can add bandwidth on existing routes via carrier supported enterprise web portals, but those are customer provisioned services, and not induced by enterprise services demand automatically.

The maintenance and recovery designs of SONET/SDH systems represent a mature technology development in which maintenance and automatic restoral were given a high place in the scheme of design. Other network-managing NEs differ in that they have control plane and bearer plane functions more formally segregated as part of their design. SONET/SDH networks are fixed connection oriented networks without a formalized control plane; they do however have a Data Communications Channel (DCC) which is sometimes called the Embedded Operations Channel (EOC). Within the section overhead a 192 kbps channel is available for Operations, Administration & Maintenance (OA&M) and control functions between section terminals. In a similar fashion the line overhead includes a separate DCC with a 576 kbps channel for line terminals. SONET/SDH network elements are able to automatically detect far end troubles and request reconfiguration, as described in Chapter 6, making those elements network-aware and network-managing elements.

Efforts have been underway for some time to formalize an optical control plane for optical networks other than SONET/SDH. ITU-T developed the Automatic Switched Optical Network (ASON) [13] control plane standard and IETF developed protocols in support of optical routing using MPLS, Generalized Multi-Protocol Label Switching (GMPLS) [14]. Deployment of optical control planes beyond SONET/SDH has been slow because of a lack of consensus on the goals and scope of the plane's responsibilities. Some of the goals of these efforts are quite ambitious. In one scenario the control plane is bridged across TDM, optical, and IP/MPLS technologies in an attempt to achieve dynamic bandwidth management, routing control, and fault management in an integrated network framework.

Network management in optical networks is presently confined to fault detection and restoral in those cases that network elements are not able to restore automatically. Optical network design is complicated by the need to account for higher risks associated with longer physical fiber routes and long transit times to enable repair and recovery. With sufficient flexibility in routing and spare capacity these networks can withstand any single route failure but can remain exposed in simplex operation for some time when remote cable faults are experienced. Undersea cable routes are the most extreme example. Reroutes on these critical facilities is dependent on sufficient advanced engineering and network analysis to verify both capacity and latency are within bounds before alternate routes are chosen and invoked.

Monitoring of optical amplifiers, repeaters, and transducers is becoming more challenging as optical modulation becomes more aggressive to meet growing demand for higher bandwidth. Narrowing tolerances require that threshold levels be set and alarmed to enable outside plant craft sufficient time to replace equipment or adjust optical levels before performance degrades. Legacy amplifiers, repeaters, and transducers are not always enabled with sufficient performance measurement and management capabilities to support the necessary monitoring. Optical EMSs may also need to be augmented with purpose built systems to monitor the elements and enable threshold alerting.

11.4.2   IP/MPLS Networks

The mechanics of OSPF, MPLS, and BGP are described in some length in Chapter 6. OSPF domains are defined by Link State Advertisement (LSA) flooding rules, which are applied within OSPF areas, shown in the diagram above. Boundaries between OSPF areas are managed by OSPF Area Border Routers (ABRs) which interconnect at internal Network to Network interfaces. Each OSPF area can be designed and managed as a separate network, with particular care and attention paid to Area 0 and to the ABRs.

MPLS domains do not necessarily map into either optical or OSPF domains directly; they are defined by the Label Switch Paths (LSPs) within the MPLS network whose domain is bounded at the PE-CE UNI. As a newer technology MPLS exposed some weaknesses in the management regime. MPLS black holes opened because of incomplete LSP designs or because of failures in label distribution. There was a lack of safeguards and metrics within the protocol and standards that caused MPLS packets to be discarded when they arrived with a tag that could not be resolved to an outgoing link.

11.4.3 Other Domains

Domains of management are implicit for a range of other groups and services within the Network.

- DSL and PON distribution networks are bounded by administrative and physical network assignments.
- Voice over Internet Protocol (VoIP) networks are bounded by the soft switch and feature server placement and homing. Carriers can choose a centralize design using very large domains or design more distributed and compartmentalized ones.
- Internet Protocol Television (IPTV) networks are usually organized into metropolitan areas, with a Video Home Office (VHO) serving the metro. IP addressing, channel lineups, and other attributes are unique with each domain.
- Mobility network domains are defined by the packet and voice cores that serve them. Global System for Mobile Communication (GSM) and Universal Mobile Telecommunications System (UMTS) have a domain separate from Long Term Evolution (LTE) by technology and possibly by serving area. LTE voice is IP based, so the packet core becomes the voice core as devices and networks move that way over the next several years. Mobility network domains are further defined by the RAN, either Base Station Controller (BSC), NodeB or eNodeB serving areas as described in Chapter 5.

Domains can be defined for consumer broadband and other services as well; the principles are the same.

11.5 Network Management and the Virtuous Cycle

Recalling the Virtuous Cycle shown in Figure 11.10, let's examine how work groups and management systems interact in the processes of fault notification, sectionalization, and verification of service restoral.

11.5.1 Notifications

When failures occur management systems are designed and configured to route notifications based on the nature, severity, and scope of the reported failure. NMSs are the primary systems used in the notification process. Alarm notifications usually generate failure tickets automatically which are associated with the alarm but tracked separately. Alarms can come and go but once a trouble ticket is created it is a permanent record of the trouble and how it is resolved. Since one failure can generate multiple alarm notifications and multiple trouble tickets, one ticket is marked as a master and related tickets are subordinated to the master.

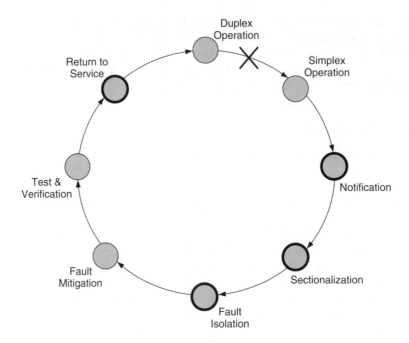

Figure 11.10 Network management and the Virtuous Cycle.

Alarm notifications can be generated by the network elements, the EMS, a telemetry system, adjacent network elements, or network probes. Trouble tickets are not ordinarily managed by an NMS but there may be links between the NMS and the ticketing system. Trouble tickets are discussed in more detail in the next chapter.

Ideally one work group is designated as having operational responsibility for resolving every customer affecting and network failure. That work group retains responsibility unless the trouble is "referred out."

- For failures of sufficient scope and severity the NOC is likely to become the responsible work group, at least initially, particularly when the failures cross domain boundaries. In general NOCs are not directly responsible for trouble resolution; their job is to diagnose problems and refer them to a work group that can resolve the problem and monitor the work group's progress. Network restoral and health is primary mission of the NOC. They bypass or otherwise mitigate failures. If they become engaged in resolving individual troubles, their mission is jeopardized.
- Transport failures, metallic, and optical are referred to a tier 2 or tier 3 Operations transport group, depending on the complexity and severity of the failure. When multiple layers, such as IP and optical are affected, transport is the logical point of referral because all layers above it are dependent. As described in Chapter 9, transport is defined as interoffice facilities and communication.
- IP/MPLS failures are managed by a separate group. That group may also have responsibility for Frame Relay and ATM networks. There is an extensive list of systems on the

control plane that affect these networks. They are generally managed by a subgroup under the IP/MPLS operations center. RRs, DNS servers, DHCP servers, firewalls, and NATs are examples of these systems. IP security may also be under this group. It is prudent to place it here because security affects how many of the systems are configured and managed. Having a security group make configuration changes independent of this group blurs operating lines of responsibility.

- Common systems failures, such as power and HVAC are referred to a separate work center. Those work centers may be regional since most failures require resolution by outside field forces.
- Voice network failures occur when systems specific to the delivery of voice services fail. A failure of a TDM transport facility that affects voice service is not a voice network failure, at least from the viewpoint of trouble resolution. It is a transport failure. Failure of STPs, SCPs, voice switching systems, SBCs and feature servers are voice network failures. A separate operations group is responsible for voice including consumer, enterprise, TDM, and mobile voice.
- Service management centers are responsible for specialized networks and systems. Examples include:
  - Consumer Broadband Service
  - IPTV
  - Video teleconferencing
  - Voice conferencing
  - Special services – video transport, telemetry, government services.

Apart from the responsible work group, other work groups may need to be informed.

11.5.2 Sectionalization

NMSs and EMSs are the primary systems used in sectionalization. Two NMS displays aid in interpreting notifications in search of the root failure: the alarm listing and a topographical display. Alarms are displayed in the NOC ordered by severity and time stamp. Multiple displays may be used for different technologies and different regions. NOC technicians are organized by technology and filter alarms at their work stations to display outstanding troubles within their purview. Topographical displays give a sense of the failure's scope. The great majority of troubles are dealt with directly by Operations tier 2 technicians. Once a trouble can be localized to a specific region or office and a specific technology, sectionalization is nearly complete. The objective is to identify a specific network element or link experiencing the failure. Once sectionalization is complete traffic re-routes may be necessary if the network does not accomplish it automatically. Notices may also be placed for other work groups. For example, if redundancy is lost in the optical network, the IP/MPLS work group needs to be made aware; they may want to delay maintenance activity or higher risk operations until the optical network is restored.

11.5.3 Fault Isolation

Once the fault is sectionalized to a network element or link, Operations must identify the specific remedy. Craft typically logs directly into the NE or possibly the EMS. CLI and TL1 are often used to identify the specific fault. All outside plant terminals and multiplexers and many central offices are unmanned. So craft will attempt to recover the failed port or device remotely first and if that fails an inside plant technician will be dispatched for troubles in a central office or data center, or an outside plant technician will be dispatched for outside plant electronic or physical systems. The Operations tier 2 analyst doesn't dispatch a technician directly, but rather uses a work force management system to enter a specific request, the trouble ticket reference, and contact information. Dispatched craft may contact the tier 2 Operations technician when the fault is repaired or may just close the work ticket with a description of the action taken. When the tier 2 Operations technician is alerted through the work force management system that the work is complete, they can test and verify the pack or device and return it to service. If successful, the trouble ticket can then be closed.

11.6   Summary

Fault management follows the Virtuous Cycle. Work groups and systems are dedicated to specific technologies and the different phases of the cycle. NOCs and NMSs have cross-domain responsibilities, monitoring networks up and down the OSI stack. NOCs primary responsibilities are for service restoral and delegation of failure response to other Operations centers. EMSs are supplier and technology specific; unlike NMSs they are dedicated to managing a specific type of network element, such as Digital Subscriber Line Access Multi-plexers (DSLAMs) and act as a central point for FCAPS management of those elements. Network elements, NMSs and EMSs use a range of interfaces to communicate with each other. Some interfaces are proprietary but SNMP and TL1 have been used for decades and continue to dominate. CORBA, XML, HTTP, and TR-069 are alternative interface technologies that have gained some traction. Proprietary CLIs still abound and are often used for provisioning and fault resolution.

Network domain boundaries are often drawn out of engineering considerations, such as address exhaust, convergence times, or geographical boundaries. They also serve to compartmentalize networks, confine outages, and prevent the spread of propagating faults. Boundaries are different for different layers of the OSI stack. Optical network boundaries may have no direct relationship to MPLS/IP boundaries.

Several operations work groups may be involved in the resolution of a single fault. Faults that result in wide spread service impacts come to the attention of the NOC; once they determine the specific network elements affected by the failure they alert the responsible work groups who perform sectionalization and fault isolation. When the trouble requires physical replacement or repair in a central office or in outside plant, craft is dispatched with a work order requesting the repair.

Trouble tickets are part of the management process. They are opened when a notification is received from the network or when a customer calls with a service complaint. Throughout its life each ticket has a responsible organization. As work groups narrow the diagnosis and identify the next work group needed to resolve the problem, the ticket is referred out to that group and the responsibility passes.

References

1. ITU-T Recommendation (2000) M.3010. *Principles for a Telecommunications Management Network*, February 2000.
2. ITU-T (1988) Z.301. *Introduction to the CCITT Man–machine Language*.
3. Telcordia (1996) Operations Applications Messages – Language for Operations Applications Messages, TR-TSY-000831 Issue 2.
4. Case, J., Fedor, M., Schoffstall, M., and Davin, J. (1988) IETF RFC 1067. *A Simple Network Management Protocol*, August 1988.
5. Waters, G. (1996) IETF RFC 1910. *User-based Security Model for SNMPv2*, February 1996.
6. Harrington, D., Presuhn, R., and Wijnen, B. (2002) IETF RFC 3411. *An Architecture for Describing Simple Network Management Protocol (SNMP) Management Frameworks*, December 2002.
7. McCloghrie, K., Perkins, D., Schoenwaelder, J. *et al.* (1999) IETF RFC 2578. *Structure of Management Information Version 2*, April 1999.
8. McCloghrie, K. and Rose, M. (1991) IETF RFC 2013. *Management Information Base for Management of TCP/IP Networks (MIB-2)*, March 1991.
9. Aklecha, V. (1999) Object-Oriented Frameworks Using C++ and CORBA, The Coriolis Group.
10. Newcomer, E. and Lomow, G. (2005) *Understanding SOA with Web Services*, Addison Wesley.
11. Kalmanek, C.R., Ge, Z., Lee, S., *et al.* (2009) *Darkstar: Using Exploratory Data Mining to Raise the Bar on Network Reliability and Performance*, AT&T Labs.
12. ITU-T (2003) G.709. *Interfaces for the Optical Transport Network*, March 2003.
13. ITU-T (2006) G.8080/Y.1304. *Architecture for the Automatically Switched Optical Network (ASON)*, 2nd edn, June 2006.
14. Mannie, E. (2004) IETF RFC 3954. *Generalized Multi-Protocol Label Switching (GMPLS) Architecture*, October 2004.

12

Support Systems

Support systems are very different than network systems in the ways they are specified and the ways they are implemented. Network systems conform to rigid functional standards, particularly at well-defined interfaces. If not, network elements will not interwork. Network systems suppliers differentiate themselves not in their approach to what network systems do, but rather how well they perform and how cost competitive they are. Conformance to standards at a detailed level is the entry bar to be in the bidding. Support systems are quite different. Many are home grown or built to replicate or interwork with legacy systems from AT&T or other carriers. So, a discussion of support systems is more illustrative than it is definitive. In this chapter systems standards, systems processes, and example systems are described.

12.1 Support Systems Standards and Design

A starting for a discussion of support systems is the Telecommunication Management Network (TMN) framework which is embodied in a set of ITU-T recommendations [1]. ITU adopted recommendations from forums and associations for standardization; in particular Telecommunications Management Forum (TMF) models and recommendations are largely reflected in ITU-T standards. ITU recommendations abstract FCAPS (Fault, Configuration, Accounting, Performance, Security) functions across a five tier logical layer architecture as shown in Figure 12.1.

The q reference points are information sharing interfaces between layered functions. The f reference points are interfaces to work stations functions and work groups. The three lower layers of the model were described in the previous chapter. This chapter focuses on the upper two layers.

ITU defined four roles for the business management layer:

1. supporting the decision-making process for investment of telecommunications resources;
2. supporting the management of OA&M related budgets;
3. supporting the supply and demand of OA&M related manpower; and
4. maintaining aggregate data about the total enterprise.

Global Networks: Engineering, Operations and Design, First Edition. G. Keith Cambron.
© 2013 John Wiley & Sons, Ltd. Published 2013 by John Wiley & Sons, Ltd.

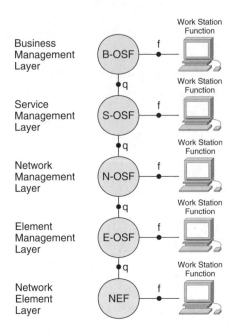

Figure 12.1 TMN functional layer model.

Business management functions do not support the day to day operation of the network or servicing of customers. Rather they support the decision process for overall management and guidance of telecommunications investments and decision making.

Service management functions include customer acquisition, order fulfillment, complaint handling, and billing.

The service management layer has the following roles:

- customer facing functions and working with other carriers;
- interaction with service providers;
- maintaining statistical data; and
- interaction between services.

The Service Management layer is responsible for working with the customer to reach a service agreement, and then insuring the agreement is met by collecting service information, provisioning network elements with the service, and tracking performance.

In addition to the functional model TMN has a business process model, and a communications model. The business process model is used to identify those processes necessary to support the business and service management layers. ITU breaks processes into two broad categories. The first category is Strategy, Infrastructure, and Product. Strategy processes cover marketing, service, and product development and management, supply chain development and management; they are executive, management, and lifecycle oriented; they are not meant to become directly involved in day to day functions.

The second category is Operations Processes. Operations Processes directly support the day to day running of the business; they are illustrated in Figure 12.2, taken from the ITU

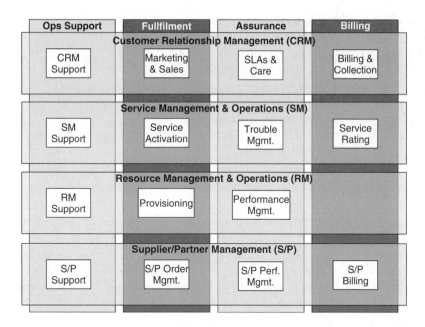

Figure 12.2 TMN operations process model.

standard M.3050 with some editing to align terms with others in the text. ITU and the TMF use the term Enhanced Telecom Operations Map (eTOM) to describe these models.

The functions defined in Figure 12.2 are self-descriptive by title. ITU and TMF do not define process flows for services but rather describe tools, frameworks, and high level class models using the Unified Modeling Language (UML).

Support systems, like their network counterparts, need to interwork with peer systems, be reliable, maintain transactional integrity, and present information and choices to the work groups they support that are succinct and intuitive. Support systems are more data driven than event driven. Transactions are batched and processing windows are liberal relative to networks systems. Support systems are built to automate business processes. Network systems engineers can design and certify a new network or service, but if the business processes to order, engineer, monitor, and manage the new equipment or service are not in place, Operations, Sales, and Care will not be able to support it. The new service will fail Operational Readiness Testing (ORT) if the systems cannot support essential processes.

An example design is presented in the following sections to illustrate the principles involved in support systems design. In practice these systems work off of common databases, share common information models, and ideally have common standards for information presentation. The example illustrates these interdependencies.

12.2 Capacity Management Systems

Capacity management (CM) is the responsibility of Engineering, as described in Chapter 1. A CM system supporting network and facility engineers must meet the following objectives:

- collect usage information daily on all of the network elements,

- collect usage information daily on all of the network facilities,
- maintain usage history and formulate usage trends,
- record engineering limits for every network resource,
- project exhaust dates for every network resource,
- notify the responsible work group of anticipated exhausts in a timely way,
- monitor capital investments and engineering efficiency,
- support modeling and engineering planning, and
- produce reports with Key Performance Indicators (KPIs) highlighting the state of network capacity and engineering management.

Automation of the CM process is dictated by the constraints facing the engineering staff:

- there are thousands of different equipment types, configurations, and versions in the network,
- equipment and facility engineers are not trained in traffic and computational theory,
- there are hundreds of thousands of ports, systems, and subsystems (e.g., memory, Central Processing Units (CPUs), file space) that can exhaust and affect customer service, and
- there are hundreds of thousands of logical resources (e.g., Private Virtual Circuits (PVCs), addresses, file descriptors) that can exhaust and affect customer service.

These constraints are in addition to the continuing need to manage staff sizes and operational costs.

A general design for a CM system is shown in Figure 12.3.

12.2.1   Work Groups

Five work groups are shown in the CM design example; others use CM data and applications as well but these are primary clients.

- **Equipment engineers** – these engineers are responsible for monitoring the capacity of individual network elements, ordering circuit packs, equipment augments, and tracking the execution of the orders. Equipment engineers are further organized by technology: voice switching engineers, data switching engineers, transport engineers DWDM multiplexers, and optical systems), microwave engineers, and other technologies. Transport engineers may be included with facility engineers described directly in the following.
- **Facility engineers** – responsible for monitoring the capacity of interoffice optical, microwave and metallic capacity, identifying routes needing augments, and preparing business cases for the augments.
- **Area planners** – area planners have responsibility for a specific metropolitan area. They consider area wide solutions, such as alternative technologies and new routing plans. Their work is a combination of technical and economic analysis. When their business case for new area plans are approved, equipment and facility engineers execute the individual augments, rearrangements, and orders necessary to achieve the plan. Area planners often design their metropolitan plans according to guidelines set by Fundamental Planners, not shown on the diagram.

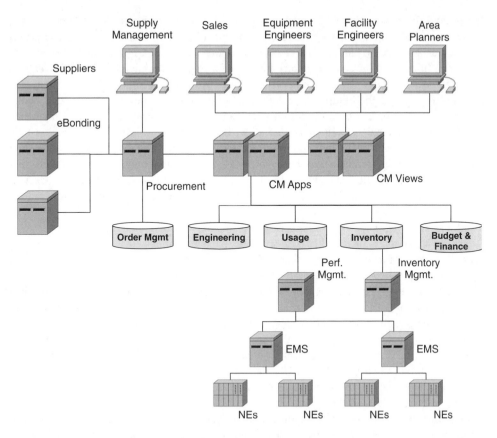

Figure 12.3   Capacity management system design.

- **Enterprise and small business sales** – CM applications identify which customer circuits and services have high occupancy and growth rates. These are ideal sales leads for customers that are sometimes too busy or lack the tools to monitor their own usage. It also enables account representatives to identify circuits that are not being used. Letting a customer know they are paying for something they are not using may result in a revenue hit in the short run, but pays dividends in customer loyalty over time.

12.2.2   Data Collection

CM and many other engineering and operations processes depend most directly on the integrity of inventory data. With thousands of network elements added and removed annually the inventory management system must be updated as part of the engineering and provisioning processes. Element Management Systems (EMSs) can aid in that process significantly. Some of them have auto discovery; if a network operator enters an Internet Protocol (IP) address range, the EMS will discover all relevant network element systems within that range.

Inventory databases should be updated daily to reflect the configuration of circuit packs, ports, processors, and other engineered elements of the systems. Some carriers use provisioning databases for inventory management, but inventories that are not based on the actual equipment in the field lack the necessary integrity for reliable decision making. *The network is the database of record; all other databases are subordinate*. Inventory systems should also maintain the state of the equipment, circuit packs, and other resources. CM system audits conducted weekly highlight those circuit packs and ports that have been in a maintenance state beyond a reasonable maintenance window. If a maintenance program in a region falls behind, it is not unusual to find thousands of ports in a craft enabled maintenance state long after the actual problem has been resolved. Engineering may find it is purchasing new equipment when all that is needed is to restore circuits marked out-of-service-manual to an in-service state. CM should track the percentages of ports in the various operations states, in-service, unequipped, out-of-service-automatic, and out-of-service-manual as a way of judging the effectiveness of engineering and maintenance teams and programs.

Performance data, maintained by performance management systems, includes usage data, discards, blocked transactions, or other off normal results. Data sets are often quite large and need to be collected at least daily. Systems that do not have persistent storage in an EMS may only hold the data for 24–96 hours before it is overwritten, making timely data collection essential. There are a lot of clients for performance data. Engineering needs it for capacity management. Operations uses it for service and fault management, and groups as diverse as Finance and Strategy use it in special studies or as a way of assessing network trends. Collecting data can be a burden to network elements or EMSs, so having one system to collect data and then provide feeds or database access for all other systems is the most reliable and practical approach, but collection must be comprehensive or a client will find a reason to collect data on their own.

Data collection designs are sometimes poorly thought out and execution can be undependable. Independent audits in one carrier showed that at times as much as 20% of the data were missing or corrupted in some implementations. The group that managed collections was unaware of the poor integrity of their system. Missing inventory data can delay order execution as provisioning systems hold other transactions until necessary equipment is in place. Missing or corrupt usage data causes inaccurate traffic reporting and invalidates forecasts used for capital investment and equipment augmentation. Data collection and routines should be treated with the same standards as network protocols and transactions. Collection protocols should be fully implemented. In particular congestion and flow control need to be included, tested, and exercised. A common problem is multiple support systems accessing a network element at the same time and executing SNMP MIB walks or bulk gets. SNMP agents can create a burden on network element processing and their thread priority may be low, causing timeouts at the SNMP manager, which creates retries exacerbating the problem. Poor collection implementations timeout and go on to the next element. Data collection metrics need to be reported daily and failures in scheduled polls should be alarmed and promptly serviced. Retrieved data should be checked for validity and consistency. Collections that are questionable should be stored but marked as suspicious with a cause code. Repeated SNMP timeouts are a sign bulk data collection should be performed using secure File Transfer Protocol (FTP); a redesign of the EMS or network element may be in order.

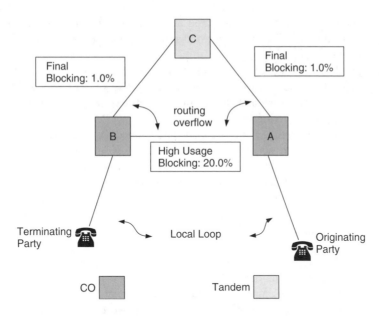

Figure 12.4   High usage engineering.

12.2.3   Engineering Rules

Capacity management is sometimes referred to as demand and capacity (D&C) engineering by network engineers. That nomenclature is incomplete; it glosses over the subjects of network objectives, traffic engineering, and forecasting.

Chapter 4 described how network objectives are derived from end to end service objectives. Figure 12.4 modifies an example in that chapter by adding an additional route, a high usage route.

In Figure 12.4, calls originating at central Office A first attempt completion over the high usage trunk group between Offices A and B, group A-B. If that trunk group is occupied, calls overflow to the final trunk group to office C, group A-C, for completion via the completing group C-B. Blocking on the high usage group is 20% in the example. Groups with high blocking rates are much more efficient than final groups and they do not take up ports on an intermediate tandem. Network engineers use different engineering rules for the two types of trunk groups and those rules have to be applied when usage data are examined. A measured blocking rate of 5% on a final trunk group is a signal for immediate augment; the trunk group is exceeding its objective by 400%. The same 5% blocking on a high usage group indicates the group is well within engineered limits.

Engineering rules and measurements for data networks are complex [2]. Core data networks are designed with a high degree of redundant routing to prevent service impacts from any single facility failure. Routes designed with a not to exceed capacity of 50% of the base bandwidth are common. Data networks also support quality of service transport; multiple traffic flows with different service objectives travel on the same ports and facilities. Unlike circuit switched networks, data networks don't block traffic; they queue it. Latency suffers when queues begin to grow. Engineers cannot use port or facility usage data directly.

Individual traffic classes have to be measured at the data switch and distinct rules applied for each class. Netflow measurements, introduced by Cisco, measure streams of IP traffic as a basis for traffic engineering [3].

Engineering rules are also set by suppliers for design specific limits. Address tables are memory dependent and can exhaust before the system or port bandwidth limits are reached. Logical resources, such as PVCs or other identifiers can also exhaust and must be incorporated into the engineering rules.

How and when usage measurements are made and applied are also part of the engineering rules base. Circuit switched network engineering rules are usually based on occupancy measurements using Erlangs or 100 call seconds, over an hour. The chosen hour can be a time consistent busy hour (TCBH), a busy season busy hour (BSBH), a bouncing busy hour (the hour with the greatest traffic), or 10 high day busy hour. IP netflow records report the duration of traffic having common source (IP address and port) and destination (IP address and port), protocol and type of service. The records are time-stamped and are created at regular intervals as long as the flow is active. Engineering rules establish how to use time dependent data. IP traffic, particularly on relatively small circuits can be very peaked. Denial of Service (DoS) attacks or one-time events can greatly skew the underlying traffic pattern. Algorithms are applied in the engineering rules to smooth and filter the traffic to reflect the measurement most closely associated with the desired network grade of service objectives.

12.2.4 Capacity Management Applications

CM applications analyze usage data and apply engineering rules to determine if engineering objectives are being met, and project when demand is expected to exceed capacity. Figure 12.5 illustrates the mechanics of the analysis. Engineers sometimes refer to the plot as a demand and capacity chart.

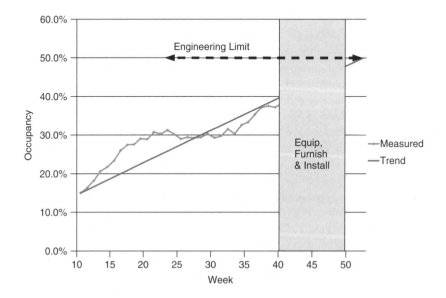

Figure 12.5 Demand and capacity chart.

In the analysis the CM application performs the following operations.

- it computes the engineering limit from rules supplied in the engineering database and any algorithms embedded into the application,
- a trend line is constructed from usage data using engineering or traffic rules in the databases or in the application,
- a lead time for equipment order, delivery, installation, and turn-up is taken from either the engineering database, procurement, or a combination of the two, and
- using all of the factors above, an order date, shown as week 40 in the diagram, is computed. Equipment engineers need to place the order on that date to forestall an under-engineered condition at week 50.

In practice small augments for infrastructure can be fully automated. Orders can be placed without human intervention.

12.2.5 Supply Chain Management

Engineering has daily interactions with Procurement. CM systems enable eBonding with suppliers and automate much of the work of Procurement. CM applications can be extended to give suppliers planning estimates in advance of actual orders, allowing them to better plan their inventories, and update their Original Equipment Manufacturer (OEM) suppliers. Suppliers report committed delivery dates which create exception reports for network engineers if they are beyond the standard intervals. Automatic reports on each supplier's delivery record can also be generated for Procurement.

12.3 Service Fulfillment

TMF and ITU-T use the term fulfillment to describe service delivery, from sales to provisioning, service activation, and procurement. Systems supporting service fulfillment are diverse and represent the largest investment and number of systems in global carrier's systems inventory. Unlike the previous CM example, there is no one model that portrays how service fulfillment works across the carrier's service offerings. Consumer and enterprise services share some functions but there are differences as well as similarities. The services, technology, and work groups supporting those markets are different and so are most of the systems. Within enterprise service fulfillment varies over a wide range. The large number of sales and orders are straightforward requests for Internet access or Private Branch Exchange (PBX) trunking services. But carriers also provide complex Virtual Private Network (VPN) services, call center services, teleconferencing, and a range of cloud services. There is no one design that accurately conveys the functions involved in all of those systems supporting service delivery across the entire range. However, a generalization of the stages of fulfillment can be used for discussion, as illustrated in Figure 12.6.

The hosts in Figure 12.6 represent processes that are usually automated by support systems, but not always. Some processes are manual. In the following sections processes and automation are described for consumer, small business, and enterprise. As used here, small businesses are those with 10 lines or less; their services align most closely with consumer services and are often served by the consumer sales organization.

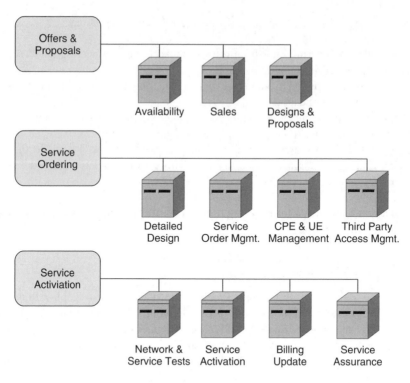

Figure 12.6   Service fulfillment.

12.3.1   Offers and Proposals

The sales process begins with marketing and advertising and concludes, when successful, with a customer order.

12.3.1.1   Consumer

Consumer and small business services are offered directly to customers on portals, through retail stores, or by toll free call center service representatives. These mass markets have a high degree of automation, necessitated by their low margins.

Availability

Service availability is automated for the majority of services. For wireline services availability starts with the customer's address and a determination of whether or not it is within a wire center served by the carrier. That is more difficult than it sounds because wire center boundaries do not readily map to political boundaries, zip codes, or any other available identifier. Address resolution has a high degree of automation and complex algorithms supporting it; it is one of the more difficult problems to solve. The problem is compounded by the quality of outside plant records and the ever changing demographics of communities. Mobile home parks can spring up with little notice and generate service demand before they are on property tax rolls or maps.

If the consumer is served by a wire center then a check for service availability is performed. For services like Digital Subscriber Line (DSL), availability checks are more complex and require a higher degree of automation. DSL service is sold by tier, based on downstream data rates. The closer the subscriber is to the serving multiplexer the more likely it is that high speed tiers are available. Telcos have loop qualification applications that use performance data gathered at least weekly to compute an estimate of the speed tiers available to inquiring consumers. Adding to the uncertainty, not all central offices are equipped with the same services. Limits to DSL reach and serving equipment arrangements often mean that one consumer on one side of a street can get a high speed service or an advanced service like Internet Protocol Television (IPTV) and their neighbor across the street cannot. That complicates the sales process and makes for a frustrated customer, or at least a potential customer.

Sales

Historically wireline consumer sales were simply order taking. When was the last time you saw an advertisement for Plain Old Telephone Service (POTS), if ever? Retail consumer sales today are dominated by wireless and broadband services like FiOS and U-Verse. Sales automation serves direct Internet sales, toll free call center sales, company retail outlet sales, and alliance sales through partners like Apple or Radio Shack. Company retail stores and call centers need a high degree of automation for promotions, availability, credit verification, and pricing. Alliance sales are restricted to specific products and automation is more limited.

Direct mail catalogs from Montgomery Ward and Sears Roebuck disappeared in 1985 and 1993 respectively, but direct mail yellow pages continue into the new millennium. I use lower case for *yellow pages* because it has been designated as a generic name for the retail advertising medium. Small businesses are the customer base that supports yellow pages advertising. Some carriers have integrated yellow pages advertising, Internet search, and a small business sales force under one business unit. As yellow pages hard copy search declines, carriers that can move that traffic to the Internet can retain, or at least offset revenue losses. Having a sales force that can visit a small business and offer phone, broadband, Internet advertising, and yellow pages advertising is also an advantage. Internet search companies have a different sales model and don't ordinarily have feet on the street. Automation for a small business sales force generally means developing mobile computing applications to aid presentation and sales. Company retail outlets also need specialists to support small business sales, or at least ready access to call centers that can respond to small business walk ins.

12.3.1.2 Enterprise

Enterprise processes and systems have to support more complex services, and support those services in territory and out of territory.

Availability

In territory service availability for IP services is usually fairly uniform across the region. Ethernet access and services have been deployed widely, improving IP service availability. Out of territory access is usually the most important factor in determining availability. Support systems use eBonding with other providers and systems with carrier pricing information to find the best design choice when responding to customer inquiries. Specialized planning systems also aid Engineering in determining where Points of Presence (POPs) should be placed out of region.

Sales, Designs, and Proposals

Account teams that work on larger accounts have technical team members and there are centralized technology teams that support all accounts. Support systems compliment the account teams' technical resources by automating complex high volume design tasks; VPN and Metro Ethernet services are examples. The goal is to simplify the design and proposal processes, enabling the sales force to respond quickly to customer inquiries and Request for Proposals (RFPs). Automated proposals have standard terms and conditions, and incorporate caveats and conditions particular to the jurisdictions included in the proposal. Proposals can move quickly to contract with the right document management system.

12.3.2 Service Ordering

Once a sale is made and a service commitment extended, the service ordering process begins. All of the components of the sale have to come together before the committed service due date. Moreover, each transaction associated with a service order is recorded as it is completed by the service order management system. Service orders that are cancelled must be able to be rolled back, returning all of the resources to their original state.

12.3.2.1 Consumer

Mass market services need systems support for complete "flow through" processing. An expression heard from Operations when new products are proposed for introduction is "no flow–no go." By that they mean they cannot support a new market product that requires human intervention in the provisioning and ordering processes.

Service Order Management

Service orders are at the center of the fulfillment process. As an example, a service order for POTS results in the following automated actions:

- a directory number is assigned from the pool of available numbers in an exchange served by the class 5 switch serving the customer's wire center,
- a copper pair on the customer's terminal is assigned, based on the customer's address,
- if needed, a cross connect order is issued for the serving area interface,
- a work order to connect the terminal to the customer's drop is generated if the drop is not connected,
- a line equipment, or switching system termination, is assigned in the serving central office,
- a work order to cross connect the line equipment to the customer pair at the main frame is generated,
- an electronic order is sent to the switching system assigning the new directory number to the line equipment, and setting all of the features, and charging class associated with the order,

- an electronic order is sent to the customer's interexchange carrier,
- the directory number is populated in the local number portability database,
- the customer's name and directory number are added to the Calling Name Service (CNAM) database along with any restrictions,
- the directory number is populated in the Line Information Database (LIDB) for third-party calls,
- the customer's name and address are added to the 911 database,
- the directory number and customer information is populated in the directory and information systems,
- the billing system is updated to reflect the customer's information and service, and
- if the customer purchased an inside wire plan, or a telco supplied station set, a work order is issued to dispatch a technician to perform the inside work at the residence on a date and time frame specified by the customer.

These steps vary somewhat depending on whether the customer has existing service or wants services beyond POTS. Flow through means that all of these steps must be completed before the service is activated. If any one of them fails, the service order is blocked until that step has been cleared.

CPE Management

Customer Premises Equipment (CPE) and User Equipment (UE) management expanded with the introduction of DSL, TV, and mobile services. Of the three, mobile device management is the most automated by far. Activation of mobile service is performed via Subscriber management system (SMS) push using over the air (OTA) provisioning. Global System for Mobile Communications (GSM's) introduction of the Subscriber Identity Module (SIM) makes CPE adoption and changes effortless and they are usually performed by the user with little intervention by Sales or Care.

Asymmetric Digital Subscriber Line (ADSL) service required craft installation when it was launched in the 1990s, but the economics made truck rolls prohibitively expensive at scale. The first attempts at customer installation only had about a 20% success rate. Today ADSL service is largely customer installed because of two initiatives. Human Factors engineers and cognitive psychologists dissected the self-install material and tools, and improved the success rate to over 85%.[1] Improvements in the Customer Management Systems (CMS) and loop qualification systems also improved success rates. Before loop qualification systems were refined it was not uncommon to sign a customer up, ship the ADSL modem, and dispatch an installer only to find out the customer's loop was too long for the service as sold.

TV services, like U-Verse and FiOS have also expanded the role of CPE management. CMS used for managing set top boxes (STBs) in the home are usually supplied by the STB manufacturer. Carriers deploying Microsoft's IPTV service rely on their subscriber management system which manages Microsoft client software and firmware from OEMs. Management of STBs in the field and inventories are performed by separate systems.

---

[1] The Human Factors group at SBC Labs achieved that improvement after a six month effort.

12.3.2.2 Enterprise

Service ordering in the enterprise market is more complex than the consumer market for several reasons.

- the services are more complex,
- there are more interdependencies. Services depend on the customer's Wide Area Network (WAN) and Local Area Network (LAN) environments and infrastructure, such as Voice over Internet Protocol (VoIP),
- multiple service providers are often involved. Most customers have global locations and use multiple service providers for reliability, access, and competition,
- there are a greater number of equipment suppliers and options for service design and configuration, and
- significant customer dependencies. Many new services require coordination with the customer to make changes and find maintenance windows when necessary transitions can be made without affecting the customer's core services.

Service Order Management

There are more similarities than differences in the service order management processes in consumer and enterprise markets. Work orders flow to work groups to implement physical cross connects, populate circuit cards, and install terminal equipment and CPE. Differences come from the degree of coordination required for the two services; consumer services are aimed at a mass market where full flow through and automation are a financial necessity.

Detailed Design

When a sale is made a detailed design review of the service is made by Engineering for more involved orders. Data associated with the order has to be validated; location information, circuit identifiers, CPE compatibility, and other data are verified as a precursor to design review. An order often involves multiple tariffed products with interdependencies.

Orders generally fall into one of three categories:

- **Standard products and offerings** – products that have been fully vetted and have full systems support can be provisioned with full flow through and apart from logistics issues, should meet committed service dates.
- **Supported products and offerings** – these are products and services that do not have full systems support but are offered as a special design. They are usually services that are in a controlled introduction, where the network implementation is complete but systems support is lacking, or they are services that several customers have bought through RFPs or on special request and there are operations workarounds that support them. If demand becomes great enough, systems are upgraded to support the service as a standard product.
- **Custom offerings** – enterprise customers issue RFPs that represent substantial opportunities, but they include an absolute requirement for some service or offering that has not been standardized by the carrier. If the revenue at jeopardy is substantial, the carrier will agree to the service without the necessary systems support.

Designs are standardized through a process that includes:

- **Development of field of use guidelines** – systems engineers work with product managers to define and codify when, where and how services are provided, specifying equipment types and configurations.
- **Integration** – systems engineers analyze the new service against all existing interfaces, network elements, and support systems to verify interworking, protocol, and feature compatibility.
- **Verification testing** – systems engineers test the new service for functionality, northbound interfaces for support system integrity, charging, and test access.
- **Support systems modification and verification** – all aspects of the TMNs model have to be examined and those that apply to the new service are implemented. This is typically the activity that determines the critical path to standardization.

In the detailed design phase services that have not been designated as standard require interim solutions. They come in the form of operational workarounds, manual processes that have to be developed and approved by Operations, Care, and Engineering. Interim scripts may be written to aid provisioning, performance, and fault management.

CPE Management

The penetration of managed services and resale in enterprise are far higher as well. Both have dedicated Engineering and Operations teams to support the range of third-party CPE that make up the services. Smaller customer routers and IP PBXs are managed remotely, usually via software systems developed by the carrier. Those systems are part of the order management process.

Third Party Access Management

Enterprise services require coordination with other carriers for access or interconnection and with the customer for implementation. eBonding, ordering, and tracking systems for global access consumes a good deal of effort and resources. Global carriers operate in 200–300 countries across the globe. Tariffs, billing, regulatory compliance, service intervals, and the sophistication of automated ordering and service assurance vary widely across the incumbent service providers. Systems support is a patchwork across the globe with major carriers investing more in those countries that have the highest traffic and a systems infrastructure that supports eBonding. A good deal of third-party access management is manual and is likely to remain so.

12.3.3 Service Activation

Service activation, with the exception of home CPE installation, is completely automated for the mass market but is labor intensive in enterprise markets. Account teams work with the customer and Operations to build service activation plans. Pre-service tests are scheduled to verify each part of the order as it is fulfilled. Activation is scheduled at the convenience of the customer, generally in a maintenance window at night or on a weekend. For many services carriers provide service assurance metrics on a network wide basis and specific metrics for each customer. Web portals dedicated to enterprise customer service level monitoring are the norm for global carriers.

12.4 Design and Engineering

Support systems, particularly provisioning systems lag network systems in the services and technology process. The TMF, ITU-T, and industry have adopted current software design practices and worked to develop frameworks to speed new service introduction and improve flow through. In spite of these efforts support systems still tend to lag network systems by about a year. Part of the explanation is the complexity and diversity of carrier systems environments. No two carriers have the same collection of systems, service definitions, business practices, or network equipment. Provisioning systems for each new service have to be designed or at least modified for every carrier.

The other and possibly more tractable problem is the lack of network domain knowledge in the support systems industry. Support systems development is outsourced to a larger degree today than in the past. Software skill, not network systems knowledge is the determining factor in hiring developers within the community. The developers are often far removed from, and may have never spoken with the network systems engineers that design, specify, test, and certify the equipment and services. Invariably important aspects of the service are missed by the systems developers causing rework and sometimes complete redesign. When provisioning systems are finally released, a flow through rate of 50% is considered a success. Network systems are expected to process transactions with a 99.999% success rate when they are introduced. Without comprehensive systems engineering knowledge of the service and how it is implemented on each network element, surprises are inevitable.

12.5 Summary

Support systems are significant investments, rivaling network in carrier budgets. FCAPS functions are addressed by families of systems that cross technologies from legacy voice to mobile and IP. Service fulfillment is possibly the most complex support system function. Sales, service design, service ordering, and service activation all fall under that category. Consumer services have end to end flow though as a pre-requisite for introduction. Mass market economics won't work without it. Enterprise services are more complex, often because they have a heavier reliance on managed services that use CPE, and because of reliance on other incumbent carriers. All aspects of enterprise service fulfillment are supported by systems, but they still need support by the account teams, Engineering, and Operations.

ITU-T and the Technology Management Forum (TMF) have developed and standardized frameworks that identify support system functions by categories. Architectural practices, based on UML are also cited as examples for business and software modeling. While helpful, industry standards for support systems lack the detail and specificity found in network requirements, ITU-T, IEEE, and IETF standards. Support system implementations are different for each carrier and are likely to remain that way.

References

1. ITU-T (2000) M.3000. *Overview of TMN Recommendations*, February 2000.
2. Crovella, M. and Krishnamurthy, B. (2006) *Internet Measurement, Infrastructure, Traffic and Applications*, John Wiley & Sons, Inc., New York.
3. Claise, B. (2004) IETF RFC 3954. *Cisco Systems Netflow Services Export Version 9*, October 2004.

Part Three

Transformation

Part Three

13

Integration and Innovation

Global networks are dynamic. Growth and expansion into new regions are part of the story, but the turnover of services and technologies that underpin the competitiveness and performance of the network are the greatest forces of change. This chapter describes how the stream of new technology and services are integrated into the network and the product mix.

13.1 Technology Integration

New network technology comes from a range of industry telecommunications suppliers. Reasons for adopting new technology are varied. Emerging designs of deployed systems incorporate new component technology, faster processors, use less power, and have higher capacity. Economics or scale can make them attractive enough to consider technology refresh. New classes of network elements are adopted with evolving services and networks, like Long Term Evolution (LTE). Designs of new network systems are largely based on industry standards, but interpretation and compliance with those standards are the responsibility of the supplier. There are no national compliance laboratories in the US. Standards are also quite flexible allowing suppliers some latitude in their designs and giving carriers options in the way they incorporate systems to form networks. Integration of technology follows a process where a supplier's product is evaluated and modified to fit a specific set of applications in a carrier's network. The system must interwork with all of the existing network elements, but also has to integrate with network management systems (NMSs), operations support systems (OSSs), and possibly business support systems (BSSs). Responsibility for integration rests with Systems Engineering, Network Engineering, IT, and Operations. Technology integration is composed of distinct phases,

Technology scanning – identifying potential or needed technologies and assessing their maturity for network introduction,

Technology selection – performing economic, technical, and operational analyses that lead to system selection,

Verification testing – laboratory testing to verify functionality and compatibility with network and IT systems,

Global Networks: Engineering, Operations and Design, First Edition. G. Keith Cambron.
© 2013 John Wiley & Sons, Ltd. Published 2013 by John Wiley & Sons, Ltd.

Operational readiness and user acceptance testing – in-field tests to confirm the system can be managed and maintained,

Soak – limited field introduction to place the system under the stress and conditions found in the real environment,

Lifecycle support – activities needed for the life of the system to insure it continues to perform and adapt to new requirements as they emerge.

In this section the process, people, and documentation are described.

13.1.1 Technology Scanning

Looking for technology solutions and opportunities to improve network efficiency and improve scale is a responsibility shared by Systems Engineering and Network Engineering. Systems engineers are exposed to new ideas and technology through industry forums they attend, collaboration with universities, suppliers previewing their forthcoming products, and their own analyses of opportunities for new designs in the existing networks. Network engineers are also attuned to technology advancements, but are usually more driven by economic opportunities. Cooperation and exchange between the two groups is continuous; each gains from the other's insights and knowledge.

Scans can be product focused, initiated because a manufacture has announced end of life for an embedded system or because it becomes clear that an existing technology, like ATM, can only scale to a certain point. End of life technology replacement is often a hard sell to Finance. Nothing is as cheap as a sunk cost and the embedded systems were paid for long ago. Engineering has to enlist Operations in building a case that illustrates the operational risk, and operational costs of trying to maintain and service end of life systems. Cap and grow strategies can be economically attractive in many cases. End of life systems may not have hardware replacements available after a supplier fixed date. Procurement is faced with making an end of life buy which only compounds the problem; investing in obsolete hardware is an unpalatable solution. A common strategy is to restrict or cap the old technology by continuing to use it in a limited set of offices, regions, or applications that are not expected to experience high growth. The aging technology can then be harvested and replaced with new technology in parts of the network that are growing and need additional capacity. Harvested hardware serves as spares and over time the footprint of the old technology shrinks, ideally to the point where operational costs make it more expensive to keep than replace.

Marketing and Sales may also play a role if there are new competitive services that cannot ride the old technology. New services, like LTE, come with new architectures and complete service overbuilds. That is often an opportunity to step back and see if older technology can be replaced and the costs carried by the new service.

White papers with technical and economic assessments of new technology and potential applications may be written in conjunction with a scan. Presentations to management without a whitepaper are more common. Management sorts through the opportunities and initiates technical and economic analyses for the ones with the most promise.

13.1.2   Technology Selection

Technical analysis is a precursor to technology selection and it can be a substantial investment for complex systems, and so is more formal and more closely managed than technical scans. If time permits, economic analysis should be performed before technical analysis because examining the economics tends to be faster, less complex and can be done without dedicating time from scarce systems engineering talent. If there is no economic opportunity, there is no need for a technical analysis. Fundamental planners who have to perform economic studies don't always agree with that sequence. Ideally planners identify two to four suppliers with systems that are economically competitive.

The starting point for a technical analysis is dependent on the system being analyzed and the motivation for considering it.

13.1.2.1   Technology Refresh

Technology refresh is the replacement of existing systems with like systems that are faster, cheaper, use less energy, and are more reliable. Analysis of candidates is simplified because the existing system functionality and performance are well understood benchmarks. Network systems have an engineering and operational specification called a Field of Use (FOU) specification. That specification identifies all network and IT system interconnections, services supported, NMS, OSS, and BSS system dependencies.

Black Box Review

First pass reviews of candidate system specifications by systems engineers and operations analysts often reduce the candidates to one or two. In the first pass candidate systems are examined as black boxes. Their implementations and conformance with mandatory protocols and standards is reviewed against the FOU specification and industry standards. Live service experience with interconnections to systems listed in the FOU specification is a positive indication that the system can achieve conformance. Network protocols and standards are more rigorous and more uniform than IT interfaces and applications. Northbound interfaces to IT systems as described in Chapter 12 are likely to be different in candidate systems than the embedded system, but the task of modifying the IT systems should be manageable. Where network systems use industry standard Management Information Bases (MIBs), the nomenclature and semantics of the data should be nearly identical. Enterprise MIBs will be different and need to be assessed to understand the impact on OSS and NMS systems.

White Box Review

In the next phase of review, surviving candidates undergo a white box review. In a white box review engineers look inside the system; they examine internal bearer and control paths, look for bottlenecks, check for design redundancy, maintenance modularity, and any weak spots that can make it more difficult to operate and maintain the system. Carrier systems engineers generally do not have the hardware and software design depth of the equipment and systems suppliers. White box reviews are essential, however. They often reveal shortcuts the designers took to get to market and even misunderstandings of standards. Day-long review sessions with carrier systems engineers and supplier design engineers also give each a sense of the others understanding of the technology and networks. It either builds mutual confidence or forewarns of potential support problems in the future.

Laboratory Scan

The last review phase is a laboratory concept test. It is the most resource consuming and so ideally only one system, or two at most enter this phase. The systems under test are connected in a test bed to traffic and load test systems, protocol analyzers and other network elements they will be interworking with in the field. Test suites, likely those used with the embedded system, are run against the candidates.

Selection and Procurement

A summary report comparing the candidates is prepared by the technical team and submitted to management for selection. These reports list the systems side by side, usually with a numerical or letter grade indicating their assessed performance against critical requirements. In summary reports technical teams often place candidate systems in three categories to clarify their findings and simplify the job of final selection,

- **Best in class** – that system, or systems judged to be the best technical and operational candidates,
- **Acceptable** – those systems that may need to undergo design changes, but can meet the minimum technical and operational requirements,
- **Unacceptable** – those systems with design or implementation issues that are severe, and are not expected to meet requirements within the time window needed.

Economic and business considerations may induce the decision maker to choose systems from the second group, at times even the third. Even in that event, the time spent in the analysis is not wasted. The technical team starts working with the chosen supplier using a list of deficiencies developed during the analysis.

13.1.2.2 New Technology Introduction

Technology introduction goes through a process similar to technology refresh, except there are additional steps that have to be accomplished before products can enter a black box review. As used here, new technology introduction means the service, architecture, or design are not currently in use in the network. FOUs have to be written from scratch and there may not be concrete network designs, interface specifications, and system management designs to consult as a requirements base. Technology introduction can be limited in scope or usher in a completely new access and service network. Introduction of Reconfigurable Optical Add Drop Multiplexers (ROADMs) is illustrative of a more confined project. It changed the way optical networks were designed and managed, and so a good deal of work had to be done before the product evaluation and selection process began. Unlike ROADMs, the introduction of LTE represented a massive project, touching all parts of the network and carrier teams. A dozen new network elements were introduced, each requiring specification, evaluation, test, and network introduction. Each new element was accompanied by a set of new management systems, or at least extensive revision of existing systems.

Different companies are likely to have different names for the specifications and plans needed to begin new technology introduction, but they are likely to address the same issues. The documents in the following are listed in roughly the order they are written. Much of the writing can be done in parallel, but they build on each other and consistency matters.

- **Architectural plan** – a high level document written at a functional level, identifying network elements or functions, their responsibilities, physical interfaces, and Open Systems Interconnection (OSI) connectivity to other elements. A baseline plan for the entire network and all systems is updated regularly, identifying current systems and target (future) systems. When new technology initiatives are launched the plans are refreshed. OSI connectivity means the plans identify elements that are interconnected above the physical layer. For example, databases queried by network elements may connect over a multi-tiered IP network, but at the application layer they are closely bound. Architectural plans should be specific enough to segregate element responsibilities, the what, but not get into design choices or preferences, the how. Architects are not necessarily skilled at design.
- **Marketing Services Description (MSD)** – if new technology brings with it new services, an MSD is written. New services are described as a user would describe them, not as an engineer designs them. Customer segmentation, expected market penetration, sales channels, customer care expectations, charging and billing mechanisms, but not necessarily pricing, are defined. The sales and service expectations are described as well. For example, mobile product MSDs should describe the various sales channels (market retail store, affiliate store, on line purchase) and how the product is activated. It should also describe any aspects of sales and service that are self-service and use websites tied to back end provisioning or care systems. Cost objectives for installation and service turn-up in the business case need to be translated into a sales and installation model.
- **Technical Service Description (TSD)** – TSDs are requirements and design documents that describe all of the new elements associated with the service, the functionality, connectivity, and the protocols they use. End to end service objectives, impairment budgets for the network elements, and performance objectives should be specified as well. Message flows, control plane designs and constraints, redundancy and failover, and a list of other network attributes are described in the TSD.
- **Operations and engineering plan** – network systems functionality is usually set by standards bodies and technical forums, but management systems should conform to the operating model set by Operations. An operations plan for new technology is the roadmap for the force and systems that are needed for network and service management. Operations plans for new technology vary, but some of the topics in typical plans include,
  - **Site redundancy** – does the service warrant complete site redundancy with failover capability? Systems Engineering can help Operations analyze this question, but Operations is likely to make the choice.
  - **Network surveillance** – identifies the operations center that has primary responsibility for new networks and network elements.
  - **Centers of excellence** – service delivery and day to day operation and administration of the network and service require either augmentation of existing centers or the formation of new ones.

- **Service management** – NOCs focus on network management and recovery. Services such as IPTV have unique service delivery and quality demands outside the capability of existing operations centers. For services of that complexity a dedicated service management center with specialized monitoring and management systems is needed.
- **Capacity and performance management** – engineering teams' responsibilities for monitoring and augmenting network capacity and network performance are described in the plan. If the new technology does not fit within existing practices and systems, the future method of operation, skill, and systems capabilities are described.
- **Field service forces** – line field organizations may be affected by the new technology. Special training, contract services, sparing, and other issues are addressed in the plan.
- **Premises technicians** – if home installation or repair is required, job titles and descriptions and estimates of service times, and responsiveness have to been spelled out.

Use cases are valuable in operations planning, describing the actors, environmental assumptions and operational procedures and expected outcomes. The cases identify which existing systems used by Operations and Engineering need to be augmented to support the new technology, and which new capabilities or systems are needed.

- **Network management plan (NMP)** – the complete set of new support systems and required modifications to existing systems are spelled out in the NMP. Interconnectivity among the systems and network elements, along with protocols, databases of record, and management networks are defined in some technical detail. In most cases support systems trail network deployment, particularly during soak and controlled introduction. NMPs identify the stages of support systems introduction and workarounds used in the interim periods. All of the Fault, Configuration, Accounting, Performance, Security (FCAPS) functions are addressed in the plan, with enough detail to identify which systems support each of the functions and what the dependencies are.

Projects like LTE depend on the entire suite of plans described above if they are to be successful. Such projects are replete with technologies not previously seen or managed by the carrier and so thorough analysis and planning are prerequisite to preparing teams and the network for change on such a scale. For smaller projects the list of plans above appears daunting and cumbersome, but for those projects the plans can be one page and they can be incorporated into a few documents. If the plans are trivial, there is no harm in putting them in writing so that the conclusion is public.

System suppliers benefit a great deal by access to the plans as well. Preparing plans sparks engagement and dialog between carrier teams and suppliers, thereby setting expectations and clarifying understanding of technical functions and details. Detailed requirements flow more smoothly as does integration and verification testing. The detailed requirements and plans often constitute the technical appendices in an RFP used for supplier selection. Steps in the selection process are the same as those used in technology refresh; they include black box review, white box review, laboratory scan and selection, only differing in scale and intensity. After selection, agreement on terms, conditions and schedules, technology verification, and integration begin.

13.1.3   Network System Testing and Verification

System testing and verification is the final gate before technology enters a carrier network. When it works well, it is a cooperative effort between supplier, carrier systems engineers and developers to assure both companies succeed. It is a chance to make the product better.

13.1.3.1   Why Carriers Test

Carriers of scale often perform a rigorous verification because they simply have more resources, have dedicated professional teams and can command a more dedicated investment of talent from the suppliers. Smaller carriers have talented teams, but they may fill many roles and lack the extensive laboratories needed for in depth verification. Altogether, I spent 10 years running verification labs and teams at the working level. That does not include the eight years I headed SBC and AT&T Labs and their extensive verification and certification staff and labs. Managers and even engineers outside of the Labs and Systems Engineering community often have misconceptions about testing. In the following list some of the misconceptions are addressed.

- **Myth #1** – Equipment suppliers verify systems and so the carrier's verification is a duplication of effort and slows technology introduction.
  - Testing is part of the integration process. During testing engineers are developing engineering guidelines, writing engineering documents (such as the FOU Guidelines) interface specifications for support systems, working with Operations to write and verify installation and maintenance procedures and verifying system capacity and limits. Equipment suppliers do not operate equipment and do not provide custom documentation and interface designs associated with each carrier they serve. Without the detailed knowledge that comes with testing, it would not be possible to integrate the equipment into the family of systems the carrier relies upon.
  - Equipment suppliers design products for a wide market, not a specific network. I have experienced cases where our engineers were testing two network elements that were directly interconnected and found them incompatible. Both pieces of equipment were from the same supplier. Suppliers of scale have vertical product lines, separating their design and development teams. In the most recent case two teams from the same supplier were not working with each other and one team had a different understanding of the interface specifications than the other. The difference was not small. It delayed service introduction by eight months. Carrier testing is based on the FOU and specific systems that the equipment will face in the field. This focus without exception results in the discovery of bugs and incompatibilities equipment suppliers failed to address in the initial designs.
  - Some systems are so flawed they never emerge from carrier testing. Supplier product managers and sales teams are under pressure to meet what are often ambitious schedules imposed by carriers. If they miss the date, there is no sale. In the face of that pressure products are offered and even showcased that are not ready for service. As the supplier's delivery date approaches the system is still in design. To meet the

date the supplier skips the last steps in their development cycle, full integration and verification testing, often with the excuse that the carrier's team will perform those tests and any bugs will be corrected during that process. Some systems are so buggy when they enter the carrier laboratory that they are shipped back to the supplier within two weeks. One system stands out in memory. Our engineers worked with a major supplier for a year and a half without the system ever passing verification. The relationship became so contentious I called a design review with the supplier. We went through their board level hardware and software designs and found the problems were not in implementation but in design. The supplier's development team did not have in-depth experience in designing carrier grade systems. They agreed with the findings and withdrew the system.

- **Myth #2** – The system is already in use in a competitor's network, so we can skip verification testing; we need to get to market.
  - Competitors don't often share their experience with a particular supplier's equipment. This argument usually comes from the equipment supplier. How extensively a competitor uses equipment and the FOU matters a great deal. If your FOU includes multicasting and the competitor does not, you may be in for a surprise. It is also unlikely the two carriers use the same support systems or adjacent network systems, so the experiences in the two networks can be quite different. Faced with this argument by suppliers, test engineers often ask the supplier for a list of known troubles submitted by other users of the equipment on condition the supplier anonymize the reporting carriers. Suppliers sometimes refuse to provide such a list, citing confidentiality.
  - When equipment enters service prematurely systems engineers have not developed the expertise to properly provide tier 4 support nor has Operations built a tier 3 organization with depth of experience. Problems begin to arise immediately, often outages come in succession and carrier engineers scurry to construct a tier 4 lab to meet the crisis. Meanwhile the equipment supplier has moved their design teams on to other projects and carrier systems engineers are working with the equipment supplier's maintenance developers, not the design developers. Maintenance engineers are often not as experienced and are seldom free to change the design; they can only correct minor implementation errors. A major flaw requires the supplier to take their design teams off of other projects, if they can locate them, and put them back on the failing system.
- **Myth #3** – We'll force the suppliers to perform verification for us by contractually binding them to interoperability testing (IOT) in their labs before they deliver their systems.
  - Lab to lab IOT can be helpful, but it should precede carrier verification testing, not replace it. Unless IOT involves a single supplier's equipment the value to be gained is marginal in most cases. Suppliers are reluctant to share proprietary information with each other about their equipment's design and operation. It can be used to their disadvantage in the marketplace. Agreement on test suites is another sticking point. Unless they are written and supervised by the carrier, which practically defeats the purpose, suppliers are unlikely to agree on any tests that might disclose a weakness or feature deficit in the systems. Even sharing test results becomes a problem.
  - The issue of lack of corporate knowledge of the systems arises in an IOT scenario. If carrier tier 3 and tier 4 engineers are not trained during the test cycle, they cannot

support the system in service. If an outage or serious chronic problem arises with the systems, the negotiations and reconfigurations needed to build an intercompany tier 4 lab to lab test arrangement that is faithful to the carrier's network preclude a timely resolution to the problem. In the 1991 outages described in the next chapter Bellcore did not have a tier 4 lab and worked hard with DSC to build one after the outage onsets. In spite of their best efforts, Bellcore was only able to recreate the problem weeks after it had been identified in Pacific Bell's tier 4 lab and a patch had been designed and deployed in the network.

- Today's systems require IOT up and down the OSI stack and so the number of systems needing interconnection precludes effective testing outside of the carrier laboratories. Systems that are not physically interconnected at a link or transport layer often have logical interconnection above those layers. An LTE MME is a prime example. Here is an abbreviated list of the network systems and interfaces formally defined for an MME: HSS – S6a, S GW – S11, eNode-B S1-C, MSC – SGs, SGSN – S3, MME – S10. The list does not include interfaces to support systems.

13.1.3.2 Test Bed Design

Figure 13.1 is a diagram of a fully developed test bed. Most facilities are not this completely developed; only technologies that have a long service life and ones in which carriers that have made substantial investments warrant a build out this complexity.

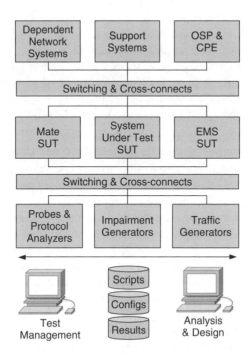

Figure 13.1 Fully developed test facility.

System Under Test (SUT)

Suppliers usually provide System Under Tests (SUTs) either on extended loan or gratis for certification to carriers of scale. SUTs are under change control and configured to match the planned field versions. Once the SUT is delivered and installed the supplier is either on site or at a minimum available remotely on demand, but does not have authority or access to modify the SUT without carrier approval. A list of known bugs accompanies the SUT and any subsequent software and firmware loads. If the SUT network design calls for a redundant mate, then the lab configuration needs to include it, as well as the Element Management System (EMS), as shown in Figure 13.1. I/O from all three systems terminates on switches or manual cross connects for flexibility. Test equipment is usually too expensive to dedicate it to one test bed; it needs to be shared.

Dependent Systems

Most SUTs need interconnection to dependent network systems for any realistic test program. Interconnection to other new technology and legacy systems may both be required. Allocating and configuring legacy systems is more straightforward; they are known and proven systems. When the SUT is interconnected to immature technology, or technology undergoing verification testing the number of unknowns begin to grow dramatically. Data capture is particularly important in those situations as a means to determine which unproven system failed to implement a protocol correctly or is otherwise at fault. Dependent network systems are instrumented as well; at a minimum alarm and message collection for those systems is activated.

NMS, OSS, and BSS system interconnection is included in the test facility as well. Provisioning and performance systems are unlikely to be ready when testing new technology; usually special test systems are scripted to emulate a minimum functional set. In some programs the IT systems lag so significantly that the interim test facility systems are placed in operational service until the IT systems mature. All FCAPS capabilities inherent in the SUT are exercised in verification testing. For example, provisioning via the EMS may be exercised via an automated script through a northbound CORBA, XML, or other interface.

Outside Plant and Customer Premises Equipment (OSP and CPE)

Some network systems, such as Digital Subscriber Line (DSL) multiplexers and Dense Wave Division Multiplexing (DWDM) optical systems operate over physical media that can vary in its characteristics and in the case of copper systems is subject to a wide range of impairments. Test equipment suppliers design and build test systems meant to simulate those media and their impairments. Copper cable simulations use lumped LRC networks intended to match the characteristic impedance of actual cable. While useful, in the end real outside plant (OSP) cable and fiber behave differently than the lumped networks used in test systems. Delay characteristics in particular affect transmission performance and lumped networks do not accurately recreate delay. Reels of optical cable are needed for the same reason. Different wavelengths of light propagate at slightly different speeds, causing delay distortion. Actual cable best recreates that phenomenon. Carrier labs have a mix of simulators and actual cable. Simulators are capable of running a large number of tests quickly and automatically documenting the results. Actual cable tests are often run after a SUT passes simulator tests. Cable tests are often more time consuming, but they are also more flexible and are faithful.

Moreover, there are only a couple of suppliers of cable simulators, so carrier engineers are reasonably sure the equipment supplier used the same simulators they are using; if the SUT failed those tests it would be surprising.

Some of the most extensive laboratories in carrier test facilities are customer premises labs. Consumer labs support all of the versions of DSL, IPTV (Internet Protocol Television), and other home services. Typical equipment includes all versions of supported modems and home gateways, different local area networks, set top boxes, and Wi-Fi access points. Like the other labs, these labs are instrumented and can be configured for a variety of consumer networks and wiring environments. Enterprise Customer Premises Equipment (CPE) labs are even more extensive. Carriers often support a range of managed services using different customer edge routers. Each new provider edge router has to be tested against the list of supported customer edge routers and each of carrier's managed services.

Environmental Chambers

A wide range of network systems are placed in cabinets, man holes, or placed on poles in the OSP without the benefit of cooling, heating, or humidity conditioning. Electronics in those environments must operate over a wide range of temperatures and humidity. Equipment has to operate in the Mojave Desert and on top of a cell tower in North Dakota in the winter. Electronic systems behave differently under this range of temperature. Critical systems are tested in environmental vaults capable of a range of temperature and humidity that replicates the most extreme conditions the equipment will face. Large systems, like multi-bay switching systems that can't fit in a chamber can be tented with thermostatically controlled heaters used to raise and lower temperatures. Results can be surprising. In one test cycle our engineers discovered that a DSL multiplexer failed heat cycle tests in an environmental chamber. As the system passed through $50\,^\circ C$ with either rising temperature or falling temperature, it failed and rebooted. An oscillator from a particular component supplier had not been tested thoroughly by the equipment supplier. That temperature is at the high end of the range for central office equipment, but not for OSP cabinets sitting in the sun, the typical environment for that multiplexer. Had that system been deployed wide scale outages could have resulted in massive service problems, followed by a costly upgrade program. On hot summer days it would have rebooted twice, once as it ascended through $50\,^\circ C$ and once when it descended below it.

Probes and Protocol Analyzers

As carriers move to an all IP services network, testing is increasingly a matter of building protocol test suites and using protocol analyzers to generate tests and automatically analyze the results. Analyzers represent an increasing investment; test systems can cost half a million dollars each and several are needed for different labs and different services. When IP analyzers came into popular use they were restricted to lower layers of protocols and test engineers crafted scripts at the session and application layers. Today manufacturers of analyzers sit on standards committees and build products as the standards evolve. Test suites for the application layers are developed early by the test equipment suppliers and used by network systems developers to test and approve their designs. Carrier test engineers purchase the same application test suites and use them to verify and certify the new systems.

There is an increasing danger in this industry direction. Effective testing is a directed statistical undertaking. When beginning a verification project, managers assess the list of functional test areas, estimate the associated risk and complexity of each area and assign time and resources accordingly. Technical leads then develop plans, define test suites, and designate individual tests for each functional area, to be reviewed by the manager. Test engineers design and develop individual tests and suites. Increasingly carrier test managers and engineers rely completely on the standard test suites shipped with the analyzer. System suppliers use conformance to the standard tests as a selling point that their systems are ready for service. When system supplier test engineers and carrier engineers develop their test plans, suites, and tests individually, the statistical sampling space is broad and deep. Working independently they explore different conformance and potential weaknesses of the system under test. Importantly their independently designed tests reflect the experience of the engineers in practice, detecting failures and weaknesses in other similar, but perhaps not identical systems. Electing to take the easy path and rely on the analyzer standard tests dramatically decreases the sampling space. There is no guarantee that analyzer engineers have the rich testing and network background of carrier engineers and so are less likely to investigate areas outside of what is written in the protocol standard.

I was aware of this trend toward standardized test suites, but the danger first crystallized for me when I visited information systems companies in 2007. The purpose of the visit was to assess the opportunity to move some verification testing to those companies, and to look at the possibility of opening our own laboratory in India. The ground rules for my visit were that I did not want extensive presentations on the capabilities of the companies, but wanted instead to visit working laboratories and have the freedom to talk to the managers and engineers. My personal experience in verification testing served as a solid base for the discussions. In some of the labs I found entry level engineers running standard tests using preprogramed analyzers; they were simply verifying that the report generated by the analyzer showed no discrepancies. When quizzed on transmission methods, types of impairments the system was likely to face, or alarm generation and fault isolation, they had limited understanding and experience. In one case I found a piece of equipment that was being verified for the equipment supplier and the same engineers, on a different project, verifying it for a carrier. In that case the engineers did not truly understand the application or design of the equipment; they simply ran through a list of settings on the analyzer, started the test, and looked at the report. That piece of equipment was not stressed in a realistic way during the process.

Switches and Cross-Connects

Switches and cross-connects enable flexible test configurations, sharing of systems and sharing of test equipment. Well managed labs do not have an expensive piece of test equipment sitting in an aisle with a dedicated connection to a piece of network equipment that is visited by an engineer for a couple of hours a week. That's useful on occasion, but the equipment should not sit there more than a few days. Test equipment bays can be segregated from the SUT. All systems terminate on either switches or cross-connects. Equipment is scheduled and connections recorded and released when the reserved test time is over.

Data Management

Automation of testing, configuration, and results analysis assures repeatability and reliability of test processes. Scripting complex tests is an upfront investment but it pays great dividends. Using scripts to capture and load systems and test equipment configurations improves efficiency but more importantly assures integrity in the test process. Actively managing failures via bug reporting systems, such as JIRA,[1] and maintaining version control using systems, such as CVS,[2] are sine qua non for test programs.

Test Fixtures

Not shown in Figure 13.1 are specialized test fixtures. These are specialized hardware and software test jigs that are constructed by test engineers for specific tests. Cable farms for DSL testing and thermal monitors used in checking for hot spots in systems are examples. For some technologies fixtures are the only way to get the needed measurements or to create realistic impairments.

13.1.3.3   Test Team Roles and Responsibilities

There are three major test phases that most network systems undergo before entering a soak or controlled introduction into a carrier network.

- **Lab testing** – these tests are the primary responsibility of Systems Engineering. They are the most complex design and functional analysis of the system and are a precursor to Operational Readiness Testings (ORTs) and User Acceptance Testings (UATs).
- **ORT** – unlike lab testing, ORT focuses not only on the new network technology; it also exercises the Operations teams, support systems, engineering systems, and documented procedures. Operations has the primary responsibility for ORT.
- **UAT** – new technology that changes the way customers use a product undergo UAT. The system or service is tested from a user's point of view. UAT is the responsibility of Marketing, Care, or human factor engineers.

Lab testing is a team effort, divided into functional areas. The Table 13.1 is typical of the test areas for a new network system. The table is a representative summary; specifics vary significantly with different technologies and systems.

Lab Manager

Effective test execution begins with lab management. Lab managers do not have testing responsibility, but they do have control of lab design, configurations, operations, access, and housekeeping. You only have to walk into a lab and stroll through the racks and test benches to determine if the manager is doing their job. Poorly run labs have components and equipment everywhere and fiber and copper test links hanging from the racks like spaghetti.

---

[1] See http://www.atlassian.com/opensource/.
[2] See http://cvsbook.red-bean.com/cvsbook.html.

Table 13.1 Lab testing functional test areas

Functional area	Description	Primary responsibility	Supporting teams
NEBS level 3 compliance	Mechanical, power, cooling, and grounding	Systems engineering	Operations
Networking	Physical layer, link layer, and network layer protocol conformance	Systems engineering	–
Security	Craft access security (e.g., SSH), database encryption, northbound interfaces (e.g., SNMP), firewalls, and session logging protections	Systems engineering	Chief security officer organization and operations
Application	Functional test of transactions and/or sessions invoked at the application layer, including error legs	Systems engineering	–
Single point faults	Fail every port and circuit pack and monitor in-service performance	Systems engineering	Operations
Fail over and fail back	For redundant systems, can the system fail over to the mate and can traffic return reliably once the failed unit is restored.	Systems engineering	Operations
Alarms and logs	Verify alarm reporting and filtering. Verify alarms are retired and restored	Systems engineering	Operations
Traffic and performance measurements	Verify counter accuracy and persistence, threshold alerts, data integrity	Systems engineering	Engineering and operations
Fault management	Check system compliance with the managed object model and correct operational and admin state management	Systems engineering	Operations
Northbound interfaces	Simulate and verify northbound provisioning, maintenance, engineering, and care interfaces	Systems engineering	IT
System and database integrity and audits	Create data integrity faults and kill processes; observe system response	Systems engineering	Operations
Configuration management	Check transaction integrity, out of bounds, or inconsistent configurations	Systems engineering	Operations

Table 13.1   (*continued*)

Functional area	Description	Primary responsibility	Supporting teams
Provisioning	Verify transaction integrity, rejection of invalid records	Systems engineering	IT
Focused overload	Create link, process and virtual resource overload; check system performance and alarming	Systems engineering	Operations
System overload	Overload system, usually via specialized software or systems	Equipment supplier	Systems engineering and operations
Documentation	Installation and configuration practices, I/O manuals, task oriented procedures	Operations	Systems engineering

NEBS: Network Equipment Building Standards; SNMP: Simple Network Management Protocol; SSH: Secure Shell.

Periodic unannounced visits by senior managers make for better lab management. Equipment inventory control is another sign of a lab's health. Labs need effective marking of equipment such as bar codes or RFID, regular audits and entrance and exit controls on equipment. Different lab managers are assigned to labs with different functions. Much like networks labs are organized along the lines described earlier. Labs supporting mobility have no need to be collocated with backbone or DSL labs. They are interconnected via facilities and trying to collocate them is counterproductive; they become unmanageable if they are too large.

Test Manager

Test managers have broad experience in system testing and serve as the responsible manager for verification of network systems. Plans for testing a new technology are developed by the manager responsible for certifying results and recommending systems for use in the network.

- Test plan objectives are stated broadly with responsibilities and all of the essential players enumerated.
- Managers work with the network system supplier to set expectations, arrange for technical support, manage joint technical reviews, and layout logistics. Tight technical coordination is needed to manage the cycle of bug discovery, confirmation, reporting and tracking, fix design and delivery, and fix confirmation.
- Teams and team leads are organized by the test manager, often on a matrix basis. Operations, Engineering, and IT may have members on teams led by systems engineers.
- Scheduling and planning for test modules falls to the test manager as does status reporting to senior management and the supplier. Selection of test suites and their construction rests with test leads.

Test Lead

The experience level of test leads is deep but often not as broad as test managers. Teams vary in size from two engineers to a dozen, depending on the technology and complexity of building and running test fixtures and the actual tests. Leads design test suites; test engineers design, and execute individual tests. Leads are responsible for the efficacy and integrity of the tests and the scope of the test suite.

Test Support

Software skills continue to rise in importance in test labs. Engineers skilled in scripting languages like PERL and Python can quickly build drivers for test equipment and systems. Test support engineers can team with web site developers to build test scheduling, config-uration, execution, and reporting applications into secure intranet sites. That enables test teams to be distributed and system developers to view configurations and results.

Lab technicians set up routine tests, run cabling, install software, and build test fixtures. Most electrical and virtually all software engineers are not skilled at many of the hands on jobs in a lab; technicians are and can keep the lab in an orderly and professional condition.

Supplier Support

Arranging for on-site technical support from the supplier is a good investment for them and for the carrier test team. Support engineers are usually systems engineers, not software developers or hardware designers. They are skilled at operating the equipment and under-stand the design intent and limitations as well. Direct access to the design team should be afforded by the supplier, through their on-site engineering team. Having a quick way of reporting a suspected problem and vetting it with the design team avoids misunderstandings on the part of the test engineers, and can also alert the design team of corner case faults or even more critical design issues. A spirit of cooperation and mutual goal of moving the equipment through the process and out into the field has to be fostered by the leadership of the carrier and the supplier. Adversarial atmospheres serve no one well.

13.1.3.4   Test Design and Execution

Test design begins with the FOU specification. The FOU identifies those systems physically and logically interconnected with the SUT. That analysis determines the configuration of the test bed(s) needed to verify the SUT. Analysis of the protocols and applications used by the SUT helps identify test equipment and any special fixtures needed. Functional analyses using a checklist like the table above are used to generate a list of test suites and test team members. Carrier requirements, industry standards, and system specifications collectively describe expected system behavior and performance. They form a basis of individual test design.

There is a hierarchy in testing design.

- **Test plans** are broad in scope, identifying functional areas, listing requirements and specifications, enumerating test teams and responsibilities, schedules, logistics, reporting structures, and points of coordination. Test plans are the responsibility of the test manager.

- **Test suites** identify individual tests by name and the functional area they address. Suites often specify a test bed configuration, or set of configurations used during testing. Test suite development is the responsibility of test leads.
- **Test scripts** are written procedures, often accompanied by scripted software that tests a specific function or capability of an SUT. Test scripts are the responsibility of test engineers, with oversight by the team lead.

An ongoing debate is centered on the sharing of documents with system suppliers. My practice has been to share test plans but not test suites and scripts. Because testing is a statistical process, coverage is the greatest when all participants design and run their own tests to verify the system. If test scripts are shared in advance, equipment suppliers may be motivated to design their tests, and even their systems, to pass the scripts. Moreover, if new scripts are developed and run without their notice, suppliers may take exception. Having each set of test engineers go about the tasks independently makes for a higher sample space and likely a more reliable system. Once a fault is discovered on a supplier's system, test results, including the message sequence or other stimuli that brought the fault to light are shared in detail with the supplier. In some cases the supplier may not have the ability to recreate the fault. In those cases the scripts and even test equipment are shared to enable the supplier to perform the necessary forensic investigation.

Engineers segregate their tests into "sunny day" and "rainy day" cases. Sunny day tests verify system functioning and performance under normal operating conditions, with no faults or impairments in the SUT or interconnected systems. Rainy day tests verify system performance under faults, impairments, and overloads. Simplex operation, fail over and fail back tests are examples of rainy day tests. Those tests are more difficult to design and more time consuming to run. Alarm reporting, system craft tests and managed object status all need verification, as well as the functional performance of the system; it has to process transactions through rainy day cases. A basic set of sunny day cases can be performed simply by interconnecting the SUT with other elements in the lab. Presuming elements in the test bed conform to industry standard protocols, the SUT should begin processing transactions once the test bed is correctly configured. Rainy day tests often have to be performed using simulators in a standalone test configuration. Production network elements, like an Ethernet switch, can't be configured to produce errors or impairments easily, and certainly not over the range needed for rainy day testing. Ethernet test sets can generate a range of impairments from invalid addresses to invalid procedures.

13.1.4 Support Systems Integration

Once a network system has been verified to conform to network requirements and proven it interworks with other network systems, it has to be integrated with the NMS, OSS, and BSS systems. Support systems integration lags network systems integration for two reasons. A solid foundation is needed before support systems can begin their integration in earnest. Network element designs can't be changing while support systems engineers design and build matching interfaces. Secondly, support systems often have release cycles that follow fixed schedules. Upgrades to existing support systems must hit a window to catch the right release cycle. As a consequence support systems releases for a new network technology usually come out in phases. Stop-gap solutions are used to support those network systems and services that have a time to market imperative.

Network systems engineers test northbound interfaces to planned support systems as part of network systems verification. They often build test simulators of those systems to enable testing in as faithful a way as possible without the actual support systems. In the process of designing tests and verifying northbound interfaces, network systems engineers enhance their understanding of how FCAPS functions operate on the SUT. Those same network test engineers often end up writing requirements for support systems developers. Test beds are devoted to support systems integration as well, with analyzers and monitors to aid those teams.

EMS undergo extensive testing prior to support systems integration. Unfortunately EMSs vary dramatically in design and functionality. Unlike network systems there is a lack of clear standards and adherence in the EMS and OSS communities, as described earlier. Without standards each integration effort is unique. Northbound interfaces vary at the session and presentation layers. In many cases the EMS is well designed, performing FCAPS functions within its domain. The other extreme is one in which network equipment suppliers furnish no EMS at all, leaving those responsibilities to third party suppliers or to NMS and OSS systems. FCAPS functions are prioritized with emphasis on fault management and accounting management at the top; carriers need to correct faults, be able to inventory equipment and bill customers. New technology often enters networks with limited implementations, workarounds or interim systems, with plans to develop full FCAPS functionality over time. That stretches the systems integration effort, and the network systems engineering support requirement over a much longer time frame than the original network equipment verification. Operations organizations can deal with poorly supported network technology during the soak and early introduction, but as the number of units in the field begins to grow, Operations can be overtaxed and scaling becomes increasingly difficult if the systems have not arrived.

13.2   Lifecycle Support

Technology integration does not end with the initial system verification and network introduction. Over the life of the technology more effort and integration investment is likely to occur in the years after integration. Here are some of the drivers of continuing integration,

- **NMS and OSS changes** – as described previously, these systems often lack full features at technology introduction. Integration support usually ramps up after the introduction of new technology. IT engineers are working to fix bugs in existing releases and develop new releases to address feature shortfalls. Interim systems may be adding features as well if the dates for the primary systems slip.
- **EMS and network element major software releases** – new features and functions, adding marketing services, improving system performance, or adding new maintenance and operations capabilities all drive new major releases from suppliers. Major releases are often accompanied by new hardware as the number of interfaces expands or additional processing and applications are added to the system.
- **Network element point releases, firmware upgrades, and patches** – on occasion these are fixes that have to be certified and quickly distributed into the in-service network.

Usually they fix minor problems and can be grouped into a single upgrade, easing the integration effort.

• **FOU changes** – not all of the features and capabilities of a system may be included in the original FOU. Marketing, Engineering, or Operations may decide they want to sell a new service or take advantage of a supported interface. That triggers network verification and new support systems development, all of which has to be integrated.

To manage a steady stream of network and system changes, supporting test beds are organized to match the release cycle of the network system and the purpose of the lab.

• Release Designation
  – **N Labs** – labs replicating the current versions of all hardware and software in the network,
  – **N+1 Labs** – labs incorporating all of the technology schedule for the next major network release,
  – **N+2 Labs** – labs with early development releases and possibly prototypes.
• Purpose
  – **Production Support Labs (PSLs)** – used for tier 4 support, to replicate in-service failures that cannot be resolved in the field, and to verify complex Method of Procedures (MOPs) (operations procedures) before they are applied in the field. Verification of point releases, firmware upgrades, and engineering change notices are also performed in the PSL. PSLs are N Labs.
  – **Network Certification Labs (NCLs)** – certification of network equipment and software planned for release in the network. Equipment and systems in the lab have been released by the supplier as ready for service but have not yet been introduced into the network. NCLs are N+1 Labs.
  – **System Integration Labs (SILs)** – NMS, OSS, and BSS design, verification and integration is supported from a SIL. Design support is necessary because without a SIL, IT system developers aren't able to test their designs until they are completed. SILs are some of the most active labs. N and N+1 Labs are used as SILs. New systems releases must be compatible with the in-service network systems, and the next set of network systems scheduled for release.
  – **Advanced Development Labs (ADLs)** – system suppliers, carriers, and third parties develop applications that build applications and mobile phone and tablet applications. ADLs are usually N+2 Labs upon or integrate with network technology. Examples are IPTV applications and mobile phone and tablet applications. ADLs are usually N+2 Labs.

Constant attention and management of these labs are needed to control investment and maintenance. Left to their own, engineers want dedicated labs and test equipment for their particular project or responsibility. Labs are shared through automated scheduling and configuration systems. Users can save and reload their own configuration automatically. Local lab support engineers check connectivity to prevent changes made by one user from affecting another. Automated cross-connects and switches assist in policing and configuring connectivity.

Systems Engineering and Operations regularly have to make judgments about the need for regression testing of changes made by network system suppliers. In the next chapter is a story of how the national telephone network was taken down by a simple assembly language change involving only a few lines of code. The change was put in service with a minimum of regression testing. A time proven strategy is to resist individual changes and put them in as large a package as possible, making regression testing more efficient. Risks associated with operational introduction of the changes also are reduced; a single maintenance interval may allow all of the changes to be made instead of taking multiple risks with a number of smaller changes introduced several maintenance windows.

Labs also have to be maintained for network systems that are not undergoing change and perhaps never will. Legacy voice switching systems are a good example. Legacy labs are needed to serve as an interface test facility for new technology that has to interwork with the older technology. New Voice over Internet Protocol (VoIP) media gateways have to interconnect with the legacy Public Switched Telephone Network (PSTN), and so the verification process uses laboratory Time Division Multiplex (TDM) switching systems for interconnection testing. Legacy labs are also used to train technicians who would otherwise learn on in-service systems.

13.3   Invention and Innovation

The title of this chapter was taken without shame or remorse from talks given by Dr. Dave Belanger, Chief Scientist at AT&T Labs and V.P. of Information and Software Systems Research [1]. Dave, Chuck Kalmanek, and Rich Cox were the three Vice Presidents who guided core research during my tenure at AT&T Labs. My distillation of the distinction Dave makes between invention and innovation is this:

- Invention is a process of discovery. It is by nature a creative exploration of a novel idea.
- Innovation is the application of invention or inventions to products, services, and processes.

Three examples of inventions that enable global high bandwidth data networks were described at the opening of Chapter 2: the laser, fiber optic cable, and the Erbium Doped Fiber Amplifier (EDFA). All three inventions had to overcome technological boundaries that had not been crossed, and their outcomes were unsure when the projects began. In the case of the laser, no one could have predicted the range of applications it has spawned; fiber optic cable was not invented until 10 years after the laser and reliable optical transmission didn't emerge until 10 years after that. Invention doesn't always have a clear application, nor should it. There is, however, a symbiotic relationship between invention and innovation. Innovations often combine several inventions in a ways that serve people. Innovations press in directions set by the marketplace, governments, or society. In the process they expose a need to solve large technological problems that stand in their way, and are beyond the intellectual grasp of innovation alone. Advances in science and technology are needed to break the boundaries. From an historical perspective, research is the most likely source of the needed breakthroughs.

13.3.1 The Role of Research

Telecommunications history is replete with examples of entirely new fields of science that were opened in search of solutions to overcome problems of scale, design and economics encountered as networks expanded and data, wireless, and video services were added.

- Traffic engineering and queuing theory [2] were both invented by A.K. Erlang, a Danish mathematician and Chief Engineer of the Copenhagen Telephone Company.
- Coaxial cable [3] was invented at AT&T in 1929 to improve voice transmission capacity over long distances.[3]
- Modern information theory was invented by Claude Shannon [4] in 1948 at Bell Labs.
- The transistor, was first demonstrated by William Shockley, John Bardeen, and Walter Brattain at Bell Labs in 1947.
- Hamming coding opened the field of error detection and correction; it was invented by Richard Hamming in 1950 at Bell Labs.
- The C Programming Language was invented by Dennis Ritchie and the Unix Operating System by Ritchie, Ken Thompson, Brian Kernighan, Douglas McIlroy, and Joe Ossanna at Bell Labs in the early 1970s.

The list of contributions by telecommunications scientists and engineers around the globe is long and storied. Technologies they invented benefit society far beyond fields of communication. Erlang's work is central to improvements in how we move people and products around the world. There are few products or services we use today that have no links to the transistor or the C programming language.

Those contributions came largely from the communications industry at a time when corporations funded industrial research as part of their corporate charter. Bob Lucky[4] recently wrote an article in IEEE Spectrum [5] in which he identified several models for funding research that have evolved in modern times.

1. **Self-funded independent labs** – Thomas Edison's lab is the example.
2. **Industrial labs** – Bell Labs, IBM's Watson Laboratory, and AT&T Labs are examples.
3. **Internet model** – Government funded collaboration with academia and industry, with DARPA and the Internet being the iconic example.
4. **MCC[5] model** – a research consortium funded by member companies.
5. **Silicon Valley model** – a mix of universities, venture capitalists, and startup entrepreneurs.

Models 1, 3, and 4 have largely disappeared from the technology landscape. In his article Bob laments the demise of the industrial lab model. While I agree it is dramatically diminished, I believe it survives in a few corporations, although basic research has decreased

---

[3] Dave Belanger accepted the Technology & Engineering Emmy on behalf of AT&T in January 2008 for the invention.

[4] Bob Lucky was the Vice President of Research at Bell Labs until 1992, and then became Vice President for Applied Research at Telcordia Technologies.

[5] Microelectronics and Computer Technology Corp.

over the past 20 years. Research, at least at AT&T Labs, can be categorized as foundational and applied. AT&T's core research labs have about 200 members of technical staff. Their work program is self-determined, led by the Research VPs, their management and senior leadership teams. Some research teams investigate problem spaces that yield technologies that may solve a wide range of problems, but are not applications or solutions in and of themselves. Examples of foundational research are investigations of technologies in visual presentation, natural language understanding, graph and network theory, software systems modeling, and database structures and design. Applied research tackles problem sets unique to global carriers, and AT&T in particularly. Those projects generally have two common denominators, unique requirements for scale and deep data mining.

Attracting and challenging premier research members of staff is dependent on supporting the right culture and environment, and having the right leadership. The senior managers must be qualified researchers in their own right to be able to build a credible program, and earn the respect of the members of staff. Apart from that, there are three essential elements to attracting and retaining premier talent,

- **A core of industry recognized researchers** – prospective researchers from academia or industry want to work with the best in the field. Having a nucleus of recognized scientist and engineers is crucial.
- **A challenging set of problems that matter** – academic studies can be, well, academic. To a practicing engineer's eyes some of the papers coming from academia just have little bearing on anything remotely related to the real world. Most, not all, researchers want to work on difficult problems whose solutions can make a difference in the way people communicate and live their lives.
- **Access to real data** – carriers of scale have massive data stores; they measure traffic flows, network topology and performance, workplace efficiency, and user behavior. Anonymized data and data summaries are used to find hidden trends and correlations, solve those difficult problems, and test the efficacy of the solutions.

Many other factors such as research community participation, mentoring of interns, and developing employees are important, but without any one of the three essential elements research may falter.

13.3.2 The Bridge to Research

The history of Xerox PARC®[6] is rich with stories of the inventions and innovations given birth there, and the degree with which Xerox was able to commercialize them [6]. What then are the keys to tapping into the inventions of an industrial research lab while maintaining the freedom of investigation? The history of research in AT&T, part of which is listed previously, is a treasure not easily replicated that bonds members of technical staff. Associating that legacy with current research members in a way that sets them apart, yet keeps them vitally connected to the business is the balance to be struck.

Technology coming out of research falls into one of the two categories described earlier, applied and foundational research. Those approaches to research are different in the ways the technologies they develop are applied to networks.

---

[6] PARC is a Xerox company. The acronym originated from Xerox's Palo Alto Research Center.

13.3.2.1 Foundational Research

Recall that foundational research programs are those not formulated to solve a specific business problem; they address broader technological challenges that can serve others in solving entire classes of problems, or they attempt to answer a scientific question of some consequence. One of the iconic manifestations of foundational research is the programming language C++. Bjarne Stroustrup,[7] an AT&T Fellow and Member of Technical Staff in AT&T Research, developed the language while at AT&T Bell Labs in Murray Hill out of a need to build better tools for his work on distributed computing in distributed networks. The list of problems solved by C++ goes far beyond what he or anyone could have envisioned. As two examples from a very long list, Microsoft adopted it in the 1990s as the base language for desktop computing with the introduction of Microsoft Foundation Classes (MFCs) and Adobe uses it extensively in their products. I doubt it is possible to count all of the applications using the language.

Graphviz and GvMap[8] are other examples. They are open source libraries developed and supported by AT&T Labs[9] that are used in a wide range of visualization and network graphing projects. Building on Graphviz and Vcodex, another AT&T Research open source library for data compression, a highly scalable web based network management system for enterprises, Vizualizer, was developed by an AT&T Research team. The project illustrates how foundational research technologies can be pulled together in a short time by highly skilled engineers to produce an applied solution.

Foundational research is a good deal like a venture capital investment. Only a fraction of projects produce something that is ultimately a useful product. But research can't be judged by a simple metric like the number of projects that end up as part of an application. The research community learns a great deal and gains insights working on projects that have no direct long term benefit. UNIX and the C Programming Language evolved from earlier work on Multics and Basic Combined Programming Language (BCPL) respectively. Multics and BCPL are long forgotten but the dividends they paid in cultivating fertile ground for UNIX and C are hard to overestimate.

A final example is the Daytona® [7] data management system. Daytona is unique in its ability to handle very large data stores. It has a proprietary query language Cymbal™ which is SQL compatible and designed to exploit the power of Daytona. One of advantages Daytona has over commercial DataBase Management Systems (DBMSs) is in data compression, again using AT&T Research's highly tunable and efficient compression libraries. Daytona is not a database itself, but a set of technologies that serve as the foundation to many AT&T databases, some of which are among the largest in the world.

Foundational research produces intellectual property that can be documented and sit on a shelf for some time, available to be called upon to solve very large problems. The model only works if the research staff is relatively stable with sufficient continuity to help in the adoption, modification, or extension of the technology in applied research or applications developed within the company.

---

[7] Dr. Stroustrup is currently a Distinguished Professor and the holder of the College of Engineering Chair in Computer Science at Texas A&M University.

[8] Formerly known as *Graphs as maps*.

[9] See http://www.research.att.com/software_tools for a complete list of AT&T Research open source tools.

13.3.2.2   Applied Research

Applied research projects are the direct engagement of research teams to solve business problems. The image of researchers as remote, unapproachable, and lacking pragmatism is far removed from fact, at least in my experience. Intellect and curiosity are not exclusive to passion, practicality, and engagement. But because research teams are a very scarce and highly valuable talent pool, their projects are carefully selected. Some of the criteria used to select projects at AT&T Labs are:

• Problems that involve deep analytics and a large corpus of data.
• Projects that need massive scale, high efficiency, and high reliability.
• Projects that process high transaction volumes with tight real time tolerances.
• Projects that depend upon a deep knowledge of networks, protocols, and communication theory.
• Projects that use different modalities, speech, text, video, and graphics in complex ways.

There are also requests that should not be entertained as research projects:

• Problems another organization in the company can solve with an acceptable solution within a reasonable time frame are not candidates.
• Lack of staffing, or lack of skills in other organizations in the company, are not reasons to call on Research.

Research in speech has yielded a range of successful projects and products over many years. Some of the work is more akin to foundational research in that it adds to the knowledge base and software libraries that have paid dividends over many years. Speech compression, recognition, natural language understanding, text to speech, and speech to text technologies are used in many AT&T products and are licensed to third parties under the name AT&T Watson™ Speech Technology.

Network planning, design, monitoring, and fault management are all areas that benefit from applied research and the projects that follow. Research teams are usually augmented with software engineers, network engineers, systems engineers, and operations research analysts to build scalable solutions. Team members from Operations and Engineering join in the design and prototyping process to validate and improve system designs. Development leaders in these groups are familiar with the range of software libraries, tools, and platforms used to build similar systems. Databases from Operations, Engineering, Sales, and Service are accessible by the teams and serve as information stores not accessible by third-party developers. That, coupled with the deep corporate knowledge held in the teams enables them to move to solutions quickly. Outsourced development operates under a waterfall model and can be no better than the requirements a systems engineer writes early in the process. Systems engineers can't always envision the art of the possible and may not have an understanding of the power of software platforms and tools that already exist.

Solutions coming from Research may need hardening and packaging. During packaging they are brought under controls and processes expected from a commercial product. Source code control, independent verification, regression testing, bug tracking, installation packages, and a host of other processes and practices have to be infused to enable projects to move out of Research and into Operations, Engineering, or IT. A dedicated incubation group can insure a high standard is set and consistent quality is enforced for all products planned for use by other groups or third parties. Most but not all applications built upon technology coming out of Research should be capable of being exported to operational groups once it has been through hardening and packaging. Six months to a year in incubation is sufficient to harden applications. At that point other software operations and maintenance teams can take ownership of the applications, just as they would for third party software. Design support and responsibility can still rest in Research for most applications. Research should not displace IT or other software groups, but rather be a resource and pipeline for technology. If transition out of Research does not take place, either the size of the incubation group grows and displaces IT and other software groups, or the application pipeline becomes stagnant. Once an application is successful it is difficult to eliminate it, so unless technology can be transferred the ability of the incubation group to assume responsibility for new technology coming from Research diminishes to zero.

13.3.2.3   University Collaboration

Effective engagement with universities develops over time and needs a commitment from managers and members of staff. The list of schools needs to be carefully screened using criteria relevant to the carrier and the qualification of the technical staff. Direct relationships with faculty in areas of interest are far preferable than ones with administrators or university development offices. Summer intern programs, symposia, visiting faculty, and guest lectures are all part of a vibrant program. Benefits of collaboration are numerous. A meaningful two way exchange of ideas and a deeper understanding of real world network technology challenges benefit all. Access to high caliber candidates and a chance to work with them before they leave academia improves the quality of recruiting. Being able to discuss, and possibly influence projects as they are formulated by university faculty can make their work more relevant and potentially pay dividends to the carrier.

There are hurdles that have to be cleared to build an effective program. Some universities and some carriers have intellectual property policies that inhibit or even prevent collaboration. Mutual ground has to be found; the carrier's legal team has to recognize that a single approach will not work with all universities, and a round of good faith discussion and compromise are likely to be needed. But the benefits far outweigh the effort.

13.4   Summary

Carrier networks have tens of thousands of network elements that comprise their transport, packet, voice, video, and wireless networks. Technology is constantly being upgraded, replace, introduced, and integrated. The responsibility for selection, verification, integration, and lifecycle technology management falls on the engineering teams, with Systems Engineering at the center of technology projects. Extensive verification and support labs are maintained to test network compatibility with other network elements, exercise failover and fault isolation, and serve as integration test beds for support systems and IT systems. Systems Engineering works closely with supplier technical teams to verify and integrate network technology, and to build relationships that are crucial when complex or widespread failures occur in the field.

Research plays multiple roles within the carrier enterprise. Carrier networks rely on scale, efficiency, and fault tolerant design. Foundational research sheds light on the limitations of existing technologies and designs, and discovers new approaches and technologies to address the difficult problems of the day. Applied research engages with other carrier teams to solve specific technology problems in the enterprise. They bring a depth of expertise and experience in the fields of analytics, network design, software design and engineering, coupled with firsthand knowledge of carrier networks and work groups. Research maintains close relationships with leaders in industry forums, standards bodies and universities, serving as a conduit for the exchange of emerging and evolving technologies.

References

1. Karpinski, R. interview with Dave Belanger (2009) Innovation and the Evolution of Bell Labs, Connected Planet, November 1, 2009.
2. Erlang, A.K. (1917) *Solution of Some Problems in the Theory of Probabilities of Significance in Automatic Telephone Exchanges*, A. K. Erlang (Published paper in 1971).
3. Espenschied, L. and Affel, H. (2012) U.S. Patent No. 1, 835, 031.
4. Shannon, C.E. (1948) A mathematical theory of communication. *Bell System Technical Journal*, **27**, 379–423, 623–656.
5. Lucky, R.W. (2011) Every New Generation Finds a Way to Innovate – Adventures in Research Funding. IEEE Spectrum (Sep 2011).
6. Gladwell, M. (2011) Creation Myth: Xerox PARC, Apple and the Truth about Innovation. New York Magazine (May 16).
7. Green, R. and AT&T Labs (1999) Daytona and the fourth generation language Cymbal. SIGMOD '99 Proceedings of the 1999 ACM SIGMOD International Conference on Management of Data.

14

Disasters and Outages

Disasters tend to create outages and sometimes outages create disasters, usually financial and public relation ones. Carriers generally think of disasters as hurricanes, tornados, earthquakes, fires, blizzards, ice storms, volcano eruptions, terrorist attacks, hazardous material spills, and floods. Usually, but not always they are acts of nature, or at least nature abetted by man. Outages are network failures of some significant scope and duration that are not formally defined by carriers, although as mentioned earlier, scope and duration are defined by the Federal Communications Commission (FCC) [1] in the US. Outages are usually caused by people, often software and hardware developers or technicians, sometimes aided by backhoes. Success at defending against disasters and outages starts with preparation, organization, and training. When you are in the midst of the problem, it is too late to prepare.

Network and system requirements and design are part of preparation. In considering one technology over another it is easy to ignore the question: how will this design ride through a disaster? One classic choice is between technology that is Network Equipment Building Standards (NEBS) compliant and commercial computing technology that operates in a data center environment. It illustrates the tradeoffs between first cost, preparedness, and operational complexity that are made in many design decisions.

The next sections examine disasters and outages, and describe ways to prepare and deal with them in the event.

14.1 Disasters

My career has been in the realms of network design and technology, not in operations; but I have watched events as disasters unfolded as a member of Pacific Bell's, SBC's and AT&T's engineering teams, and have helped design networks and systems to live through or mitigate the effects of disasters. I have witnessed the effects and responses to a number of disasters including:

- San Francisco Loma Prieta earthquake on October 17, 1989
- Oakland Hills fire on October 19, 1991

Global Networks: Engineering, Operations and Design, First Edition. G. Keith Cambron.
© 2013 John Wiley & Sons, Ltd. Published 2013 by John Wiley & Sons, Ltd.

- Northridge earthquake on January 17, 1994
- Gulf Coast Hurricane Katrina on August 29, 2005
- Greensburg Kansas Tornado on May 4, 2007
- North Bend Nebraska blizzard on April 5, 2009
- Japanese earthquake and tsunami March 11, 2011
- Southeast Missouri and Southern Illinois floods of April 2011
- Southeast US tornado storms of April 2011
- Joplin Missouri tornado on 22 May 2011.

Disasters affect different parts of the network and the carrier organization differently. Hurricanes, blizzards, floods, ice storms, and often earthquakes are regional events, affecting wide swathes of the network and the organization. For discussion these are grouped into a category of "regional disasters." Fires and tornados are usually more confined to a smaller area, although the tornado storms in the southeast US in April 2011 were more regional in nature. In the rest of the section these are grouped into a category of "site disasters." They are sometimes called "black hole" events, meaning that an explosion or tornado can turn a site into a black hole.

14.1.1   Carrier Teams

The only way to begin this section is with a sincere acknowledgment of the dedication and contributions of the operations and engineering teams that respond to disasters. Telco, cable, and power company crews work under difficult and dangerous conditions in ice storms, wild fires, floods, tornados, and earthquakes to restore service. They often do this when their own homes are at risk or damaged.[1] Thank you.

Carrier teams responding to disasters can loosely be grouped into the four categories that follow.

14.1.1.1   Disaster Recovery Teams

Carriers have dedicated teams trained and equipped to respond to site and regional disasters. Investments in these teams are significant. Disaster recovery centers are regionally located at a few sites to speed deployment and to insure against the entire team being caught up in a disaster at their site. The teams are equipped with dedicated custom trailers containing emergency modules capable of being relocated and deployed into service on very short notice. Equipment includes, but is not limited to:

- complete voice switching systems, local and tandem,
- complete wireless radio cells with fiber and microwave backhaul capabilities,
- complete IP backbone optical and routing equipment,
- a range of portable and transportable generators and power systems,
- trailers with stocks of public telephones, cell phones, and Wi-Fi access points,

---

[1] In 2007 Melissa Lucht and Ed Stauth, Greensburg's only AT&T employees, rode out a tornado in Greensburg Kansas in their basements and then headed to the AT&T Central Office (CO) which had no roof. Working under makeshift tarpaulins they restored service.

- support trailers for tools, IT support, administration, and
- trailers for logistical support.

The recovery teams are specially trained and have only one purpose, respond to disasters. Just as their trailer farm is a miniature version of the complete network, so is the team. Power is the most common need in disasters, so power systems and technicians constitute a critical part of the team. Team members are also trained in voice, mobility, transport, and IP technologies.

14.1.1.2 Regional Teams

As valuable as the dedicated recovery teams are, they are small compared to the size of regional disasters. Recovering from most regional disasters takes hundreds of thousands of craft hours. Power and connectivity drive the majority of the jobs that have to be undertaken. Power crews, construction and engineering, fiber and copper splicers, premises technicians, and many other crafts in the region train and prepare for regional disasters in addition to their regular jobs.

In large regional disasters it is not only the regional team in the disaster area that responds. Carriers bring in critical teams from other regions to speed the recovery. Out of region teams quickly join the recovery effort without the need for specialized training because their skills and work practices are consistent with the local region's standards.

14.1.1.3 Operations and Administration Centers

Operations centers outside the area invoke emergency procedures and dedicate staff in regional disasters. Network management controls are often invoked in disasters. Disasters that hit with no warning and are regional, like earthquakes, can quickly generate network congestion, typically inbound. In addition to degrading overall service, congestion prevents emergency responders in the region from communicating. Network management controls can rate limit inbound demand from areas not directly associated with the outage, and network operators can directionalize network traffic improving the ability of those at the disaster site to reach others.

Where management links are still in place to network management systems, important information arriving at Network Operations Centers (NOCs) and other operations centers provide valuable maps to the size and scope of the disaster and almost always are the first indication of disasters that occur without warning. Effects of the Japanese tsunami in 2011 were seen by network monitoring systems in AT&T's Global Network Operations Center (GNOC) in Bedminster NJ long before it appeared in the media; undersea cable routes in the affected area alarmed and alternate routing through other Pacific routes went into effect immediately. As a regional disaster hits, NOCs can see the loss of facilities, power and systems as it moves across the area. NOCs, working with regional command centers can quickly begin to asses and triage sites for repair and restoral.

In regional disasters data centers in the area may be directly affected, either from direct damage or a loss of power or connectivity. Alternate administrative centers must assume additional responsibilities. Network and IT systems must be rehomed to the alternate sites and technicians and support systems have to adjust to the redirection as well. When the primary sites recover, they may have to rebuild their databases or at least update them with changes that were made during the outage.

14.1.1.4   Operations Tier 3 and Engineering Organizations

Network management controls cannot mitigate structural damage inflicted in a disaster. Regional repair teams work with all possible speed, but even they take hours or sometimes days before critical routes and systems can be restored. Operations and Engineering work together to asses which routes and systems are still in place, and what rerouting and traffic policies should be modified. Engineering may need to augment or reassign capacity to routes outside the affected area to handle rerouted traffic. Operations tier 3 quickly builds the necessary procedures, verifies them with tier 4 if needed, and begins to put the changes in place.

14.1.2   Disaster Response

Flexibility, preparation, and reserves are the keys to disaster response.

- Flexibility must be built into network designs, data center, operations center designs and placement, and force management structures.
- Preparation, as described in team formation above doesn't assure success, but without it failure is assured. Business Continuity Plans (BCPs) have to be prepared, updated, and practiced across the business, not just by first responders.
- Investments in reserves of personnel, their training, equipment, and facilities have to be made in good times so that when bad times arrive, they are there.

14.1.3   Engineering and Design

Engineering alternatives have dramatic differences in their ability to withstand or recover from disasters. Two examples are described below that illustrate practices and decisions that affect a carrier's ability to respond in a disaster

14.1.3.1   T1 Augments

Consider facility investment rules governing T1 and fiber placement. Demand tends to grow incrementally and so when line engineers look at a particular route that is nearing exhaust they have to choose the best way to augment it. If the route has several T1s in it and spare copper, they are likely to choose more T1s. But T1s are electronically repeatered. Every 6000 ft or so a line powered repeater receives and regenerates the signal. If part of that route is in a flood plain or an area subject to 50 year floods, survival is problematic in a disaster. If water penetrates the T1 repeater it can fail, particularly if electrolytes, like chemicals from a nearby field mix with the water. An additional consideration is route capacity; if the next T1 augment raises the fill in the copper cable to 80% or higher and there is no indication demand will drop in the future, is a T1 augment the right decision?

Fiber placement is an expensive proposition, but if conduit exists and fiber can be pulled the cost will be far more manageable. T1s serving Time Division Multiplex (TDM) voice will require terminal multiplexing equipment at both ends if fiber replaces copper transport, increasing the cost. If the route is being used for IP connectivity, such as a mobility cell site, terminal equipment performing inverse multiplexing can be eliminated.

Fundamental planners in Engineering develop engineering and investment rules that guide line engineers in making these decisions. Cost and operational considerations are always primary considerations, but the ability to withstand regional disasters or to quickly recover from them need to be part of the decision criteria.

14.1.3.2   Consumer Broadband Networks

Cable MSOs (Multiple System Operators) and telcos make strategic decisions on how to continue to deliver more bandwidth to consumers as demand continues to grow at double digit rates each year. For over two decades both industries have debated and wrestled with the question of how deep to push fiber. Bandwidth and signal quality improve for both as they push fiber deeper and serve smaller node sizes, but there are some differences in the operations planning for disaster recovery.

MSOs replace trunk amplified routes with fiber when they place fiber nodes in a neighborhood to increase available bandwidth. That eliminates electronics in the trunk plant, but adds electronics in the neighborhood that needs local powering. Amplified trunk routes are powered from a headend over a hardline coaxial cable and need no local power. If the trunk cable is intact, the amplifiers should be receiving power, assuming there is power at the launch point or headend. Signal quality and reliability are affected by the number of amplifiers on the route. As fiber pushes deeper more amplifiers are eliminated, signal quality improves, and the reliability of the trunk route increases. The motivation for MSOs to push fiber deeper is to share the bandwidth at the node over fewer homes. However, as fiber pushes deeper more nodes are needed; the number of cabinets with local power distribution and batteries increases inversely with the number of homes served by the nodes. That is, cut the average node size in half and you double the number of nodes. When local power is lost to those nodes, local batteries assume the load; they are usually engineered to power the nodes for 4–8 hours. Beyond that local generators have to be deployed to serve the node and recharge the batteries. An alternative used in some limited deployments is that MSOs can leave the trunk cable route in place and use it only for powering. The solution is limited because the power demand of the nodes is likely to exceed the carrying capacity of the trunk cable. Moreover, the MSO needs to continue to maintain the coaxial trunk route to maintain backup powering.

A different set of circumstances faces telcos when they push fiber deeper. Copper pairs are passive and so when fiber nodes are introduced it is difficult to match the reliability of the passive circuits they replace. However the reliability of remote terminals, managed by telcos, has proven to be able to meet high standards of availability, even during regional disasters. Those fiber nodes have similar power designs to the ones used by MSOs, except telcos do not have an option of using existing copper pairs to power an entire node in most practical designs. Like MSOs, telcos deploy nodes with battery backup and external alternate power connections so that portable generators can service them if needed. The motivation for pushing fiber deeper for telcos is somewhat different from MSOs. Telco designs are unicast designs so they don't share bandwidth on the cable in the same way as MSOs. Digital Subscriber Line (DSL) technologies used by telcos are distance sensitive however, unlike coaxial cable. The further a customer is from the node, the lower the effective bandwidth. Telcos push fiber deeper to reduce copper cable length, and thereby increase bandwidth. The calculus is a bit different for node counts with telcos than it is for MSOs. Cable length, not node size is the independent variable used to determine bandwidth for telcos. As the length of the copper cable in the design is halved, the number of nodes grows by four. Node counts increase geometrically with the reduction in loop length, not arithmetically. Fiber to the curb is the extreme example, stopping short of fiber to the home. In a typical 30 000 line office, if curbside terminals serve an average of 10 customers, 3000 terminals have to be powered. There are powering designs to place power nodes that feed about 10 terminals. Using that design, 300 power nodes are needed to serve a single central office. Fiber from the central office is delivered to the terminals along with copper cable from the power node. In the Loma Prieta earthquake in California in 1989 110 central offices were on emergency power immediately after the quake. Commercial power was restored to many of them within hours, but if 100 central offices with curb terminals had to be serviced in a regional disaster in which commercial power was lost beyond the life of local batteries, 30 000 power nodes would need to be served via portable generators.

Portable generators work well for both telcos and MSOs in localized power outages, but regional disasters, like hurricanes and ice storms, can create power outages lasting a week or more over wide areas. In those events multiple crews with portable generators are called upon to "hopscotch" from node to node charging the batteries until power companies can restore service. As the number of nodes increase more crews and generators have to be called out in a regional disaster.

14.2   Outages

In this section network outages of scale are addressed. Scale outages are ones that affect at least a city and sometimes an entire region or even a country.

14.2.1   Anatomy of an Outage

This case study illustrates common mechanisms scale outages use and how Engineering and Operations can work together to contain damage, pursue the underlying cause, and restore the network. In 1987 US telcos began deployment of SS7 to tandem switching offices to enable portable 800 Service, in part to meet FCC mandates. In the same year conversion to SS7 was extended to end offices to support Integrated Services Digital Network User Part (ISUP) common channel trunk signaling and interconnection. Services relying on SS7 in early 1991 were 800 Service and Automatic Billing Service (ABS) or Calling Card Service and trunk signaling. Service signaling volumes were light compared to the SS7 message traffic generated by SS7 trunk signaling. Similar conversions were in process in the other Regional Bell Operating Companies (RBOCs) in the late 1980s and early 1990s. In Pacific Bell 200 000 trunks had been converted from multi-frequency (MF) signaling to SS7 by mid-1991. Conversion teams worked during non-busy hours, often late in the evening or early in the mornings to swing trunk groups from MF signaling to SS7 over a period of two years. Work was underway, but not complete to interconnect Pacific Bell's SS7 network

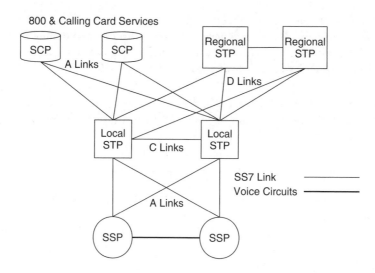

Figure 14.1   SS7 networks circa 1991.

to AT&T, MCI, US Sprint, and other US Interexchange Carriers (IXCs). Figure 14.1 is a simplified view of the state of the network in June 1991.[2]

Outages that began in June 1991 would ultimately affect two Signaling Transfer Point (STP) pairs in Northern California and two pairs in Southern California. The Northern California (NC) STP pair and Southern California (SC) pair are regional STP pairs. Local STP pairs affected were the Los Angeles (LA) STP pair and the San Francisco (SF) STP pair. Local pairs provide signaling service within a Local Access Transport Area (LATA). Regional pairs interconnect all of the LATAs for services like toll free service, but also serve some end offices and tandems directly. Mated STP pairs are interconnected by C links. Regional and local STP pairs are interconnected by D links.

Outages also hit Bell Atlantic during the same period. The outages were widespread; measured by the number of customers affected and the duration of the outages, the impact was even more severe than those experienced by Pacific Bell. During the period June 10 to July 5, SS7 outages affected over 13 million customers. Washington DC was one of the areas that experienced several hours of service outages. The scope of the outages and complaints and concern of the public caused the FCC and the US Congress to hold hearings early in July and the after effects of the outages were felt in the industry for years.

14.2.1.1   Chronology

The new network element carrying the signaling load was an STP. Under the interpreted rules of the Modified Final Judgment (MFJ) Pacific Bell had placed a mated pair of STPs in each of the 11 California LATAs and interconnected them with two regional pairs [2].

---

[2] Switching offices that are SS7 capable are called Service Signaling Points.

Each signaling point (end office, operator system, and tandem) converted to SS7 trunk signaling was completely reliant on connectivity to at least one functioning STP to be able to complete inter-office calls. If SS7 message transfer was lost, only calls between lines terminating on the same switching system could be completed.

- **Early June 1991** – STP stability problems began to appear. In the San Francisco bay area one of the two STPs went into congestion and then recovered. Service was not affected.
- **June 10, 8:55 a.m.** – One of the LA STPs generated a high volume of message routing errors. It went into congestion and was removed from the network and congestion subsided. The mate LA STP experienced congestion over the C links. About the same time one of the SC STPs began to experience congestion over the C links to its mate. The problem was quickly escalated to Pacific Bell's Electronic Systems Assurance Center (ESAC), the highest tier of technical support. ESAC decided C link congestion was enabling a potential total outage and decided to remove one of each of the mated pairs in the LA and SC STPs, and rely on a single STP in each area for service. That strategy worked and stability returned about 10:30 a.m.
- **June 10, 12:00 p.m.** – An attempt was made to restore the two out of service STPs in the greater LA area. When links were restored between the LA and SC STPs, called D links, both of the SC STPs experienced congestion over the connecting links and were removed from service, isolating 31 end offices. The outage lasted until 1:30 p.m. when one of the SC STPs was restored to service. 3200 customer complaints were received because of the outage. Technical escalation to the STP supplier, DSC Communications, led to the discovery of a corrupted database in one of the LA STPs caused by a new software load inserted on June 9th. Pacific Bell engineers accepted the explanation, but were far more concerned about the propagation of a problem in one STP to other STPs. No plausible explanation for the contagion across C and D links was offered. DSC and Pacific Bell technical teams worked together to develop a plan to isolate and understand the propagation phenomena.
- **June 26, 10:40 a.m.** – A routine audit in one of the LA STPs failed and the erred controller and some associated signaling links were removed from service. The automated link removals caused congestion in one of the SC STPs. Congestion quickly spread via C links and D links and all four STPs in the greater LA area went into congestion and were removed from service. Inter-office service was lost to 103 central offices in the area. Three million subscribers were affected. In addition, 800 and calling card service failed to all of Southern California. Emergency 911, Operator Services and Long Distance, none of which had yet converted to SS7 signaling, were not affected. ESAC quickly adopted a strategy to bring up a single STP in each of the pairs, stabilize the network, then bring up the mate STP, and finally bring up the interconnecting C and D links which had propagated the failures. Service was restored shortly before 1:00 p.m., an outage of about 3 hours.
- **June 26** – Bell Atlantic had outages in Washington D.C., Maryland, West Virginia, and Virginia very similar to those in Pacific Bell. Those outages affected 6.3 million subscribers and lasted up to 8 hours. Disruptions began before noon and service was not restored until 7:20 p.m. SS7 failures were no longer a west coast problem; it was a national problem. The common denominator in all of the failures was a DSC STP. Pacific Bell had chosen DSC as the supplier after the initial choice, the ITT 1240 was withdrawn

from the market in the mid-1980s. Bell Atlantic and Ameritech had also chosen DSC as their STP supplier. Other STP suppliers like AT&T and Nortel had STPs in service in other regions, but none had experienced failures of this magnitude. Pacific Bell, Bell Atlantic, and DSC agreed on a joint forum for sharing information and plans in pursuit of a solution.

- **June 27** – Front page articles describing the extent of the outages and the effect it was having on businesses and communities were published nationwide in the *Wall Street Journal*, *USA Today*, the *New York Times*, and regional papers.
- **June 28** – Pacific Bell Network Systems Engineering (NSE) reconfigured their tier 4 SS7 laboratories in Walnut Creek and Concord California to match the configurations of the failed STPs and begin writing test scripts to drive network loads and recreate the problems experienced in the network using the Network Services Test System (NSTS) [3]. Equipped with NSTS and captive STPs and Service Signaling Points (SSPs), Pacific Bell had one of the few facilities capable of recreating failures of the magnitude being experienced in the network. Bellcore, the technical services organization serving the seven RBOCs began efforts to interconnect their labs with DSC.
- **July 1, 10:30 a.m.** – The SF STP lost synchronization with office Building Integrated Timing Supply (BITS) clock and the ensuing switchover to the local synchronization source caused multiple link failures and the STP went into congestion. ESAC recognized the problem immediately, isolated the failing STP and prevented the problem from propagating to other STPs in the network. Service was not affected.
- **July 1, 11:15 a.m.** – Pittsburgh, Pennsylvania, in Bell Atlantic territory experienced an outage that lasted until 5:15 p.m. One million customers in western Pennsylvania were affected.
- **July 1** – Pacific Bell NSE engineers wrote and ran NSTS test scripts in the Walnut Creek Lab that replicated the congestion and failure of the STPs seen in the field. The problem could be recreated reliably at will. They also recognized that the SS7 congestion procedure, Transfer Controlled (TFC) was not being invoked when they created congestion onset to a routable destination. In a conference call with DSC they found DSC was pursuing a line of investigation along a similar path, but had not discovered the failure of the TFC procedure. DSC was in the process of issuing a software patch, designated the EX patch, but that patch did not correct the TFC problem. The EX patch did discard messages once the STP experienced congestion onset, which was a significant improvement over the in-service load.
- **July 2, 10:30 a.m.** – Pittsburgh experienced another outage lasting until 12:30 p.m.
- **July 2, 1:14 p.m.** – A 4ESS tandem switch in Santa Clara California caused a half second interruption on SS7 links homed on the NC STP pair. One of the pair went into congestion and the congestion quickly propagated to one of the SF STPs. ESAC reacted quickly, isolating the failing STPs and prevented further propagation of the problem. Service was minimally affected and all STPs were restored by 3:00 p.m.
- **July 4** – Pacific Bell released a formal Memorandum For File to all affected parties, Bell Atlantic, Bellcore, and DSC, that described the fault condition, the TFC failure, and how to recreate the problem in a laboratory [4]. Test logs were also made available. Pacific Bell and DSC described the results on conference calls with Bell Atlantic and Bellcore, but Bellcore in particular remained skeptical that the problem sited by Pacific Bell and DSC were the cause of the outages.

- **July 5** – Pacific Bell chose to load the EX patch, recognizing it was a stop gap, while waiting for a solution that provided both discard under congestion and properly implement the TFC procedure. Pacific Bell's General Manager of Network Services, Ross Ireland, notified the FCC that the problem had been recreated in Pacific Bell's tier 4 lab and a software patch to stabilize the network was being loaded.
- **July 17** – DSC delivered a new generic load, 03.07, with patches that addressed all of the outstanding problems identified by Pacific Bell and Bell Atlantic. Pacific Bell NSE ran full regression tests against the new release in the Pacific Bell Tier 4 lab, including tests designed to specifically create the congestion failure, and with minor issues the release passed the tests and was released to ESAC for loading in the network. Networks were stable and the crisis was over.
- **July 22** – Pacific Bell submitted a formal report to the FCC, California Public Utility Commission (PUC) and other regulatory agencies describing the events and chronology, along the lines described in this section [5].

14.2.1.2 The Failure

Figure 14.2 is a simplified drawing of the DSC STP design.

In this simplified model there are three components that came into play. LP is a link processor, called a CCLK by DSC. RP is a route processor, called a CCD by DSC. The fabric is the core of the switch that acts as an SS7 packet router. Link processors are responsible for the link layer of the SS7 protocol and the route processors are responsible for the network, or routing layer. In Figure 14.2 SS7 messages move from Signaling Point (SP) A, an end office, to SP B, another end office. If there is any disruption (shown by a + sign), even for a few milliseconds on the link between the model STP and SP B, traffic can be temporarily queued in the LPs facing SP B. Because SS7 has a window size of 127, up

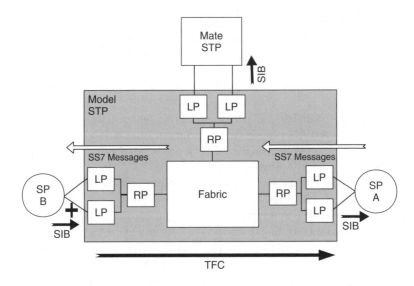

Figure 14.2   DSC STP logical model.

to 127 messages can be queued before the LP can no longer accept messages from the route processor. The link layer of the SS7 protocol is a bit like Transmission Control Protocol (TCP), requiring positive acknowledgments, having a window size, and using retransmission to insure delivery. Without acknowledgment, messages sit the LP. If acknowledgments are slower than the incoming stream, again the LP will fill to capacity. That is what happened in the failures in 1991.

In the outages that occurred in June 1991 different triggering events caused some delays in the delivery and acknowledgment of messages on a single link. The operation specified by the SS7 protocol is that once the LP reaches a high water mark in the transmit queue, as an example 80 messages, the LP should notify its RP that a state of congestion exists toward SP B. That RP in turn notifies the other RPs in the system and the administrative processor of the congestion condition. Messages arriving from SP A or the mate STP addressed to SP B should be discarded by their respective RPs, never reaching the STP fabric, much less the RP and LPs serving SP B. In addition, a congestion message, called a TFC should be returned to the originating SP each time an SS7 message is received from SP A or the mate STP informing them that the route is congested. Those signaling points should then stop sending and enter a periodic test mode to see if the congestion condition has eased.

14.2.2 Congestion Onset

The software load that caused the outages had implementation and design defects that came together to create the massive outages that were experienced nationwide. The implementation failure enabled small failures to trigger STP congestion. The reason the LP did not realize it was in a transmit congestion state is that a single line of assembly code had been mistyped. Rather than monitor the size of the transmit queue, the assembly code error caused the LP transmit queue monitor to look at a null value, indicating the transmit queue was empty. So it could never recognize a congestion situation and notify the RP. When messages arrived at the RP serving SP B and it tried to write to the LP buffer, the write blocked but the message was not discarded. It remained in the RP queue. That blocked the RP transmit buffer to not only the affected link, but all links served by the RP. Without a TFC return notification SP A continued to transmit messages to SP B. As other messages arrived from SP A and the mate STP their RPs could not write to the fabric because fabric writes became blocked, unable to write to SP B's RP. The processes responsible for managing the fabric did not discard messages but rather queued them in the system. In other words, a single blocked buffer could bring the whole system into a state of congestion with no way of recovery. Even though the mate STP had very little traffic to SP B, it didn't matter. It was a question of time before even small streams of traffic caused congestion back into the receiving links.

14.2.3 Congestion Propagation

Had the LP recognized congestion and returned TFCs to the source signaling points, the failures might not have propagated. Without TFCs all directly connected SPs and STPs continued to send traffic. The receiving LP, like the one facing SP A kept acknowledging traffic until the receive buffer in the LP serving SP A filled. When receive buffers fill a different SS7 procedure, called Status Information Busy (SIB) is invoked. A TFC indicates a route to a particular destination is busy. An SIB indicates a link is busy and asks the sender to wait briefly in the expectation the condition will clear. So SIBs could be returned to the mate STP and other signaling points when STPs began to experience congestion on a single outbound link. But an SIB is a triggering event, causing the transmitting LP at the mate to congest and begin the same failure sequence all over again in another STP.

14.2.3.1 Causal Analysis

In examining systems in the aftermath of failures of this order it is useful to segregate errors and omissions into three categories, triggering events, contributing causes, and root cause. It's a bit like investigating armed robbery. There may be a getaway driver and a bag man, but there is at least one person with a gun. The person with the gun is the root cause. The root cause is often a design mistake, not a software, hardware, or procedural error. Like armed robbery prosecutions, sometimes the gunman goes free and the getaway driver receives public scorn and retribution.

14.2.4 Root Cause

The root cause of the DSC STP failure was the lack of a design principle and discipline that recognized the importance of message discard under congestion within the RPs and Fabric. Under no circumstances can those resources be allowed to block. Recall the characteristics of a well-behaved system described in Chapter 3 and shown below with an additional line showing discarded traffic (see Figure 14.3).

When a blocked RP or fabric was discovered by a process the only practical design choices are to either begin discarding messages or dump the buffers in the RP and possibly the fabric. Major or even critical alarms can be raised simultaneously, but blocking in the hope that things will recover somehow assures complete system failure. The consequence

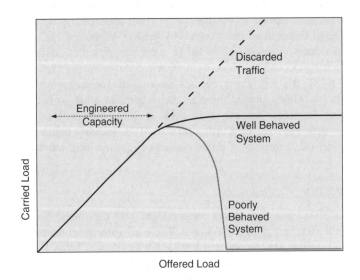

Figure 14.3 Traffic discard in a well-behaved system.

of dumping messages is that the individual transactions associated with those messages will fail. So a customer making a phone call would have the call fail and be routed to an announcement or reorder. That is not the first time that has happened, and it is an acceptable treatment under these circumstances.

This design failure was the gunman. When a system's fabric is in congestion discard is the only solution. It can be phased in increasing levels of discard or it can be invoked immediately but doing nothing is not an option.

14.2.5 Contributing Cause

The mistyped assembly code in the LP was a contributing cause and should not have gotten past integration and acceptance testing. But as noted in Chapter 3, testing is a statistical process and it has to stop at some point. When it stops and systems are placed in service, you can be sure there are still bugs in the system. Bugs like this one should do limited harm and soon be discovered and corrected as the system sees more service time, and so matures. That is, unless there is a root design deficiency that is triggered by an event associated with a contributing cause. When that happens the result is so massive that clues to the contributing cause are hidden.

Although it received great press as the culprit, the assembly code error was the getaway driver.

14.2.6 Triggering Events

A wide range of events triggered the temporary busy condition on links to the DSC STPs, including:

- A corrupted database localized to a single route processor escalated when that route processor was taken out of service.
- Loss of BITS synchronization and the ensuing switch to the STP's internal clock caused an outage.
- A 500 ms 4ESS link busy condition caused an outage at the serving STP.

These are common events that occur every day in a live network and so should not cause service disruption. In crises triggering events are sometimes labeled as the root cause. The list of triggering events when a design or implementation is flawed is endless. Triggering events cannot be prevented. They are part of the operational landscape.

Triggering events were the customers in the store when the robbery occurred.

14.2.7 Teams in an Outage

The teams responding to Pacific Bell's SS7 outage are similar to the ones used by AT&T today. Other models can be effective as well, but are likely to have these elements.

14.2.7.1 Operations Team

The primary task of Operations is network restoral and stability. Causal analysis is secondary. In the 1991 outages Pacific Bell's ESAC organization was able to identify signs of instability in STPs before they failed and develop mitigation procedures they could rapidly apply to prevent complete failure, and more importantly propagation to other STPs. The approach is akin to treating symptoms if you are not sure of the underlying disease. As time progressed Pacific Bell's outages grew shorter and fewer, not because we had loaded any software fixes but because ESAC and other Operations teams developed better surveillance and operational responses to signs of instability. Although the networks and serving areas were different, Pacific Bell outages averaged about 2 h, compared to outages of 4–6 h being experienced on the east coast. Some of that difference may be just luck, but I credit Pacific Bell's ESAC and Operations teams.

Forensic analyses of alarm logs and log files in general are operations tasks that need to be organized quickly and assigned to experienced team members early in an outage. Analysis is not restricted to logs generated by the failing network element; connected elements are likely to give a different view of events up to the failure than logs from the failed element. Working network elements can continue to log messages through a failure and recovery. Scripting filters help but there is no substitute for experienced eyes reviewing the logs. The search is not for a root cause event, but rather patterns that repeat before an outage and at the onset.

Network monitoring and surveillance may need reconfiguration or need to be expanded during extended and repeated outages. This has to be done without disturbing or burdening fragile network elements. In the 1991 outages Pacific Bell was using a Bellcore SS7 management system named Engineering Admin Data Acquisition System (EADAS) to administer and monitor the SS7 network. Among other features it had an impressive topological map of the network that displayed alarms and the service state of the elements and links. Like some of the SS7 network systems it was not yet mature. In the outages of June and July 1991 the system alarm and log files backed up with over an hour's worth of data. The alarm panel would indicate a failure one or even 2 hours after the event. Operations quickly changed their attention to the No. 4ESS maintenance terminal. The 4ESS was proven and the alarms were informative and generated in near real time. By watching the 4ESS Operations was able to judge the health of the STPs. The STP maintenance terminals were watched carefully as well.

14.2.7.2   Systems Engineering

Systems Engineering can play an important role in an outage investigation if they have a tier 4 lab that can closely replicate the field deployments, and if the engineers have been trained on the suspect network elements. Carriers that use Systems Engineering for verification testing of network systems and have a rigorous process of certification and approval for use are able to put those engineers directly into the effort of identifying and recreating failures. The engineers have an in-depth understanding of the design and behavior of the system under a variety of the conditions. Moreover, they have the necessary test equipment to load traffic on the network element, drive it into focused overloads and capture data not available in the field to analyze the system's reaction to stress and failover. Operations tier 3 engineers works in the lab with Systems Engineers and serve as liaison to the Operations Team.

The primary task of Systems Engineering is to recreate the failures seen in the field. They are not there to find the line of code or the faulty hardware component causing the failure; they only need to recreate the failure consistently with a set of conditions, traffic stimuli, or craft commands. Working with Operations, that often includes examining differences between two network elements, one of which fails and another that does not, and developing a map of differences that may point to a vulnerability.

Another important task of Systems Engineering is to quickly assess patches or other fixes recommended by the vendor or by Operations. At times isolating and fixing a problem comes down to trial and error. Timers and buffers may be changed and software and firmware patches may be recommended by the vendor. Systems engineers have to quickly evaluate those changes in a tier 4 lab to insure, at the least, that they do no harm.

14.2.7.3   Vendor Teams

Equipment vendors usually have counterparts to telco Operations and Systems Engineering organizations. They work with each other on a regular basis, but during outages contact becomes more frequent and more intense. Vendor Technical Assistance Centers (TACs) sometimes take direct control of a network element from Operations and are responsible for monitoring and restoring them. That's rare but it does happen in some situations. Certain debug tools, such as the setting of real time software traps on in-service network elements may be restricted to the vendor; they have the tools and experience to gather the needed information and stay out of trouble in the debug mode.

Systems engineering teams can be more effective during an outage if they are able to triage and cooperate on test cases to more quickly eliminate suspected causal conditions. They also verify each other's work. It is easy for a telco engineer to misinterpret how the network system should behave under a given set of conditions. The vendor knows not only the system, they have access to the developers and can ferret out expected system behavior, and design intent, what the developer had in mind. Telco engineers can share exactly how the network is configured, and importantly how all of the interconnected systems behave and what parameters they use. Vendors seldom have a rich environment with competing vendors' equipment in their labs. Working together the two systems engineering teams are far more effective than they are working independently.

14.2.7.4   Operations Program Management

Rather than have the outage focused teams report up through their normal management chains, a program manager with operations and systems engineering experience should be put in place with authority from the leadership of the company to call on any resource in the company for support. External companies seeking information work through that office; vendors are directed to work directly with the program management office as well until the failures are diagnosed and the networks are back to normal operation. The Program Management office not only coordinates activities, but shields the teams attacking the problems from outside distractions. In the 1991 outage Eva Low assumed that role, coordinating efforts across teams and communicating with external carriers and companies. Eva has experience in ESAC and has managed systems engineering teams. She was the ideal candidate for the job.

14.2.8   Press and External Affairs

It may seem odd to have a section devoted to public affairs in an engineering text, but I assure you it is relevant as the history of this outage demonstrates. Senior management of the companies involved and company spokesmen are in a difficult position in the midst of a wide scale outage. They come under intense scrutiny by the public, regulators, politicians, the media, stockholders, the investment community, industry pundits, their employees, and partners. Their ability to respond, create, and sustain an atmosphere that enables technical teams to focus on the problem and cooperate with each other is a severe test of their leadership. Two of the most common mistakes made in statements to the press and to Congress in their hearings were trying to deflect responsibility for the outages, and making public statements that had no technical basis. The latter sin, public speculation in a crisis, is human. Under extreme pressure people want to be seen as knowledgeable and in control of the situation; they want to reassure the public and customers that all will be well. The former sin can be mortal. Executives that claim others are at fault, only to be proven wrong later do damage to themselves and their company.

The following quotes are from publications chronicling the events of June and July 1991.

New York Times, June 27 – Michael Daley, a spokesman for Bell Atlantic, was quoted as saying the failures on the east coast were "totally unrelated and purely coincidental" to the problem in Los Angeles. He said the California problem stemmed from a mechanical problem rather than a software problem.

San Francisco Chronicle, June 27 – Pacific Bell spokesman Michael Runzler said the blackout occurred as new software being installed in the company's main computer in downtown Los Angles. The reason for the failure was unclear last night. The 911 number was one of the few that remained in service during the blackout.

Pacific Bell, Los Angeles Media Relations Release, July 1 – Chuck Johnston, Vice President and General Manager of the Los Angeles Region was quoted as saying, "We know what caused the problem. Our engineers are working around the clock to identify all of the issues surrounding this problem. We are aggressively pursuing and implementing any and all actions to ensure a similar situation does not reoccur."

News Roundup, a publication of the *Washington Post*, July 3 – Terry Adams, spokesperson for DSC was quoted as saying "There has not been an identification of the problem ... We can't point to a software issue or hardware issue ... He noted that the equipment was built to industry standards and that [the standards] might be flawed."

Wall Street Journal, July 8 – "On Friday, DSC, Bell Atlantic, and Pacific Telesis Group said in a joint statement they had duplicated the crisis in a laboratory and DSC had developed a software change which had been installed, to prevent the problem from recurring." Later in the article are the following quotes, "Still, DSC isn't taking any responsibilities for the troubles that have the telecommunications industry in an uproar." and "the company [DSC] made it clear that it wasn't willing to take the rap. DSC, angered that Bell Atlantic officials indicated DSC was at fault, insisted on issuing a joint news release that telephone companies and all their suppliers were working together to resolve the problem."

Philadelphia Inquirer, July 10 – "[DSC Vice President] Perpiglia said the software flaw appeared to be the 'root cause' of failures of DSC-built computers called signal-transfer points that have been at the heart of the investigation. He added that in networks as a whole, in which equipment made by many different companies is linked together, other causes might be found as well."

The Baltimore Sun, July 11 – Reporting on testimony before the US Congress, the *Sun* published the following, "DSC Communications Corp., whose software has been linked to recent phone outages, knew a year ago that there were potential problems with the new signaling systems used by the 'Baby Bell' phone companies to handle calls, a DSC official told a House subcommittee yesterday. ... Mr. Perpiglia said that SS7 has an inherent problem. That problem, he said, is this: When engineered to the precise specifications of Bellcore and other standard-setting organizations, SS7 has no way to disregard maintenance messages."

Telephony, July 15 – DSC Vice President Perpiglia was quoted as saying "This has been a blow to us as a company. We must be forthright and accept our responsibility. Our equipment was without question a contributor to the outages."

Newsweek, September 1 – "DSC Communications declared that flaws in its own software had triggered the plague of busy signals and dial tones."

Readers can judge for themselves how the various companies handled the extreme pressure of the crisis from the quotes previously. But the leaders that represent models of how to respond aren't always mentioned in the press. From my vantage point within the technical teams working to isolate and solve the problems, we were fortunate to have the stewardship of Pacific Bell leaders like Ross Ireland, Jack Hancock, John Neal, and the executive team at Pacific Bell and Pacific Telesis. They did not seek the press and responded with facts and transparency when approached in public and private forums. Pacific Bell's public statements from Michael Runzler and Chuck Johnson were responsible and factual as well. We in the technical team updated the leadership team on a daily basis, and they did not place undue pressure on us nor did they instruct us to work in a way that supported a public stance different from the technical facts.

Other carriers and suppliers in the industry also showed restraint and privately offered to provide any assistance necessary. AT&T had experienced an SS7 nation-wide outage a year earlier, and so understood what it feels like to be in a crisis. To my knowledge, STP suppliers AT&T, Nortel, and others made no public statements in an attempt to take advantage of the situation.

14.3 The Vicious Cycle

The antithesis of the Virtuous Cycle is the Vicious Cycle. Figure 14.4 illustrates the cycle.

Begin with System A at the top of the cycle, in-service with normal operation. Working in a clockwise direction, a trigger is detected. The exact trigger is unimportant; the design defect already exists in the system and eventually it will be triggered by a failed circuit pack, a bouncing link, or simply a surge of traffic. Failure onset is induced; the spread and signature of the failure is dependent on the nature of the defect and system design. Because the defect is embedded in the core of the system, it is unlikely the root cause is properly alarmed. Instead, a storm of other alarms are generated. Triggers are propagated to other systems, usually via control plane messages or traffic reroutes, creating surges in the network. As shown by the dashed line, the cycle starts all over with System B. What is not shown is that there may well be Systems C, D, E, and so on. Failures following the Vicious Cycle have the potential to expand geometrically through a network in short order. Technicians in the NOC watch with disbelief as node after node turns red on the system maps. Thankfully Vicious Cycle failures occur rarely in networks of scale.

Vicious Cycles are not limited to networks. Data center applications that operate with server groups using common software in a pooling design can also experience these failures. Application developers may fail to design well behaved systems, relying on load sharing as

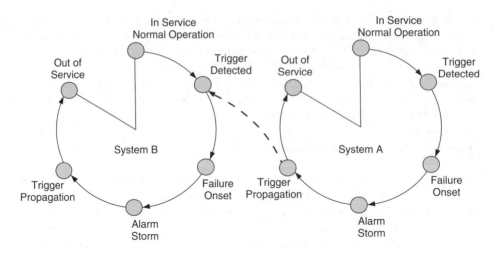

Figure 14.4   The Vicious Cycle.

a guarantee their design will never have to run above their design transaction limit. In those designs failure of a subset of servers because of a power or communications fault shifts the entire load to the remaining servers. That surge may fail other server groups, particularly if the load sharing mechanism under failure is not carefully designed and verified. Cascading traffic can bring down the entire complex. If you want to see the latest data confirming this thesis, perform an Internet search on "website crash." My search while writing this text yielded 120 million hits.

As a final note there is one community that has extensive expertise in the Vicious Cycle, Internet hackers. Standard Query Language (SQL) Slammer can be viewed as a triggering event that exploited a design defect in Microsoft's SQL Server database. To Microsoft's credit they had issued a patch six months before the worm infected servers on the Internet, but unfortunately many, including Microsoft, had not fully installed the patch.

14.3.1 Engineering and Operational Defense

In some sense this entire text is meant as a roadmap for practices that improve network performance and availability, and minimize the probability of a Vicious Cycle outage. Network integration practices described in Chapter 13 are foundational to vetting network systems and system upgrades, and are repeated here for emphasis.

- System verification and certification is the last opportunity to detect and correct design faults before they enter a network.
- As demonstrated by the 1991 SS7 outages, small changes to firmware and software should not be introduced into in-service systems without thorough regression and verification testing. Carriers must refuse the upgrades and insist they be grouped into larger packages that warrant full verification and certification.
- Operational Readiness Tests (ORTs) are essential to verifying the Virtuous Cycle.

- Introducing new systems or services with the promise that surveillance and performance management systems will follow is folly. Those systems are integral to the Virtuous Cycle and ORTs cannot realistically be successfully completed without them.
- New software loads and new hardware systems should not be deployed directly after they complete ORT. Soak periods with a few systems is mandatory. The extent of the soak and controlled introduction should be commensurate with the risk. Losing a single system or a few systems is embarrassing. Losing a network affects the reputation and brand of the carrier.
- Once an outage begins, there is no substitute for tier 4 labs and capable Systems Engineering and Operations tier 3 organizations.

An insidious characteristic of Vicious Cycle outages is that they trump Virtuous Cycle response and standard operational practices for network system restoral. Operations has to recognize they are in the midst of a Vicious Cycle outage and use dramatically different procedures on a network scale rather than focus on faulty network elements. When redundant network systems fail, the standard response of Operations is to restore it because the network is in simplex mode. Another fault in the remaining system can cause a service outage. As Pacific Bell's ESAC quickly figured out in 1991, in a Vicious Cycle outage that is exactly the wrong thing to do. Rather than focus on root cause investigation, Operations must break or at least contain the propagation mechanism. They are not going to find the root cause in the midst of the emergency and they will be distracted and outages will last far longer than necessary.

The probability of generating and propagating a triggering event is directly proportional to the number of elements and links operating in the network. That is why ESAC took a single STP in a mated pair out of service when it exhibit signs of instability. They also brought A links up before they brought up C and D links. A links could not propagate triggers to end offices because end offices had not shown a vulnerability to the triggers, only DSC STPs were at risk. Gradual reintroduction of load on essential systems lowered the probability of a triggering event. SS7 links and IP links usually have a mechanism to bring the link into service without allowing traffic on it. Stabilizing the link and monitoring it for an extended prove-in time prevents the generation of erred traffic, another trigger source. Load can slowly be applied to a recovering system. Bouncing links that go in and out of service must be placed in an out of service manual state. They generate instability in the systems and network.

Dramatic steps may be needed on a network scale in a Vicious Cycle outage. The design principles of compartmentalization and coupling were described in Chapter 6 in the section on the IP core. Naval combatant ships have water tight doors between all compartments. When they go to battle stations all of those doors are sealed shut, not be opened without direct orders. In that way any damage done to one or a few compartments on the ship cannot extend to other compartments. In the same way segments of the network may have to be isolated to prevent a complete outage. That is a hard choice to make and brings with it a whole set of risks. But if Operations sees failures beginning in one region and propagating within the region, isolating the region may be the best of a set of bad choices. Control plane faults in particular may be candidates for this defense. Network systems, like routers cache forwarding information; loss of a route reflector means that updates won't be made. In the midst of a network outage that means messages will be misrouted but isolating a suspect reflector may save the network.

14.4 Summary

Carriers prepare for disasters and outages and work on their response processes and systems every day. Dedicated disaster recovery teams are staffed and equipped, able to move anywhere natural or man-made disasters strike. When a local disaster, like a tornado strikes, disaster recovery teams arrive at the site within 48 hours and reestablish site connectivity to mobile, voice, and core IP networks. They have the ability to replace entire central offices with essential services until emergency repairs can be made. In regional outages, such as hurricanes disaster recovery teams restore critical backbone connectivity, but local recovery teams, augmented by craft from other carrier regions do the hard work of service restoral street by street.

Outages often accompany disasters but most outages are caused by a design defect, triggered by human error, or a routine failure. During outages of significant scale and scope carriers bring multiple teams to bear on the problem. Some Operations teams are dedicated to network stabilization and restoral. Others gather data and work with Systems Engineering and vendor teams on root cause analysis. Systems Engineering laboratories are reconfigured to replicate the conditions present in the network failures and try to recreate the problems and isolate the triggers and propagation mechanisms. Emergency patches and workarounds are tested in the labs before they are introduced into the networks. Widespread outages and ones that recur or last more than a few minutes receive a good deal of media attention and can affect the carrier's brand. Communication with the press, government agencies, network peers, and partners is an important part of outage recovery management and mitigation; it is most effective when it is clear, factual, and solution-focused.

References

1. FCC (2009) FCC Network Outage Reporting System User Manual, Version 6, April 9, 2009.
2. Cambron, G.K. (1992) Signaling system no. 7 local exchange deployment. *Proceedings of the IEEE*, **4**, 628–636.
3. Cambron, K. (1986) *Testing the Intelligent Network*, Bellcore Communications Research Exchange.
4. Cambron, G.K. and Ratta, B.A. (1991) An Analysis of June 1991 SS7 Network Failures – Case 240.000, Pacific Bell Memorandum for File, July 4, 1991.
5. Ireland, R.K. (1991) Pacific Bell's Response to the Federal Communications Commission Regarding the June/July 1991 SS7 Network Outage, General Manager Network Services, July 22, 1991.

15

Technologies that Matter

Networks are shaped by the marketplace and by emerging technologies. The global growth of the Internet and the speed with which video and sound were introduced, accompanied by social networking show how quickly networks need to adjust to changing tides. Significant improvements in mobile network capacity and bandwidth efficiencies built upon the Internet phenomena created growth rates in mobile traffic exceeding 1000% per year. In this chapter several technologies are profiled that are likely to produce changes in global networks in the next decade in ways difficult to foresee. In an effort to better appreciate the role technologies may play in the future, let's look at the past to see how they can combine in unexpected ways to change the course of network evolution.

15.1 Convergence or Conspiracy?

Before deregulation changed the regulatory landscape, franchise carriers around the world worked through standards organizations like ITU, to define the shape of new network technology and services. Vertically integrated systems suppliers developed equipment and applications according to those standards and carriers incorporated them into their networks. Convergence was a dominant theme; the operating model of carriers and standards bodies called for new technologies and services to fit within the existing network in a predictable and controlled way. Integrated Services Digital Network (ISDN) exemplified the standards based move to convergence. Data and voice were carried over the same Time Division Multiplex (TDM) network. Routing and signaling for voice and data used a common protocol, SS7, and common support systems. While the goals of network reuse and incremental capital investment were assured with the model, new services and applications were compromised; they had to work within the limitations of the existing network. Organizations that created the new services were still wedded to the embedded network, and so the process for developing new services followed the old. A few years after deregulation that all began to change.

Global Networks: Engineering, Operations and Design, First Edition. G. Keith Cambron.
© 2013 John Wiley & Sons, Ltd. Published 2013 by John Wiley & Sons, Ltd.

15.1.1   Enter the World Wide Web

By the early 1990s Internet Service Providers (ISPs) were connecting their customers with the web over the TDM voice network using voice band modems. The rate of growth of dial up subscriptions to the Internet was typified by America OnLine, later named AOL; it had 10 million subscribers in June of 1995. AOL was formed only six years earlier from Quantum Computer Services, a company working with Apple to promote AppleLink, an Apple based data service. Working at Pacific Bell during this period, I watched the local phone network change in ways never expected. At one point we were adding thousands of trunks each night to local networks to increase traffic capacity on routes to local ISPs. Those ISPs had added thousands of voice band modems as a flood of subscribers seeking to access the Internet came to their doorstep. My job took me to our lab in the Walnut Creek, California central office (CO) on a regular basis. I watched collocation cages being built for ISPs with hundreds of voice band modems. Some were screwed into plywood attached to the walls; others were sitting on the floors or chairs. Apart from the fact that some of the ISPs were being run on a shoestring out of garages, early in the period there simply weren't concentrators that housed multiple modems on a single circuit card. Stand-alone modems were all that were available. Telcos had invested billions in ISDN, but it was voice band modems accessing the Internet through the switched voice network that captured the market and dominated network traffic. How did this happen?

15.1.2   Silicon Valley – A Silent Partner

Silicon Valley and their kin around the globe had always been viewed as a force for advancement of technologies furthering the progress of the public networks and by association the telcos. After all, the latest advance in local switching, the No. 5 Electronic Switching System (ESS) used the Intel 8086 processor initially and later the Motorola 68000 family throughout its design. The adoption of UNIX and Reduced Instructed Set Computing (RISC) computing by Sun Microsystems and others ushered in the move to cost effective mid-range computers, replacing costly main frames. Tandem Computer's design of commercially available fault tolerant systems meant network elements, like Service Control Points could be developed using commercially available computers rather than more expensive and less capable proprietary designs. The mindset of many in the telco community was that they and the standards bodies would continue to shape the networks and the services they offered. They would take advantage of the benefits of Silicon Valley as they arrived, but there was little serious interest in trying to understand, predict, or influence the directions technology evolving there might take. Differences between Silicon Valley engineers, the "netheads," and telecommunications engineers, the "bellheads," were not grounded in intellect, training, or scientific approach. Both communities understood networks and software. Their differences were more profound; they were cultural [1].

15.1.3   US Telecommunication Policy

Telco's ability to determine the direction technology would take to channel the flow of information across networks was overestimated by governments as well. In one of the most remarkable pages in US telecommunications history the executive branch of the government relinquished its rightful and historical role in regulating US telecommunications for almost two decades. In 1978 Judge Harold H. Greene was handed the responsibility for presiding over an antitrust case, US versus AT&T, on his first day on the bench as a newly appointed Judge for the U.S. District Court for the District of Columbia. An agreement to terms and conditions for the breakup of the AT&T monopoly was reached between the US Department of Justice and AT&T and submitted to Judge Greene for his approval. His decision, rendered on August 11, 1982 was not a simple endorsement. His 178 page decision began to prescribe specific conditions on AT&T and the spun off Regional Bell Operating Companies (RBOCs). From that day until the passage of the 1996 Telecommunications Act AT&T, RBOCs, and the rest of the industry looked to Judge Greene for US telecommunications policy. During this period Judge Greene took up the question of information services and became an advocate of France Telecom's Minitel system. Videotex was the technology used in Minitel; Videotex was offered in the US by AT&T and by Prodigy, a joint venture between IBM, CBS, and Sears, Roebuck and Co. In the US service the partners created the content and Prodigy ran the network. Minitel did not use the "walled garden" approach of Prodigy, but rather invited businesses of all forms to join the network and as a result Minitel was a success. France Telecom gave subscribers a terminal when they signed up for the service. That and the open network nature of Minitel appealed to Judge Greene's ideas about the role telcos should play in information services.

15.1.4   The Conspiracy – A Confluence of Events

In the space of three years, from 1994 to 1997 a set of technologies came together to create congestion daily in local switched networks, something that had only been experienced in isolated offices previously. Each technology had little effect on the voice network on its own, but when combined they brought consumers and producers together in new ways, setting off a storm of demand that led directly through the voice network. It was as if technology itself was conspiring, with an advance in one technology sparking an idea and an advance in the other. What set of events and technologies came together to derail Videotex, sideline ISDN, and create congestion in the public voice network?

- **Voice band modems** – The state of the art in 1984 was the 1200 baud Bell 212 modem, boasting a downstream bit rate of 1.2 kbps. By 1995 V.34 modems were readily available with speeds of 33.6 kbps, a 2800% improvement in about 10 years.
- **Personal computer (PC) operating system (OS) advances** – Apple broke ground with the release of Mac OS in 1984 and set the standard for a desktop that everyone could use. Microsoft followed with Windows 3.0 introduced in 1990 and Windows 95 in 1995, important steps in making the PC more reliable and more usable. The introduction of plug and play meant users with a bit of technical know how could upgrade their voice band modems on their own.
- **Internet browsers** – Netscape's Navigator, following on the success of Mosaic, was released in 1994. Internet Explorer was released a year later and was subsequently included as part of Windows 95.
- **Web servers** – An open source web server, Apache 1.0 was released on December 1, 1995. It quickly became the most popular server on the Internet.

- **Web sites** – Yahoo!® had 100 000 unique visitors by the fall of 1994. AltaVista®, a search engine developed by Digital Equipment Corporation, was launched in December of 1995 and received 300 000 hits on its first day.
- **Equipment collocation** – Regulatory changes that enabled the ISPs and Competitive Local Exchange Carriers (CLECs) to collocate modem farms in COs dramatically reduced the cost of entry into the market.
- **CLEC interconnection** – In some jurisdictions, like Pacific Bell, charges for interconnection were paid by the originating carrier to the terminating carrier. Recognizing the opportunity, CLECs became ISPs or encouraged ISPs to locate in their networks. The longer subscribers stayed on the Internet, the more money the CLEC made. Compuserve had charged $10 per hour for Internet service at one time. By mid-1995 some ISPs were giving unlimited service with no charge because the CLECs were collecting usage charges from the Incumbent Local Exchange Carriers (ILECs); the CLECs were in turn paying ISPs for usage.

The flexibility and freedom of choice offered by the Internet, coupled with the ubiquity of the public voice network quickly spelled the end of walled garden networks like Videotex and later triggered significant subscriber migration from AOL. Voice band modems, desktop computing, and plug and play teamed to obviate the need for a complex service like basic rate ISDN. Why call the Telephone Company, schedule an installation and purchase expensive adapters when you can go to Egghead software, purchase it for $20, install it and be on line that afternoon?

In the end basic rate ISDN saw limited adoption because of these conspiring technologies. Ironically Egghead facilitated its own demise. By selling ever improving and ever cheaper voice band modems it served as a key enabler of the Internet. At one time there were over 200 Egghead stores but by 1998 revenues were declining. Certainly the introduction of the Internet and increasingly sophisticated web based applications cut into the over the counter software market. Online sales of software with web downloads also made for a much more cost effective distribution network, bypassing brick and mortar retailers like Egghead.

15.1.5 Local Phone Service in Jeopardy

Engineering rules for local switching systems had been in place for decades and the underlying traffic assumptions were taken as axiomatic. Switching systems in downtown areas experienced heavy call volumes during the day and were engineered according to one set of rules; switches serving residential areas were engineered more sparingly since the long historical record was one of light call volumes, particularly during the day.

- There busiest hour for residential switching systems was around 5 p.m. when kids were coming home from school and commuters arrived home from work.
- The average holding time for residential calls was three CCS[1] or 5 minutes.
- Call volumes varied between one and two calls in the busy hour. One incoming call and one outgoing call in the busy hour was a typical estimate.

---

[1] CCS is a 100 call seconds. There are 36 CCS in an Erlang.

- The toll connect trunk group serving all traffic headed outbound from the switch was engineered for a probability of blocking of 1% during the busy hour.
- A typical line to trunk ratio for residential offices was $10:1$. For a 30 000 line residential switch there were about 3000 trunks. Half were originating and half were terminating.
- Approximately half of the originating and terminating traffic was intra-office, one line on the local switching system calling another line, reflecting a high community of interest. For these calls no outgoing trunks were needed.
- Residential switch fabrics were engineered for $6:1$ or $8:1$ concentration ratios. Local switching systems do not necessarily have full connectivity. The way in which switch matrices are connected is not a complete field, meaning originating lines are organized in modules and do not necessarily have access to all intermediate cross connect fields.

Silicon Valley sits in the San Francisco Bay area and Pacific Bell was the franchise local exchange carrier. To say that those residential communities are early adopters of technology is to state the obvious. In effect they served as an early warning system for the changes ushered in by the confluence of events in the 1990s. In the crises created by Internet dial up service the average holding time on residential calls in the evening jumped from 5 to 30 minutes. An analysis by Pacific Bell's Systems Engineering group showed that some calls were lasting more than 24 hours. Customers were dialing up ISPs that had no connection limits or timeouts and simply never hung up. In a typical 30 000 line CO if just 5% of the customers stayed on line in the busy hour and stayed connected for 30 minutes or more, they occupied 1500 trunks, the entire outgoing capacity. All other calls were blocked, including public services. The 911 service was not affected because that is a dedicated network, but a police station would have difficulty calling a hospital emergency room in such congestion.

There is a good account of the congestion reaching Contra Costa County in the east bay of San Francisco on January 6, 1997 [2]. The sale of PCs with built in modems during Christmas 1996 is identified as one the culprits pushing the network into congestion on that day.

The immediate response of Pacific Bell was to begin adding more outgoing trunks to these offices. Tens of thousands of trunks were added over a period of about two years. Another effect of the confluence of events was a dramatic demand for second, third, or even fourth line for residences. The additional lines became the Internet dial up lines. For Pacific Bell that created two problems. As described in Chapter 5, distribution networks, the cables passing your home in the street, are sized for 1.2–1.5 lines per home in a typical neighborhood. If the Internet craze caught on in a neighborhood there was a real chance there were insufficient pairs to serve the increased demand for Plain Old Telephone Service (POTS) lines. The second problem created by adding these lines was that usage increased even more. When households shared one or two lines, there was some pressure in the family to hang up the Internet connection so they could receive incoming calls. With dedicated Internet lines that pressure disappeared.

The economics of this period have similarities to the present period. Dramatic growth in traffic led to significant capital investment with no compensating revenues. There was no reimbursement for connection to many of the ISPs and in fact because of the misguided separation rules Pacific Bell was paying CLECs to terminate the traffic. Busy hour blocking after 5 p.m. in these neighborhoods increased, blocking revenue bearing calls. Additional trunks in affected offices meant not only increased facility costs, new frames had to be added in local and tandem offices to terminate the trunks. Neighborhood distribution networks had to be augmented with either more cable or pair gain systems, both expensive investments that take time to deploy. Pair gain systems help with feeder congestion but not necessarily with congestion in the distribution plant. Another irony is that the additional lines deployed by ILECs in the crises were new inventory for leasing under unbundling rules. The ILECs were in effect, enabling competitors who they believed could use their investments with an insufficient rate of return for the ILECs. Regulatory rules computed leased rates using long term incremental accounting, ignoring significant capital outlays required under these circumstances. The combination of misguided regulatory policies and a confluence of technologies created a drag on the economic health of ILECs.

15.1.6   Technologies in Response

ILECs looked to the telecommunications supplier community for solutions and they responded with a range of solutions. ILECs were faced with congestion in three places,

- **The distribution network** – ILECs were running out of available pairs in the cables that pass people's homes,
- **Class 5 switching systems** – the network fabric of some of these systems had concentration ratios of up to 8 : 1; if voice band modem traffic continued to grow the switch would have to be de-loaded, meaning new switching equipment would have to be installed and lines would have to be redistributed in the existing switch to lower the concentration ratio,
- **Interoffice trunking** – capacity had to be added to meet the demand for outgoing connections to the ISPs.

Technical solutions were offered to solve all three problems. In some cases the technologies were backward looking, assuming trends of the current crises would extend indefinitely, with ever increasing numbers of dial up modems being deployed in homes with a commensurate increase in dedicated lines. Other technologies that truly changed the way consumers used the network, like Digital Subscriber Line (DSL) received a boost that helped speed deployment.

15.1.6.1   Digitally Added Main Lines (DAML)

Lack of pairs in the distribution plant can drive a lot of costs. Pair recovery and reuse generates truck roles and often ends up adding bridge taps and costs. In effect outside plant technicians end up moving any spare capacity from one place to another. Outside plant operational costs are lowest when pairs can be dedicated to living units. Digitally Added Main Line (DAML) is an electronic technology that places a transceiver at the side of the home and another in the CO or remote terminal. It uses digital processing to provide 2 equivalent 3 kHz lines over one copper pair. The two most popular line transmission technologies used for DAML were ISDN-BRI (Basic Rate Interface) and HDSL (High Bitrate Digital Subscriber Line). Approximately 3 million lines of DAML were deployed in the US [3]. Different versions of DAML have different limitations, but they sometimes limit voice band modems to the range of 26 kbps. DAML is also a dead end. xDSL technologies cannot operate over a DAML line, and the two common transmission technologies used for DAML are interferers with xDSL technologies, as described in Chapter 5. When DSL service demand took off a lot of DAML had to be removed.

15.1.6.2   Modem Concentrators

Voice band modem technology and Digital Signal Processors (DSPs) were a perfect fit for collapsing all of the standalone modems into a compact multiplexer. Equipment like Nortel Network's Versalar Concentrators came with a range of port counts and backhaul options. The most direct application of concentrators was by the ISPs. That only made the ILECs problem worse. With expanded capacity calls were less likely to be blocked at the ISP and so more traffic demand was seen by the local switching system.

Businesses and a limited number of multiple dwelling units used concentrators to put voice band traffic directly on T1 lines, bypassing the distribution plant and the local switching system, but that was a small market. One variant of modem concentrators was designed for COs to bypass the voice network. Nortel's Internet Thruway product tapped onto POTS lines at the main distributing frame (MDF) before they reached the class 5 switching system. It monitored numbers dialed by subscribers and when the unit detected a voice band modem call it used onboard modems to connect to the subscriber modem, bypassing POTS terminations on the switch. One problem was knowing which lines were the most likely to use voice band modems. ISPs had to buy into the solution and have a data connection to the Thruway multiplexer for the concept to be viable. It was deployed briefly in SBC's network but later removed.

15.1.6.3   ADSL1

ADSL is a technology that arrived in time and was an ideal solution for congestion in the distribution and CO. As described in Chapter 5, ADSL uses an expanded band of the copper loop's spectrum to serve voice and data. One copper pair carries both services. The first scale deployments of ADSL began early in 1999 [4]. By the middle of 2000 over 1 million homes in the US had DSL service and by mid-2002 that number had grown to over 30 million. ADSL was deployed aggressively in COs for loop lengths of 14 000 ft or less of 26 American Wire Guage (AWG) cable. Areas further from the CO were served by remote terminals, fed by fiber. Adoption of ADSL dwarfed ISDN BRI in its first year.

15.1.6.4   ATM

ADSL1 was standardized at a time when Asynchronous Transfer Mode (ATM) was the emerging standard for layer 2 transport and switching. Adoption of ATM by the ADSL Forum as the framing and interconnection standard created an immediate demand for ATM switches and networks. ATM networks were built to connect the ADSL DSLAMS (Digital Subscriber Line Access Multiplex System), CO and RT based, to ISPs using virtual path and virtual circuit connections. ATM was already gaining market share in the enterprise space as a bearer of frame relay and Internet Protocol (IP) services.

The enthusiasm for ATM was carried too far with ATM to the desktop and Voice over Asynchronous Transfer Mode (VoATM). Some carriers and suppliers believed the fabric of local and tandem switches could be replaced by ATM switches. Nortel and Lucent developed

VoATM systems and a few were deployed but the economics did not prove out; in practice there was no obvious advantage over the embedded digital switching systems.

15.2 Technologies Beyond 2012

The lesson of the 1990s is that technological advances need to be considered together as part of a larger picture, not as individual actors and trends. Particular attention needs to be given to those technologies that can play off against each other as processing and software have done over the years, one pushing the other. Conspiring technologies, ones that come together in unexpected ways to create societal changes in how people communicate, entertain, and conduct business are likely to have the most profound impact on global networks. While we are likely to be surprised by the form it takes, we can pick out some of the conspirators and speculate about the outcomes. The following is a list of technologies that are poised to change the way network operates, and some potential scenarios of how they may work to create convergence or conspiracies.

15.2.1   IPv6

This is a movie that's been 10 years in the making, but it's coming to a theater near you soon. The International Address Numbering Authority (IANA) ran out of IPv4 addresses in February of 2011. IANA allocated blocks to regional authorities and when they reach their last /8 they reach what is called Stage 3 and begin allocating blocks no greater than /22s, or 1024 addresses at a time. Here's a projection made in February of 2012[2] of the exhaust dates of the world regions (see Table 15.1).

IPv6 expands the available IP addresses from 2^{32} to 2^{128}. There are more IPv6 addresses than there are grains of sand on the earth. More than just address expansion, IPv6 is an elegant design that significantly improves the ability to scale the Internet. The designers of IPv6 streamlined packet routing and forwarding and yet provided flexibility for end point differentiation and future service expansion. There is no mechanism in IPv6 to allow direct interworking with IPv4. As a consequence dual stack operation will be around for a very long time. In dual stack operation the two protocols, IPv4 and IPv6, exist side by side but have no

Table 15.1   Projected IPv4 address exhaust

Regional authority	Exhaust date	Remaining addresses (/8s)
APNIC: Asia Pacific	April 19, 2011	1.2
RIPENCC: Europe and Middle East	July 27, 2012	3.0
ARIN: North America	June 19, 2013	5.7
LACNIC: Latin and Central America	January 29, 2014	3.9
AFRINIC: Africa	October 27, 2014	4.4

[2] Courtesy of http://www.potaroo.net/tools/ipv4/index.html.

native ability to interwork directly. Transition technologies have been developed to enable interworking between IPv4 and IPv6; these technologies were not a primary consideration in the design of IPv6 so interworking comes with risks and limitations. Information Technology (IT) professionals and network engineers must choose from a variety of techniques to bridge operation from an IPv4 world into a dual stack world, but there is not a single well-marked map through this transition that can be adopted. Each enterprise and network has to develop their own transition plans. Unlike Y2K this is not solely an IT or network technology program. This is a transformation of the enterprise and consumer Internet and electronic market place. It touches all parts of the business, government, and the public at large. Apart from address expansion there are changes that accompany IPv6 introduction that are likely to have profound impacts on the ways network operate and the ways they are designed.

15.2.1.1   Elimination of NATs

Network Address Translation (NAT) is used widely in consumer Local Area Networks (LANs) and enterprise networks. It enables reuse of address space, usually private addresses from one of the following ranges designated by IANA 10.x.x.x, 172.16.x.x, or 192.168.x.x. The greatly expanded range of IPv6 means reuse of private addresses no longer makes sense. NATs (Network Address Translations), however, have taken on a dual role in IPv4, translation, and security. The identity of a host machine accessing the Internet from behind a NAT is hidden. The NAT uses a public IPv4 address from a pool or an assigned gateway address. Multiple hosts can use the same public address by being assigned different ports for their outbound sessions. While NATs are not a security device in the strict sense, the anonymity they provide serves as a barrier for certain classes of intrusion. There are no plans in the vendor community at present to build an IPv6 NAT. IPv6 to IPv4 NATs, called NAT64s, are being built to enable IPv6 hosts to access an IPv4 Internet.

15.2.1.2   Harvesting IPv4 Address

In 2011 Microsoft purchased a block of 666 624 IPv4 addresses from bankrupt Nortel, paying $7.5 million or $11.25 per IPv4 address. Enterprises and network providers able to harvest IPv4 addresses, particularly in significant blocks, have an incentive to convert to IPv6 early and pay for the transition with harvested public IPv4 addresses. At the Microsoft rate, a public class B or /16 address is worth $737 000. A class A or /8 address is worth $188 million. As ARIN and EMEA exhaust their inventories, the price may even increase.

15.2.1.3   Transition Technologies

There are three categories of technologies that introduce features of IPv6 while preserving existing IPv4 functionality.

- **Dual stack operation** – hosts, networks, or systems (e.g., Domain Name System (DNS)) that support both protocols directly. This is always the first choice since it avoids inter-working and tunneling.
- **Tunneling** – enabling compatible hosts (e.g., IPv6) to communicate over a different connecting network (e.g., IPv4). It avoids interworking, but may be difficult to scale.

Static tunneling is more secure and more easily managed. Automatic tunneling is more flexible and scalable, but is more problematic for security and network management.

- **Translation** – enabling incompatible hosts (e.g., IPv6 and IPv4) to communicate through an intermediary and translating the protocols by acting as a proxy or Application Level Gateway (ALG).

Over the next several years, global carriers, enterprises, and consumers must decide how to use these technologies in concert, over time, to transition networks from an all IPv4 to dual stack operation. In planning for a specific design network engineers work from the list in Appendix A or a similar list; each solution has drawbacks or implementation issues so solutions will vary.

Global Carriers are affected in a variety of ways by IPv6 introduction. Each of the affected access and core networks has distinct migration plans. In examining each network and service there are really three modes of operations to consider. These modes are covered in depth along with a range of other IPv6 adoption topics in texts from Cisco [5] and O'Reilly [6].

- **IPv4 access to IPv6 services** – the great majority of LANs will continue to host IPv4 clients for some time. At some point they will need connectivity to IPv6 services.
- **IPv6 access to IPv4 services** – some locations, starting in Asia and Europe, are converting their LANs to IPv6 addressing. They need to access IPv4 services in intranets and the Internet on day 1.
- **Dual stack designs** – both protocols run over the same physical network.

Key considerations for each network are described in the following.

15.2.1.4   Backbone Networks

Carrier backbones have largely migrated to Multiprotocol Label Switching (MPLS) which is protocol agnostic and so core label switch routing is largely unaffected in the near term, providing the edge routers support dual stack operation. Core routers can remain IPv4 MPLS routers. This arrangement is called 6PE. Provider Edge (PE) routers can serve compatible Customer Edge (CE) IPv6 routers; however, both routers have to implement exterior border gateway protocol (eBGP) in a way that supports IPv6 routing updates. MP-eBGP supports IPv6 as an address family and so is a likely choice. MP-eBGP operates independently of the transport protocol, so MP-eBGP can distribute IPv6 routing information over an IPv4 network. In a similar fashion MP-iBGP (interior border gateway protocol) can serve as the internal label distribution protocol (LDP), associating addresses with label switch paths (LSPs) and distributing route information to PE routers over the internal IPv4 network.

IPv6 address administration, control plane design, route reflectors, ICMP6, firewalls, DNS, DHCP6 (Dynamic Host Control Protocol), multicasting, and a host of other issues have to be dealt with in addition to the basic routing described above. Fortunately network system suppliers anticipated this well in advance and most carrier networks support dual stack operation today.

IPv6 traffic is very light in the US today and demand has been low, possibly due to the economic recession of the last few years. There are open questions about the efficiency and engineering of high volume IPv6 networks, but those questions are likely to be answered in Asia or possibly Europe before they are faced in the US. While the size of the address field is greatly expanded with IPv6, there are efficiencies gained by having fixed header sizes and flow information contained in the header that may make it as, or more efficient than IPv4 forwarding. Routing table sizing and the processing cost of updating both IPv4 and IPv6 in dual stack routers isn't well understood at this time, but barring a steep rise in IPv6 adoptions carriers should be able to learn as service demand increases.

Investments to enable the transition to IPv6 are in the billions of dollars for global carriers, with much of that devoted to IT systems that support the network. In addition to the initial outlays there are continuing costs for testing and supporting both protocols with new services and systems, and the ongoing operational cost of managing interworking with customers. On the positive side it represents a revenue opportunity for managed services in enterprise markets.

15.2.1.5   DSL Networks

IPv6 will generally run on any layer 2 technology that supports IPv4. However implementation compatibility depends on whether the network equipment can terminate the layer 2 link into an IPv6 routing domain. For DSL that means customer modems and gateways must be dual stack capable and the DSL multiplexer, as well as the backhaul transport from the multiplexer to the Broadband Remote Access Server (BRAS) must be dual stack capable. Legacy ADSL1 multiplexers and legacy ADSL1 modems do not support dual stack operation. Most ADSL1 deployments use Point to Point Protocol over ATM (PPPoA). An ATM Private Virtual Circuit (PVC) is assigned and provisioned in the ATM network, the Customer Premises Equipment (CPE) gateway and the ISP BRAS. It is unlikely in the extreme that carriers will choose a wholesale upgrade of their ADSL1 networks to natively support IPv6. They are more likely to selectively introduce IPv6 Internet access using IPv6 Rapid Deployment (6rd) or other translation technologies. If customer demand develops for IPv6 services in areas only served by legacy ADSL1, customer modems, and gateways can be replaced with dual stack capable units on a selective basis. The customer LAN is likely to operate in a dual stack mode, using IPv6 to connect the host with the residential gateway. From the gateway an IPv6 tunnel can be established over the ADSL IPv4 and BRAS aggregation networks to reach a centralized 6rd server. IPv6 routing is enabled from the 6rd server to the targeted IPv6 site.

The Broadband Forum has thought through IPv6 over Point to Point Protocol (PPP) and defined a standard [7]. However the standard was issued in 2010, so equipment manufactured prior to 2011 is unlikely to comply. ADSL2+ and VDSL (Very High Bit Rate Digital Subscriber Line) have Ethernet framing options and use Point to Point Protocol over Ethernet (PPPoE). PPPoE is more likely to adapt to a native IPv6 implementation than PPPoA. So, the potential for upgrading existing VDSL and ADSL2+ networks to native dual stack operation is far greater than it is for ADSL1. Carriers are likely to use a translation technology, like 6rd until true IPv6 demand develops. At that point they can decide whether they upgrade ADSL2+ and VDSL networks to IPv6 or replace them with newer technology, like Fiber to the Home (FTTH).

Carriers will also have to deal with IPv4 address exhaust in legacy DSL networks. As IPv4 address pools diminish, double NATs, or NAT444s are likely to be deployed. Using two

NATs, one at the subscriber's gateway and another in the network, carriers can improve the utilization of the pool of IPv4 addresses. Subscribers use one private address domain in their home. A second private address domain is used in the carrier's access network until the carrier NAT is reached; there a public IPv4 address is dynamically assigned when the subscriber accesses the Internet. The carrier NAT is a stateful NAT; traffic needs to return to that NAT so the dynamically assigned public address can be associated with the private address. Certain messaging and peer to peer services may fail in a double NAT network design.

15.2.1.6 MSO Cable Networks

Cable networks that have deployed DOCSIS 3.0 can support native dual stack operation. Native dual stack operation does not require tunneling for either protocol. When a host queries the cable DNS with a Uniform Resource Identifier (URI), a AAAA record is returned with IPv4 and IPv6 addresses. The host can resolve the addresses and connect natively to the best resolved site. Native IPv6 is supported in compliant CPE and the Cable Modem Termination System (CMTS). Operators have to consider legacy set top boxes and cable modems which may only support IPv4 when transitioning. A likely deployment scenario is to deploy DOCSIS 3.0 CMTS with initial IPv4 addressing, move to dual stack and then finally to an all IPv6 network. Dual stack operation is likely to last a very long time.

Operators operating on DOCSIS 2.0 or DOCSIS 1.1 are likely to use tunneling or translation technologies, such as 6rd to reach the IPv6 Internet. There are extensions to DOCSIS 2.0 available from some suppliers that add IPv6 support.[3] Some MSOs are facing exhaust of private address space because of the way subscriber domains are defined and the unexpected growth of cable devices in the home. They may be forced to convert to IPv6 and DOCSIS 3.0.

Deployment of DOCSIS 3.0 CMTS in the affected network is no small chore, but the larger issues are likely to rest in the support systems. Subscriber management systems, provisioning systems, network management systems (NMSs), DNS, and DHCP are all affected by a move to dual stack operation.

15.2.1.7 FTTH

Ethernet Passive Optical Network (EPON) (IEEE 802.3ah Ethernet in the First Mile, EPON) and Gigabit PON (ITU-T G.984, GPON), 10 GigE EPON (IEEE 802.3av, 10GEPON), and 10 GigE PON (ITU-T G.987, XG-PON) standards support IPv6 but few carriers have deployed dual stack service. Testing is underway in the US[4] and Asia. Dual stack native service is expected sometime 2012.

15.2.1.8 LTE

Long Term Evolution (LTE) is one sure trigger for scale implementation of IPv6. IP Multimedia Subsystem (IMS) in Third Generation Partnership Project (3GPP) Release 6 makes

---

[3] See CableLabs standard DOCSIS 2.0+IPv6.
[4] Verizon has field trials of IPv6 on FiOS underway in late 2011.

IPv6 mandatory, but allows IPv4 to be used in early implementations and deployments, without defining what early means and specifying a definitive transition strategy. The forgiveness may be moot in any event. The number of devices and increasing usage, or the "always on" syndrome will exhaust the carriers' pools of IPv4 addresses before there is wide adoption of devices. Carriers of scale are using private addressing and Port Address Translation (PAT) to conserve IPv4 public addresses for IP services. But addresses are being consumed by the steady stream of devices and increasing use of addresses by applications. The build out of LTE networks creates new demand for IPv4 although much of LTE mandates IPv6. The reason is that LTE systems must be dual stack in most cases because not all transport systems that interconnect the network elements, and certainly not all of the Operations Support System (OSS) and NMS systems will be capable of operating in a full IPv6 mode.

A likely sequence for the introduction of IPv6 into mobility networks [8] is as follows;

- IPv6 network infrastructure – IPv6 capable DNS, DHCPv6, and associated management systems are placed first,
- network transport and routing – dual stack router operation is proven and can be converted in mobility networks well in advanced. Some transport systems, particularly those that are ATM based, will be replaced or capped; they can continue to support Global System for Mobile Communications (GSM) voice, GSM EDGE Radio Access Network (GERAN), legacy OSS, and NMS systems that may never migrate to IPv6,
- deployment of IMS as the core IPv6 application,
- build out of IPv6 proxies and ALGs to reach the IPv4 Internet from IPv6 sessions,
- enablement of native IPv6 peer to peer and push services, and
- deployment of 6rd or a similar technology to enable IPv6 Internet access for IPv4 devices.

15.2.1.9 Over the Top Mobility

IPv6 has an interesting capability not found in IPv4, a provision for layer 3 mobility that is built into the protocol. Two approaches to enabling IPv6 mobility are a host approach and a network based approach.

The host approach was proffered by two IETF RFCs (Request for Comments) that defined IPv6 mobility procedures [9, 10]. Devices that use IPv6 are called mobile nodes which are normally resident behind a gateway that acts as an IPv6 home agent. Peers that communicate with the mobile node are called correspondent nodes. The home agent acts a bit like an Home Location Register (HLR). Mobile nodes can notify the home agent when they roam into a visited network and send a care of address to the home agent for forwarding. Two procedures can then be used to forward messages to the mobile node while it is served in the visited network. If the peer does not support IPv6 mobility a network tunneling approach is used; in that case traffic from a correspondent node hair pins through the home agent to the mobile node, wherever it is on the globe. If the peer supports IPv6 mobility the care of address is passed to the correspondent node and peering takes place over a more direct route.

The network based approach uses an IPv6 mobile proxy that enables IPv6 network based mobility [11]. In effect the mobile proxy assumes the responsibilities of the mobile node when a mobile node visits the access network managed by the proxy. It has the advantage that the host does not have to be equipped with the mobile node capability. Network administrators will probably prefer the mobile proxy approach since they have more control over roaming and there is less host administration.

There are two forms of mobility anticipated by IPv6 mobility protocols, local mobility, and global mobility [12]. Local mobility manages connectivity across a campus arrangement where a mobile node moves between access points and even between access gateways within a limited geographical area, but retains its IP address. Global mobility supports movement across wide areas where a mobile node changes its IP address.

It's too early to tell if IPv6 mobility will be adopted as the Over The Top (OTT) answer to LTE mobility and peering, but the amount of thought and effort that has gone in to incorporating it into the protocol is substantial. It has an advantage over wireless mobility in that it is access agnostic. Mobility is managed at the IP layer, not at the transport layer as it is in wireless networks. While IPv6 mobility is appealing in an all IPv6 or a dual stack network, the barriers to wide adoption seem high during the extended transition away from IPv4 networks.

15.2.1.10   Trends and Outcomes

The next year will see the exhaust of IPv4 addresses in all of the Regional Internet Registries (RIRs) around the world, having started in Asia in 2011, moving to Europe in 2012, and to the Americas in 2013. Progress of IPv6 adoption worldwide is difficult to judge. There are no general guidelines on how to measure it but some trends can be measured and are being tracked.[5,6,7] Here are some statistics taken from the referenced sources in early 2012.

- 85.9% of top level domains support IPv6.
- 48% of the Internet's devices have an installed IPv6 stack.
- 39% of the Internet's transit networks appear to be dual stack capable.
- 12.7% of autonomous systems (networks) are running IPv6.
- 1% of the top web sites are accessible via IPv6.
- less than 1% of global IP traffic is IPv6.
- 0.4% of traffic reaching Google could use native IPv6. While the number is small, it doubled from a year earlier.

Trends in IPv4 address allocation are also telling [13] (see Table 15.2).

APNIC stopped allocating IPv4 addresses in blocks larger than 1024 in April of 2011. So the rise in allocations was all experienced in the first four months of the year, likely indicating a last minute effort to stock the dwindling supply. It doesn't appear a last minute surge in allocations will be experienced in RIPE NCC or ARIN at present, but those regions are not growing at the rate of APNIC. Looking at the available statistics widely, Asia and Europe are leading the way with adoption, Asia out of necessity and Europe as a policy adopted by the regulatory agencies.

---

[5] Hurricane Electric provides regular reports at www.bgp.he.net/ipv6-progress-report.cgi.
[6] Geoff Huston provides regular reports at www.potaroo.net.
[7] Google tracks end to end IPv6 availability at www.google.com/intl/en/ipv6/statistics/.

Table 15.2   IPv4 address allocation in millions by year in each RIR

RIR	2005 (%)	2006 (%)	2007 (%)	2008 (%)	2009 (%)	2010 (%)	2011 (%)
APNIC	31	31	34	44	46	48	53
RIPE NCC	35	33	31	22	23	22	22
ARIN	27	28	26	28	22	18	11
LACNIC	6	7	7	6	6	6	10
AFRINIC	1	2	3	1	3	3	5

There are two paths that open after the exhaust of IPv4 addresses. One path embraces IPv6 and moves steadily, if not swiftly to adoption. The other path is to delay and invent increasingly arcane and complex workarounds to reuse and conserve IPv4 addresses [14, 15].

The Internet Society is leading the effort for adoption. Two notable events were World IPv6 Day on 8 June 2011 and World IPv6 Launch on 6 June 2012. The first event was a demonstration of IPv6 and the second is a commitment by ISPs, equipment providers, and web services companies to make IPv6 commercially available on a permanent basis. The following ISPs committed that IPv6 will be available automatically as the normal course of business for a significant portion of their subscribers: AT&T, Comcast, Free Telecom, Internode, KDDI, Time Warner Cable, and XS4ALL. Two equipment providers, Cisco and D-Link, committed to enable their home routers for IPv6 as the default. Four web service providers, Google, Facebook, Microsoft Bing, and Yahoo agreed to permanently enable their web sites for IPv6.

The second path, building mechanisms to extend the life of IPv4, not only forestalls IPv6, it reduces the utility and interworking of the existing IPv4 network. Solutions fall into three general categories, IPv6 tunneling over IPv4 access networks, ALGs, and IPv4 conservation. Tunneling generally means the client side establishes a connection to a host server over an IPv4 access network. It severely restricts opportunities for peer to peer or push services since tunnels are established from the clients to a restricted community of servers. Moreover there is no agreement on which type of tunneling should be universally adopted. IPv6 traffic being delivered via tunnels like Teredo constituted over half of all IPv6 traffic in 2010. It is now insignificant. Tunneling also defeats local IPv4 security, since by definition firewalls are opened for the tunnel with no ability to monitor what passes through them.

ALGs allow interworking by performing protocol conversion between IPv4 and IPv6. They are preferable to tunneling because they can be made more secure, but are expensive and require regular administration and engineering. ALGs are not universal. Different protocols, such as Session Initiation Protocol (SIP) and http require different solutions.

IPv4 address conservation is generally accomplished with PAT or double NATs (NAT444). Both approaches obscure a reachable IPv4 address associated with the client host. Some existing services, like peer to peer, SIP, and home monitoring are likely to break under those arrangements with no obvious means of mitigation. Services like home monitoring deal with DHCP assigned public addresses in home routers today by using dynamic DNS. A client side application, like ddclient logs into a dynamic DNS server and updates the WAN IP (Wide Area Network) address of the host. That will fail with NAT444. Web sites will experience multiple hosts sharing a single IP address and a single host with multiple IP addresses. Security becomes more difficult under those arrangements. Simple services like ping and trace route no longer work.

Opening up a secondary market for IPv4 addresses has its own share of consequences which may not be obvious. If ISPs decide to sell IP address services they may come in three categories:

- **Static IPv4 addresses** – the most valuable,
- **Single NAT IPv4 addresses** – the second most valuable because dynamic DNS can be used to maintain accessibility,
- **Double NAT IPv4 addresses** – of little value other than access to, not from, the Internet.

Signs to watch for the adoption of IPv6 are the emergence of IPv6 enabled services and subscriber migration away from networks deploying PAT and double NATs. For example, if home monitoring or other services that rely on network access into the home start to fail, consumers may switch ISPs to ones that support IPv6 or at least static IPv4 addressing, which will hasten the exhaust of IPv4. In the end, carriers must see an economic advantage to conversion over address reuse and conservation.

15.2.2 Invisible Computing

Invisible computing is the expansion of Internet connected devices into appliances and products to the point where it is commonplace. Products you are likely to recognize as being IP connected are home computers, printers, smart phones, eTablets, eBooks, security cameras, thermostats, Digital Video Disc (DVD) players, TVs, set top boxes, MP3 players, and games consoles.

Some current day examples of products you may not recognize as IP enabled are listed in the following.

- **Home appliances** – LG Electronics introduced Wi-Fi Internet enabled washing machines, clothes dryers, refrigerators, microwave ovens, and vacuum cleaners under the brand Smart ThinQ at the Consumer Electronics Show in 2011.
- **Pets** – PetsCELL™ and AT&T offer Global Positioning System (GPS) enabled pet collars that use mobile networks to allow owners to track their pets.
- **Light bulbs** – Netherlands-based NXP has developed IPv6 addressable light bulbs that operate on a 6LoWPAN (Low power Wireless Personal Area Network) (IEEE 802.15.4) wireless network.
- **Athletic shoes** – Adidas sells soccer shoes that are equipped with ANT™+TM protocol, ANT FS (2.4 GHz) transceivers to monitor stride and performance. If popular the trend can move to IPv6 LTE and even GPS. Do you know where your $300 shoes are?
- **Utility smart meters** – water meters, gas meters, and electric meters all have IP enabled versions.
- **Planes, trains, and automobiles**[8] – airlines and cruise ships offer Wi-Fi on board. BART, AmTrak, and other rail systems offer Wi-Fi. In addition to new models that are Wi-Fi equipped, companies like AutoNet™ Mobile will retrofit your automobile with

---

[8] This is not a reference to the classic Paramount Film with Steve Martin and John Candy.

a Wi-Fi network. Siemens and others offer GPS tracking of vehicles and freight with their Vicos CTmobile product.

- **Pallets** – American Security Logistics (ASL) and AT&T have a joint venture for tracking high value pallets and other containers.
- **Wrist watches** – the Lok8u GPS Child Locator is a wrist watch with GPS and GSM capability.

Virtually all of these developments have one thing in common; in addition to being IP enabled, they are wirelessly enabled. They rely on Wi-Fi, GSM, LTE, or 6LoWPAN to reach a LAN or the Internet. Missing from the wireless list are ZigBee®, ANT, and Bluetooth (IEEE Standard 802.15.1-2005). Bluetooth is IP compatible, ZigBee and ANT are not, but ZigBee and ANT gateways can enable IP connectivity.

Most of the progress in invisible computing is being made in the wireless world. Telematics is the term coined by the industry for technology that merges automated monitoring with wireless communication. UPS has been a leader in the adoption of telematics to manage their fleet of delivery vehicles. They have about 47 000 vehicles equipped with telematics that monitor the location, speed, braking, engine, and drive trains. Other companies like FedEx, Ryder, and Service Master have also adopted the technology.

Caterpillar, John Deere, Komatsu, and Volvo formed the Association of Equipment Management Professionals (AEMPs) and developed a telematics standard using Extensible Markup Language (XML) that defines the semantics for information on the identity, operation, location, performance, and location of construction equipment [16]. Equipment sensors and GPS location systems monitor equipment continuously and send the data back to the manufacturer where it is collected and made available to fleet owners using the standard and via web services offered by the manufacturers. Cat Product Link is an example of the service. The benefits of using telematics for high value equipment is obvious and a clear win for the manufacturers and their customers.

The cost of insurance has always been linked to usage, but it relied on the customer to honestly report their annual mileage. Telematics is automating the collection of usage and other information about the performance of the automobile. Companies like KORE marry their telematics technology with analytics companies like Agnik to collect, process and present fleet information for insurers and automobile rental companies. Commercial services consumers will recognize are Progressive Insurance's Pay As You Drive® product and GM's OnStar™ product.

Progress in home automation has been steady but is highly segmented by a lack of industry consensus around automation and communication technologies. The Digital Living Network Alliance (DLNA®) was founded by Sony in 2003 and it focuses on audio visual equipment. It has broad support across equipment suppliers and carriers and has developed unifying standards under the trademark Universal Plug and Play (UPnP). Adoption by entertainment systems is well established and growing but appliances, home security, and monitoring devices are almost non-existent. Home Automation Networks (HANs) are supplied by automation companies like HAI and service providers like ADT, Verizon, and AT&T (with the purchase of Xanboo). Phillips and ActiveCare have services that specialize in monitoring seniors or the disabled. A wide range of products based on Zigbee and X10 are available for home monitoring self-installs from D-Link, Motorola, and others.

Enterprise and commercial advances in energy management, fleet management, and logistics that are enabled by embedded computing and wireless networks are likely to continue to thrive and surprise. Automation in the home appears stalled or at least destined to advance incrementally. Unless the wireline access networks match the speed and ease of connectivity coming with LTE and IPv6, automation and remote management will lag in those networks.

15.2.3 Beyond 400G

Three decades ago archaeologists discovered there was a dramatic change in the earth's population of dinosaurs at the end of the Cretaceous Period and the beginning of the Tertiary Period, about 65 million years ago. A layer of ash, discernible worldwide, marks the boundary between those periods. Below the layer dinosaur fossils are found in abundance. Above it, no dinosaurs, except for avian fossils are found. Scientists named the divide the KT Boundary. While the debate goes on about the cause of the mass extinction, called the KT event, it is clear that nature had one model for life prior to the boundary and one afterward. Before the KT event species prospered by growing ever larger. After the mass extinction event nature took a new course and smaller species prospered. In some way the environment changed and the model of ever increasing size was no longer sustainable. Smaller and more prolific species prevailed in the new world.

The voice telecommunications world experienced its own KT event around the year 2000. In the latter half of the twentieth century switching systems grew in scale and size with the evolution of technologies like large scale integrated circuits and third generation languages. Switching systems, particularly class 5 switches grew ever larger. Nortel's Digital Multiplex System (DMS) 100 and Lucent's 5 ESS led the way in North America, each capable of hosting over 100 000 subscriber lines. Today there are still thousands of those systems in service, but they have stood still in time for a decade. Typical COs serve 30 000 subscriber lines and the switching systems often occupy 20 bays or more. Prior to 2000 class 5 switching systems received new software loads about every 18 months and new equipment often accompanied the new releases. Equipment suppliers had a steady stream of income and maintained substantial engineering staffs to support the systems and develop the next generation of software and hardware. The subscriber base for wireline service began to flatten in the mid-1990s from simultaneous inroads by cable, mobile, and the move to DSL broadband, displacing dial up Internet. Orders for new systems, hardware upgrades, and new software came to an abrupt halt, the TDM switching systems' KT event. In response equipment suppliers looked for ways to evolve their embedded base of large systems into a form suited for the new environment. Designs were undertaken that changed the core of those systems from TDM to ATM and voice over Internet Protocol (VoIP) but those evolutionary directions ignored the rules of the new environment. New services, based on broadband, were delivered by DSL multiplexers that connect to subscriber loops at the MDF, bypassing entirely class 5 switching systems. No amount of reengineering the old model in an attempt to evolve it worked. The family of TDM switching systems is becoming extinct, replaced by smaller and more agile systems. In the new communication environment broadband and IP systems flourish, delivering services like video, messaging, and VoIP, pushing aside their TDM predecessors.

Two questions emerge from the story:

- What will become of the embedded switching systems of the Public Switched Telephone Network (PSTN)?
- Can the same fate fall upon today's Internet backbone routers?

The first question is addressed in the next chapter. The second question is moot on the surface, but worthy of discussion.

Network dislocations are driven by changes in technology, changes in user behavior or more often both. TDM switching suffered from the move to broadband at the same time VoIP and wireless appeared. It is unlikely in the extreme that the IP network will be replaced with some other form of transport in the next two decades. But it is possible that the way IP traffic is switched on the backbone will change; MPLS is a recent lesson on how fundamental technology changes reshape backbone routing. Without MPLS backbones, cores would need to implement IPv4, IPv6, and Ethernet control planes for user data, placing a significant burden on processing and route management. MPLS came along at the right time and eliminated one barrier to the never ending demand for more capacity at lower costs.

The most obvious barriers and enablers, to the annual increase in backbone bandwidth demand of 40–50% are optics and switching. 40G technology deployment in backbone networks began about 2007. 100G backbone deployment began about 2010[9] and will see significant deployment over five years as the economics become more compelling and the routing infrastructure to support those rates expands. Long haul optical transmission is moving from SONET/SDH (Synchronous Optical Network/Synchronous Digital Hierarchy) to OTN (Optical Transport Network). OC-768 may be the last synchronous optical standard that will see wide spread deployment. The leap from 40G to 100G was made possible by incorporating coherent detection and DSP processing. Long haul and regional fiber networks in North America were largely deployed in the 1990s. The physical standards for optical cable were designed to support optical transmission technology at that time, with some foresight for future systems. Impairments introduced by chromatic dispersion and polarization mode dispersion inherent in long haul fiber routes can be corrected using coherent reception coupled with advanced signal processing; both are used in OTN 100G designs.

Optical engineers are generally optimistic that 400G transport can be achieved with the existing fiber base. Router engineers also feel confident they can continue to increase line speeds and network fabrics to switch the traffic at 4X today's rates. That takes us to the end of the decade, but moving beyond may take a different model. The physics and economics of ever higher optical data rates and ever larger core routers may force other solutions. Some candidate technologies are wavelength switching, highly distributed content (much like Akamai's model today), and smaller more numerous routers using new generations of blade center systems. To take advantage of these technologies backbone networks may become less hierarchical. Mesh networking is used in backbones today, but a lot of traffic still travels through area 0 transit routers. While there are no signs that IP networks will suffer the fate of TDM switching systems, they will continue to evolve, but the form and direction of the evolution beyond 400G is uncertain. Optical switching, expressing routing, and highly distributed content are all likely candidates for new backbone architectures.

15.3 HTML5 and WEBRTC

Hyper Text Markup Language 5 (HTML5) is an emerging standard for web applications. Not all features have been implemented by all browsers, but developers are committed to the standard. At the end of 2011 here's a list of browsers implementing HTML5 in order

---

[9] Verizon announced deployment in their European backbone in December 2009.

of the most complete implementations first: Google Chrome (14.0), Mozilla Firefox (7.0), Apple Safari (5.1), Opera (11.5), and Microsoft IE (9.0).

Here's a simple summary for non-programmers of what the HTML5 and Cascading Style Sheet 3 (CSS3) bring to web applications.

- Web pages have even more multi-media content and it is integrated in a more consistent and predictable way. Video and audio play more seamlessly.
- Users don't have to download plug-ins for different media types and be stuck with the plug-in player and its controls to view media. A lot, but not all, of the capabilities of Adobe® Flash® can be replaced by native HTML5 working with JavaScript.
- Web pages have drag and drop capabilities and the ability to draw interactively, and so they start to acquire some of the features of desktop applications without the need to load Java applets or ActiveX components.
- Web applications can detect screen orientation and size, so they can work as well with smart phones and tablets as they do with a desktop.
- Web applications have client side (browser) data cache, so when users enter data they don't have to wait on a server response. The browser can hold the data until the time is convenient to forward it to a server side database. It can also improve reliability since loss of a server side session doesn't mean the client side has lost all the data the user entered.
- Web pages can use client side communication tools to interact directly with other HTML5 applications and users.

For programmers, here are a few additional comments: HTML5 has audio and video elements just as HTML4 has an img element, and like the image element they allow linking via a reference without redirection. Unlike HTML4 the browser does not have to search for a plug-in suitable for the referenced media type. Drag and drop are also enabled with HTML5 with the element *draggable* and a JavaScript Application Programming Interface (API). A new element, *canvas*, enables drawing much like Microsoft Foundation Classes or Java AWT/Swing. Client side threading and sockets are added via the WHATWG specification, not formally part of HTML5 but arriving about the same time and compatible with it. By making web applications perform much more like native applications HTML5 has the potential to cut into the mobile and desktop applications markets. Apple iOS does not support Adobe Flash™ but Android™ does. If a developer writes their application in HTML5, it doesn't matter. It will run on both OSs as well as Linux and Windows desktops. Moreover, there's no application charge from the OS or device supplier because the application is accessed through the browser.

The Web Real-Time Communication Group (WebRTC) was formed in May 2011 under the auspices of W3C to develop client side (browser) APIs for real time communication. A companion working group was formed at IETF in 2011. The charter of the IETF working group states that the goal is to produce the necessary communications protocol standards for the transfer of media between web clients (browsers). The standards support IPv4, IPv6, and dual stack operation and include provisions for security and NAT traversal. The working group will also consider compatibility with legacy VoIP equipment. Google® is a sponsor of WebRTC. The SIP protocol is not specifically mentioned in the charter but a review of the APIs and FAQs shows it does interwork with SIP. WebRTC relies on HTML5 and JavaScript. WebRTC can enable peer to peer web based applications that support voice and video as well as client server applications. The specific focus on NAT traversal indicates that peer to peer is an important mode for the working group.

HTML5 and WebRTC have the potential to move more applications off of the device and into the cloud. Improved graphics and drag and drop make HTML5 competitive with device based applications from a usability viewpoint. However, HTML5 is dependent on a reliable and high speed communication link to the servers in the cloud. For those applications that are latency sensitive, such as gaming, IMS is a possible technological solution. Quality of service (QoS), authentication, policy management, high speed communication, and network caching can be provided by IMS. However IMS becomes yet another platform for applications in an already crowded space. Developers are faced with building applications for desktops (Mac and Windows), mobile devices (iOS, Android, Blackberry OS, etc.), and the World Wide Web. Choosing an HTML5 enabled application on IMS might be the best choice for a mobile application, but will the same application work on IMS platforms provided by different mobile carriers? Can the application be ported to the Internet to serve non-IMS clients without a re-write? IMS and HTML5 seem to be logical partners, but these questions have to be answered.

15.3.1   Video Evolution

In 2010 video content accounted for about 40% of the traffic across AT&T's backbone and it was growing at an annual rate of 75%. Video compression technology using H.264 or MPEG-4/AVC (Motion Pictures Expert Group; AVC, Advanced Video Coding) has reached the point where a High Definition Television (HDTV) stream that needs 16 Mbps using MPEG-2 can be compressed to 6 Mbps using MPEG-4, depending on the content. Future gains are harder to achieve, particularly with live content. Trends in video evolution matter a great deal for backbone and access networks. For backbone networks it requires significant capital investments each year to keep up with the demand. Access networks are an even greater challenge. Higher adoption of more bandwidth intensive forms of video services can tax access network technologies beyond their physical limits. Unlike the backbone, access networks can only add capacity with incremental investments to a point within the limits of the technology. To go further requires a significant upgrade of technology or a move to a completely different one. The following sections describe trends in video technology without choosing one over another.

15.3.1.1   Ultra HDTV

Ultra-High Definition Television (UHDTV) is a video technology that can deliver up to 16 times higher resolution than today's HDTV. UHDTV has two formats, a 2160 format at $4\times$ HDTV's 1080p, and a 4320p format at $16\times$ 1080p. Sharp Electronics and NHK, Japan Broadcasting, demonstrated the technology using an 85 in. monitor in May of 2011. BBC is promoting the technology as part of the 2012 London Olympics. That resolution needs a large screen like the 85 in. model used in the demonstration to distinguish itself from HDTV's 1080p. UHDTV has been standardized by ITU [17] and SMPTE [18]. NHK is considering satellite broadcast trials in 2020. Since this technology can't be displayed on today's HDTV sets it requires a change out in consumer sets much like the introduction of HDTV. In addition, studios, cameras, and access networks will require modification if not replacement.

Table 15.3 Streaming video bandwidth requirements

Quality	VuDu	Netflix
SD	1–2 Mbps	2.4 Mbps
HD	2.25–4.5 Mbps	5.2 Mbps
HDX	4.5–9 Mbps	–

15.3.1.2   Streaming Video Services

Hulu, VuDu, Netflix, Major League Baseball (MLB), and other sites offer streaming video services. The bandwidth estimates in Table 15.3 were taken from their web sites and may change with time. The services are offered on a variety of devices including PCs, Blu-Ray players, game players, eTablets, Roku™, and other purpose-built appliances.

15.3.1.3   3DTV

Three-dimensional video can stream over most TV access networks today. Comcast and other MSOs offer 3D service and cable channels, such as ESPN and 3net, a joint venture with Discovery, Sony, and IMAX started with event related services and have expanded to full 3D channel service. Three-dimensional video uses 30–40% more bandwidth than an equivalent 2D stream.

15.3.1.4   User Generated Video

During the 2011 Super Bowl in Cowboys Stadium in Dallas AT&T installed Distributed Antenna Systems (DAS) in anticipation of the dramatic increase in demand from fans using their 3G devices. What few expected was the directionality of the traffic. More bandwidth was used leaving the stadium than entering it. Fans were using their mobile devices to take pictures and videos and sending them to friends and relatives.

High Definition (HD) and night time infrared video have advanced into the home security market. 3G surveillance cameras are now available that can be left anywhere there is 3G coverage. Apple's FaceTime® application enables iPhone4® users to set up video chat sessions with the iPhone®, iPod touch®, iPad 2®, iPhone 4, the Mac®. Skype™, Google®, Yahoo!®, and others have competing services. Mobile video resolution is moving from 240p to 720p and even 1080p as devices come with larger screens and better screen technology.

15.3.2 High Definition Voice

The idea of HD voice,[10] also called wideband audio, has been around a long time. ITU standardized G.722 in 1988 [19]. The PSTN uses G.711 which has a 3.1 kHz bandwidth and uses 64 kbps for transmission; G.722 has a 7 kHz bandwidth and supports coding at rates of 48, 56, and 64 kbps. 3GPP also supports wideband audio, but they adopted the Adaptive

---

[10] Polycom has a trademark for "Polycom HD Voice"™ but the generic term "HD Voice" does not appear on their list of trademarks.

MultiRate WideBand (AMR-WB) ITU standard, G.722.2 which has a 7 kHz bandwidth but supports coding at rates from 6.6 to 23.85 kbps; 12.65 kbps or higher is needed for a competitive grade of service. Standardization by ITU and 3GPP on G.722.2 is significant. Wireless and wireline networks that share a common standard, working with the device industry can create a market where wideband voice works end to end, and becomes the norm, like HD TV, rather than the exception. But to realize the widespread adoption of wideband audio there are three roadblocks that have to be overcome:

- **Networks** – PSTN and PLMN (Public Land Mobile Network) 2G/3G switching systems are primarily based on G.711.
- **TDM interworking** – when an originating AMR-WB device initiates a call to another AMR-WB capable device, more often than not it is handed off to the PSTN where it is converted to G.711 for completion. Although both ends of the call are AMR-WB capable, the originating network sends the call to the PSTN by default.
- **Devices** – legacy devices have 3 kHz filters or have audio response limited to the 3 kHz range. The benefits of an AMR-WB analog terminal adaptor (ATA) are defeated by legacy devices.

All of the limitations can be overcome with the growing penetration of VoIP, broadband, and LTE.

- **Networks** – IMS is the wireline and wireless standard for global carriers. 3GPP specified AMR-WB as a requirement for IMS, which translates to an LTE requirement. Over the top competitors can adopt AMR-WB as well as long as their networks are SIP compliant.
- **TDM interworking** – unless the called number is on the originating VoIP network, it will probably complete via the TDM network even though all wireless phones are IP capable and 28 countries have broadband penetration rates of 50% or higher.[11] The problem is not one of connectivity but rather awareness. The originating network does not know the IP address of the called party. E.164 Number Mapping (ENUM), which has been addressed in earlier chapters, can resolve E.164 numbers into URIs and DNS can resolve the URIs into IP addresses. ENUM has been trialed in over 50 countries, but adoption is moving at a glacial pace.
- **Devices** – Avaya, Cisco, VoiceAge, Nokia, and others have AMR-WB compatible SIP phones.

While the technology is ready, it is likely that HD voice will be confined to closed networks for some time. That can change if the industry or regulatory agencies drive universal implementation of ENUM and align behind adoption of G.722.2 as a universal standard. Without broad support or a mandate HD voice will continue to be a niche service. There is an obvious irony when voice advancement is compared to video. Voice service began in the later part of the nineteenth century. Video service, TV, began in the middle of the twentieth century. Video services abandoned analog formats for HD digital formats in the 1990s, gaining broad support from regulators and the industry. It is impossible to purchase a new analog TV and any TV purchased today provides stunning HD video whether it is connected to a cable company, a telephone company, a satellite company, the Internet, or

---

[11] OECD Broadband statistics (see http://oecd.org/sti/ict/broadband).

uses an antenna for off air reception. Not so for a VoIP phone; there are over 25 different codecs used in VoIP, many of which are proprietary.

The lack of international alignment for VoIP interworking blocks some of the advantages of voice transmission that have been designed into IMS and LTE. Consider the case of a call that originates on an IMS wireline network and terminates on a wireless IMS network. Assuming the two networks are operated by different carriers, without ENUM the call traverses the PSTN. The call is transcoded at least twice and possibly four times depending on how the IMS networks are designed and how they use SBCs. Each time the call is transcoded from one audio format, AMR, for example, to another, G.711, up to three impairments are introduced. Quantizing distortion occurs when converting from one companding/compression format to another. Latency is increased by the need for receive and transmit buffering to manage jitter and accommodate differing coding rates; increases of 10–40 ms are common. Jitter can be increased as well. Jitter can be traded off with latency; increased buffering increases latency but can lower jitter. The ITU objective for one way voice delay is 150 ms [20] but mobility networks generally accept one way end to end delays up to 300 ms. QoS and Radio Access Network (RAN) improvements were designed into LTE with a view toward reducing latency. Those gains are jeopardized by transcoding impairments. If the move from mobile TDM transit to Voice over Long Term Evolution (VoLTE) increases the likelihood of transcoding, latency may increase, not decrease. Finally, the use of TDM transit eliminates the possibility of making full use of SIP or IMS. Features are restricted to those available for SS7 PSTN networks.

Alignment behind ENUM and standards for interconnection would enable end to end SIP interconnection with the following benefits.

- reduction of traffic on the PSTN enabling better conservation of an aging technology,
- reduction of the numbers of media gateways and possibly SBCs,
- reduction of transport costs by eliminating the need to detour to the PSTN and use 64 kbps G.711 transmission,
- reduction and rationalization of echo cancellation equipment,
- improved voice quality by a reduction in quantization distortion, latency, and jitter,
- an opportunity to provide HD voice.

15.4 Summary

While technology trends are often viewed as isolated threads, historically they have come together in ways that surprise network engineers and operators. In the 1990s carriers scrambled to add voice capacity to local networks to serve the surge in voice band modem traffic, only to see Internet access move to DSL and cable modems in a few years. Today there are a range of technologies, each of which has the potential to change the nature of traffic on the Network. Moreover, those technologies may come together in unforeseen ways and create extreme demand in segments of the network that exceed design capacity. In that event carriers will respond the way they did in the 1990s and in the recent triple digit growth in wireless data demand; they will add capacity as quickly as possible and look for engineering solutions that use network assets in new and more effective ways.

References

1. Ash, G.R. (2006) *Traffic Engineering and QoS Optimization of Integrated Voice and Data Networks*, Morgan Kaufmann Publishers.
2. Goralski, W. (1998) *ADSL and DSL Technologies*, McGraw-Hill Series on Computer Communications, McGraw-Hill Associated, Inc.
3. Vaidya, A. (2002) *ADSL over ISDN, DAML and Long Loops*, Whitepaper, Vice President and Chief Technology Officer, Charles Industries, Ltd.
4. Starr, T., Sorbara, M., Cioffi, J.M., and Silverman, P.J. (2003) *DSL Advances*, Prentice Hall.
5. Popviciu, C., Levy-Abegnoli, E., and Grossetete, P. (2006) *Deploying IPv6 Networks*, Cisco Press.
6. Hagen, S. (2006) *IPv6 Essentials*, O'Reilly Media Inc.
7. TR-187 (2010) *IPv6 for PPP Broadband Access*, Broadband Forum, Issue 1, May 2010.
8. Transitioning to IPv6 (2008) Whitepaper, 3G Americas, March 2008.
9. Johnson, D., Perkins, C., and Arkko, J. (2004) IETF RFC 3775. *Mobility Support in IPv6*, June 2004.
10. Arkko, J., Devarapall, V., and DuPont, F. (2004) IETF RFC 3776. *Using IPsec to Protect Signaling between Mobile Nodes and Home Agents*, June 2004.
11. Gundavelli, S., Devarapall, V., Chowdhury, K., *et al*. (2008) IETF RFC 5213. *Proxy Mobile IPv6*, August 2008.
12. Kempf, J. (2007) IETF RFC 4830. *Problem Statement for Network-Based Localized Mobility Management (NETLMM)*, April 2007.
13. Huston, G. (2012) Addressing 2011 – One Down, Four to Go, ISP Column, January 2012.
14. Claffy, K.C. (2011) IPv6 evolution: data we have and data we need. *ACM SIGCOMM Computer Communication Review*, **41** (3), 48–48.
15. Huston, G. (2011) IPv4 Address Exhaustion: A Progress Report, Plenary, RIPE 63, Vienna, Austria, 2nd November 2011.
16. AEMP (2011) AEMP Telematics Data Standard V 1.2, March 15, 2011.
17. ITU-R Recommendation (1995–2004) BT.1201-1.
18. SMPTE 2036-1-2009 (n.d.) Ultra High Definition Television.
19. ITU-T (1988) G.722. *General Aspects of Digital Transmission Systems, 7 kHz Audio Coding within 64 Kbits/S*, 1988.
20. ITU-T (2003) G.114. *International Telephone Connections and Circuits – General Recommendations on the Transmission Quality for an Entire International Telephone Connection*, May 2003.

16

Carriers Transformed

Markets, technology, and regulation all have the ability to change the course and form of networks. Carrier transformations occur over a span of 15–25 years and are typified by broad technology deployments that dominate capital and operational demands on carriers, and open up new services. Periods of transformation overlap, with new technologies gaining momentum as others wane.

16.1 Historical Transformations

Having lived and worked through these transformations my point of view is shaped by events in the US, which serve as mileposts. But the US had and has no monopoly on innovation and invention; these trends were underway in the rest of the developed world.

   As engineers, we see transformations as logical adoptions of superior technology. But engineers don't shoulder the burdens of financial proof and fiduciary responsibilities born by CEOs, CFOs, and corporate boards. Transformations take 5–20 years and tax the treasuries of carriers, justified by the belief in a future payoff. But market trends and technologies can pivot more quickly than capital programs, and several years into a transition the technology and market assumptions that underpin the original business case may fade or be entirely reversed. Two examples cited earlier, Asynchronous Transfer Mode (ATM) and Integrated Services Digital Network (ISDN), are technologies that garnered substantial financial commitments, were years in formulation, but had short lives that failed to live up to the claims. Few executives that launch these broad capital intensive technology transformations are still at the helm when the ship arrives at the destination, or flounders in transit. So executives that have the foresight and confidence to embark on transformation, recognizing their administration will bear the costs and scrutiny, and their successors will receive any benefits, deserve recognition alongside the engineers that forge the technology.

16.1.1 Stored Program Control Switching 1965–1985

The era of computer controlled switching began with the introduction of the No. 1 Electronic Switching System (ESS) in 1965 [1]. Capital investment over the next 20 years emphasized

Global Networks: Engineering, Operations and Design, First Edition. G. Keith Cambron.
© 2013 John Wiley & Sons, Ltd. Published 2013 by John Wiley & Sons, Ltd.

the evolution from electromechanical systems to stored program control (SPC) switching systems with electromechanical space division switch fabrics. SPC systems had higher call capacities than the systems they replaced and made possible a wide range of services, such as three-way calling, call waiting, automatic call distribution (ACD), and speed dialing. These systems significantly improved the ability to offer advanced business services under the Centrex and Essex offerings. But the No. 1 ESS was introduced using a growth strategy, not as a wholesale replacement of existing systems. Development and upgrades of the No. 5 and No. 5A Crossbar systems continued well into the 1970s; the last system was manufactured in 1976. The success of the No. 1 ESS led to a family of SPC space division systems, the No. 2B, No.3 ESS, 10A Remote Switching System (RSS), and Dimension Private Branch Exchange (PBX) are examples. The 10A RSS and Dimension PBX had solid state network fabrics but were not time division switching systems like the generation that followed.

16.1.2   Digital Wireline Communications 1975–2000

The SPC era was almost entirely driven by switching, but the era of digital wireline communications spans switching, signaling, and transmission.

16.1.2.1   Switching

The No. 4 ESS toll switching system was introduced in 1976 as the first time division switching system in the US Bell System. It was designed to fill the need for a large metropolitan toll office capable of terminating more than 40 000 trunks and completing more than 300 000 busy hour calls. It eventually exceeded those goals, terminating 100 000 trunks and completing up to 550 000 busy hour calls. What sets the No. 4 ESS apart from earlier SPC systems is the network fabric. Unlike the space division fabrics used by electromechanical and other SPC systems, the No. 4 ESS introduced digital multiplex switching using a time slot interchange (TSI) and time multiplex switching (TMS). The combination is sometimes referred to as a time-space-time interchange. Technology advances that made the No. 4 ESS possible were the No. 1A processor, developed for the No. 1 ESS, small scale and medium scale integrated circuits, and pulse code modulation (PCM). All emerged in the early 1970s, with G.711 being accepted as the worldwide standard for PCM. Local switching systems, like the DMS-10, DMS-100, DMS-200, and No. 5 ESS followed the path of the No. 4 ESS. By the mid-1980s these systems were favored over their space division predecessors for new systems growth and for replacement of electromechanical systems. They benefitted from an aggressive deployment program sponsored by unlikely partners, the US Department of Justice and Judge Harold H. Greene. As a condition of closing the anti-trust suit against AT&T, equal access signaling also known as Feature Group D signaling, was developed to allow customers to choose different long distance carriers; AT&T was no longer the default. Older switching systems, the No. 4A Crossbar, Step by Step, No. 5 Crossbar[1] and some of the smaller ESS systems were either incapable of supporting Feature Group D, or the economics of upgrading them were inferior to replacement. Some local carriers deployed third party modifications to Step by Step and other systems to forestall

---

[1] The No. 5 Crossbar with an Electronic Translator System (ETS) did support Feature Group D.

replacement; vendors implemented Feature Group D as an adjunct, but in general they were Rube Goldberg designs that were short lived.

16.1.2.2 Signaling

Prior to 1976 interoffice signaling was either dial pulse or in-band tone signaling. Both used Channel Associated Signaling (CAS), where the same circuit that carried the voice channel carried the signaling. In 1976 a CCITT digital signaling standard, Signaling System No. 6 (SS6[2]), was introduced in the US. A single channel could carry all of the signaling information between two switching systems; multiple channels were used for redundancy. While successful, SS6 lacked the ability to scale sufficiently to serve local offices and so CCITT developed SS7, which was introduced in the US in 1987. It also enjoyed a boost from Judge Greene. SS7 was needed for portable 800 phone numbers and speeded call set up for Feature Group D calls. SS7 was quickly deployed in the late 1980s and early 1990s.

16.1.2.3 Transmission

Digital transmission began in earnest with the introduction of T carrier in the US in the early 1960s and in Europe with E carrier in the early 1970s. They opened the door for digital switching; time division switching systems were the perfect partners for digital carrier systems. Carrier systems terminated directly on the switching systems, eliminating expensive carrier channel banks used to convert between analog and digital formats. While T and E carrier systems digitized local transmission it wasn't until the 1980s when SONET/SDH (Synchronous Optical Network/Synchronous Digital Hierarchy) systems were deployed that optical systems achieved end to end digital quality.

The development of voice codecs was already mentioned, but in 1996 another important coding standard was released, Motion Pictures Expert Group (MPEG−2). Digital video transmission became economic for satellite links, and for distribution into homes equipped with digital set top boxes. Digital transmission over coaxial cable to the home was matched by the introduction of Asymmetric Digital Subscriber Line (ADSL) on copper pairs in the same time frame.

In the space of 25 years the digital transformation from the home into the core of the network, encompassing switching, signaling, and transmission was completed.

16.1.3 *Digital Wireless Communication 1990−Onwards*

As the successors to analog wireless and digital standards,[3] Global System for Mobile Communications (GSM) and CDMA2000 began deployment in the early 1990s. There are now over 5 billion subscribers worldwide and over 30 major carriers. Over 220 countries have GSM coverage and over 30 countries have CDMA2000 coverage. Capital investments to deploy these networks are in the hundreds of billions of US dollars. Spectrum auctions alone in the UK garnered $34 billion in 2000 [2] for 3G spectrum and US auctions totaled

---

[2] SS6 was often referred to as Common Channel Interoffice Signaling (CCIS) in the Bell System.
[3] DCS1800 and IS-95 were predecessors to CDMA2000 and GSM respectively.

$19.6 billion for Personal Communication Service (PCS) spectrum in 1994–1996 and $19.7 billion in receipts in 2008 for 700 MHz spectrum.

In retrospect, apart from the speed and ubiquity of adoption the greatest surprise is that what started out as mobile phone networks quickly evolved to the point that voice is the third most popular service. Internet access uses more bandwidth than voice, and subscribers average about twice as many text messages daily as voice calls; the average is highly skewed by younger subscribers who average 100 text messages each day.

The physical infrastructure created for mobile networks includes radio access networks, core voice, and core data networks. The latter two are quickly converging with the wireline counterparts. While growing at triple digit rates, mobile data still accounts for less than 10% of the data carried on common backbones. Mobile voice traffic relies largely on wireline tandem networks for completion; the amount of traffic on wireline tandem networks has grown with the adoption of mobile voice. Long Term Evolution (LTE) in particular has also accelerated the deployment of Ethernet over fiber as it is being deployed to about 100 000 cell sites in the US.

16.2   Regulation and Investment

To set the stage for the conclusion of this last chapter a discussion and set of assumptions about the directions regulation and investment may take in the coming decade is in order. The reason is simple. Technology, markets, and economics determine how carriers should invest capital, what businesses carriers should enter and which ones they should cap or divest. But governments determine what obligations are placed on the carriers, what businesses they are allowed to enter, what businesses they are allowed to expand, and possibly which ones they must serve at a loss.

It may seem odd to include investment in a section on regulation, but I'm in good company with my pairing. Germany's Federal Ministry of Economics and Technology sets policy for both. Regulation has the ability to drive investment shifts in dramatic fashion. It trumps technology, markets, and economics. Regulatory choices can advance or retard investments in a nation's infrastructure. In the US communications policy is split among the Justice Department, Department of Commerce, State Department, and the Federal Communications Commission (FCC). Policies can change dramatically when administrations change, causing significant shifts in investment. The 1996 Telecommunications Act is an illustration of the point. It attempted to convert a natural economic monopoly, local access, into a competitive market by pricing Incumbent Local Exchange Carrier (ILEC) local loops below cost under wholesale network sharing. The regulation was struck down by the US Appeals Court in 2004, but for eight years ILECs made investment decisions differently because returns were uncertain in local markets.

16.2.1   Regulation

The relationship between governments and carriers has a long history and it continues to become more complex with the passage of time. Markets and technology tend to pull carriers into the future; regulation tends to anchor them to the present, if not the past. Without changing my career from engineering to law (both would suffer), I am enumerating some principles and assumptions that are used in the remainder of the chapter for analyzing where technology and markets may take carriers.

- As franchise carriers, ILECs have special obligations in the markets they serve. In addition to being the carrier of last resort, their tandem Time Division Multiplex (TDM) networks are common transport networks for connectivity between other carriers.
- Regulators are reluctant to allow franchise carriers to drop services offered to the public no matter how small the market. Small constituencies can have loud voices and regulators often yield to those groups unless there is a counter balance in the public discourse.
- Most US voice services are offered as interstate services, subject to FCC regulation, and as intrastate services, subject to state public utility commission (PUC) regulation. Changing those services requires approval of both regulatory bodies.
- Further industry consolidation will be marginal in the US, except when forced by regulatory bodies.

16.2.2   Investment

Network transformations are funded by legacy networks and services. The SPC transformation was funded by cash generated by electromechanical systems. Wireless transformation was funded by cash generated by wireline voice. For many carriers wireless and legacy broadband services will provide the cash for the next transformation. The following section describes economic and market trends that are forcing a decade long transformation in telco wireline networks. That transformation has been underway for 10 years for some carriers, and others are in the process of forming their strategies.

Transformation may not be desirable for some carriers. Economics may force them to divest businesses and franchises. If wireline loss trends continue until the end of the decade, legacy US voice franchises that lack high speed competitive broadband networks will become cash flow negative, but regulators will expect carriers to fund those networks out of profits from other parts of the business. Carriers may seek to separate wireline franchises from wireless and successful broadband franchises. There are precedents for this strategy. In 1994 Pacific Telesis, under regulation of the most activist led PUC in the nation chose to spin off Air Touch Communications, their wireless enterprise, and thereby avoid cross subsidization of their wireline Plain Old Telephone Service (POTS) base. Air Touch merged with Vodafone and became part of the US Verizon Wireless network; the financial wisdom of that divestiture has been substantiated.

16.3   Consumer Wireline Networks and Services

16.3.1   Market Trends

Current market trends in the US are:

- consumer switched voice (TDM) connections are declining at an annual rate of about 12%,
- ADSL broadband connections are declining at an annual rate of 3–4%,
- fiber to the node (FTTN) (Very High Bit Rate Digital Subscriber Line: VDSL) and Fiber To The Home: FTTH; Broadband Passive Optical Network: BPON and Gigabit Passive

Optical Network: GPON) broadband connections are increasing at an annual rate of 20–30%,
- cable broadband connections are increasing at an annual rate of 7%,
- cable voice connections are increasing at an annual rate of 6–8%.

These rates have to be considered against the base populations used for the estimates. AT&T's and Verizon's 2011 financial reports were used for the first three statistics. Comcast's and Time Warner Cable's 2011 annual reports were used for the last two statistics; their results are consistent with industry results published by NCTA.

Additional statistics shown here illustrate the size and penetration of competing technologies in the US. All statistics shown are approximate, taken from the sources cited above, the NCTA, USTA, the FCC Broadband Plan, and other sources and adjusted to a baseline of 2011.

- There are 23 million ADSL broadband connections,
- there are 4 million VDSL broadband connections and 30 million homes passed,
- there are 5 million FTTH connections and 17 million homes passed,
- there are 46 million cable broadband connections,
- cable MSOs pass 130 million homes, of those 85 million have access to DOCSIS 3.0 (Data Over Cable Service Interface Specification),
- cable MSOs have a penetration rate of 45% for basic video service,
- cable has 25 million voice subscribers.

In 2011 Internet data rates are between 10 and 20 Mbps for standard service on DOCSIS 3.0, FTTN, and FTTH systems. Premium data rates of 50–300 Mbps are available on DOCSIS 3.0 and FTTH systems today. US data trends are consistent with Europe; current estimates by Cable Europe [3] are that 50% of homes will receive downstream data rates of 30 Mbps by 2013 and 69% will receive 100 Mbps by 2020.

DOCSIS 3.0 and FTTH also support native IPv6 access, which addresses Europe's exhaust of IPv4 in 2012 and the anticipated exhaust in the US in 2013. Given LTE's adoption of IPv6, there is greater opportunity of service interworking and convergence of wireless and wireline access networks when both support native IPv6 access. Interim technologies, such as tunneling and 6rd can ease the transition, but by 2020 full dual stack support, including native IPv6 access will be the norm for competitive services.

16.3.2 Technology

Central offices and the wire centers they support are the largest assets carried on the books of global wireline carriers. Consumer and small business wireline voice networks have not undergone a technology refresh in over a decade. Many of the TDM switching systems entered service over two decades ago. Entire product lines, like some of the NORTEL systems are manufacture discontinued. However, there is a robust third party market that supports these systems. Software loads are stable with no anticipated upgrades for the remaining life of most of the systems. Some systems still receive occasional patches in those cases where the supplier maintains active support. Overall the systems are capable of operating through at least 2020. If wireline voice subscriber losses continue at an annual rate of 8–10%, systems can be cannibalized to maintain circuit pack inventories.

Operating costs in wire centers that are in decline are a more difficult issue than switching system modernization. Copper cable in the outside plant has a life span of 20–50 years, depending on location and placement; every year 2–4% of the cable plant should be placed or rehabilitated. Aerial cable and pulp cable tend to have shorter life spans. Underground Polyethylene Insulated Cable (PIC) has longer life spans. Distribution areas with ADSL only, including those served by remote terminals, are likely to see accelerated losses above the 3–5% experienced annually today unless new broadband technology is deployed. Remote terminals in these distribution areas require regular servicing of their battery and power equipment and occasional dispatches for customer troubles. The costs of rehabilitating and maintaining copper cable routes and maintaining remote voice and ADSL terminals are likely to rise as revenues are declining in these distribution areas. Wireline carriers are faced with a choice of modernizing these wire centers or selling them.

Carriers are approaching the question of wire center modernization at different paces with different solutions.

- South Korea and Japan are committed to complete fiber deployments and currently have FTTH/FTTB (Fiber To The Building) penetration rates of greater than 40%.
- By selling some wire centers and deploying FiOS, Verizon will soon have half of their wireline subscriber base served by FTTH.
- Australia has undertaken a national broadband network that plans to cover 93% of their subscribers by 2021 with FTTH/FTTB.
- France Telecom is deploying FTTH with an innovative business model, partnering with competitors to build and share infrastructure; their goal is to reach 10 million subscribers by 2015.
- British Telecom (BT) in the UK is rolling out a mixed network, 75% FTTN (ADSL2+) and 25% FTTH by the end of 2014.
- Germany announced a national broadband strategy in early 2009 setting a goal of 75% coverage of households at a minimum downstream speed of 75 Mbps by 2014 [4]. A mixed technology approach is being implemented, using VDSL2, cable, and FTTB in metropolitan areas and some deployments of FTTH as well.

In those markets where DOCSIS 3.0 is being deployed, wireline carriers have a limited window to deploy a competitive strategy. History shows that consumer wireline services are a natural monopoly or duopoly when two providers have competitive offers. In the US 78% of households are served by two providers.[4] Only 13% are served by a single provider, and the remaining 9% are equally divided, having either none or more than two. Access network upgrades to VDSL2 or FTTH/FTTB take a minimum of four years when done at scale. Construction, permits, logistics, and technology adoption all take time. To meet a goal of 100 Mbps by 2020 technology choices and plans need to be in place by 2015. There are limited technology choices, illustrated by the carrier broadband plans listed above.

---

[4] Data taken from US Telecom statistics published at http://www.ustelecom.org.

- ADSL2+ lacks the bandwidth and reach to provide 100 Mbps. Using pair bonding, where two pairs are used to double total bandwidth, ADSL2+ can consistently deliver 30 Mbps downstream and 1.5 Mbps upstream at 3 kft of copper.[5] But even bonded ADSL2+ is not suited for speeds of 50 Mbps at practical loop lengths.
- VDSL2 can deliver 50 Mbps at 3 kft when two pairs are bonded. VDSL is most effective below 1.5 kft. When used as an in-building distribution technology in conjunction with FTTB it can deliver speeds approaching 100 Mbps.
- FTTH/FTTB with GPON (ITU-T G.984) and Ethernet Passive Optical Network (EPON) can deliver downstream rates well in excess of 100 Mbps.

Japan is a pioneer in the deployment of FTTH and their experience yields some interesting insights. Digital Subscriber Line (DSL) service peaked in 2007. In 2011 FTTH served 34% of Japan's households; DSL served 20%, and Community Antenna Television (CATV) served 8% [5]. There are 187 providers of FTTH services. FTTB accounts for over half of the service base. Characteristics that support FTTH/FTTB economics in Japan are;

- High household density – as of 1990 there were 327 people per km^2 in Japan [6], 432 people per km^2 in South Korea, and 27 people per km^2 in the US.
- A favorable infrastructure – aerial plant and conduit fed underground are common. Fiber cable can be placed in either without the need for trenching.
- High per capita income and a technology oriented consumer.
- A national policy in place since the late 1990s fostering the adoption of high speed Internet access.

These advantages largely exist in South Korea and Taiwan, both of which also have high adoption rates of FTTH/FTTB. The lesson for carriers around the world is where these conditions exist, minus perhaps enlightened regulation, an opportunity for FTTH/FTTB also exists. When planning a multi-year modernization program, begin with areas with high population density and aerial or conduit access.

- These are areas most vulnerable to competitors. In Japan and in some US metropolitan areas FTTH Competitive Local Exchange Carriers (CLECs) have emerged,
- they are the most cost effective,
- they are the fastest to deploy,
- in the latter stages of deployment cash flow from high density areas can help fund less attractive builds.

VDSL2 is a good bridge to the future, but is unlikely to survive past 2020 where loop lengths are above 1.5 kft. In the US there is no convenient interconnection point between the Serving Area Interface (SAI) and the home. The curb, where terminals serve 12–20 homes, is the next logical connection point, but fiber to the curb has not seen widespread adoption because of powering and deployment issues. Carriers have concluded that if you are willing to go to the curb, you might as well go all the way to the home with fiber. Europe and other parts of the world may have distribution networks that have logical

---

[5] All cable lengths cited are 26 American Wire Guage (AWG) whose transmission characteristics are comparable to 0.4 mm cable.

points of interconnection within 1.5 kft of the residence. In those cases VDSL2 should be competitive into the next decade.

Cable MSOs should be able to continue on DOCSIS for the foreseeable future. Bandwidth demand may cause them to split distribution areas and reduce node sizes but they do not need to change their fundamental technology choice. They may begin to use FTTH technology for greenfield starts.

16.4   Wireless Networks and Services

LTE deployments began this decade and will become the dominant mobile technology by the end of the decade, with likely retirements of 2G technologies and spectrum reassignments.

16.4.1   Market Trends

Devices and applications will continue to surprise.

- iTablets are likely to continue to take market share from laptops; they have decimated the netbook market along the way. iTablets in the enterprise are likely to continue to be integrated with enterprise cloud applications, again displacing laptops in the field.
- Automotive mobile technology and applications have advanced quickly. Car makers around the globe are incorporating mobile technology into their autos and building smart phone applications that interwork with the automobile technology. Examples of applications are:
  - car finders which enable your smart device to locate your car on a map of the local area,
  - mileage, gas range, braking habits, and other driving statistics are available on your smart device,
  - automobile vital signs are sent directly to the manufacturer and in some cases insurers,
  - mobile social networking to let your friends know where you are,
  - calendar items and Short Message Service (SMS) texts display on the automobile's console,
  - mobile entertainment (presumably for passengers) including streaming video, games,
  - finance, news, and anything else that's available on the Internet.
- High Definition (HD) Voice Adaptive Multi-Rate-Wideband (AMR-WB) was launched by Orange in 2009. In 2011, a number of other carriers launched. If Voice over long term evolution (VoLTE) spurs Voice over internet protocol (VoIP) adoption and E.164 Number Mapping (ENUM) becomes more widely deployed, HD voice could gain traction.
- Instrumentation will continue to grow. Smart meters and smart grids, Global Positioning System (GPS) tracking, medical monitors, surveillance systems, construction equipment, and so on.

Pricing is also changing. Carriers are more aggressive in implementing caps and data usage charges for consumers; they will look for ways to do the same with content and application providers. LTE brings quality of service (QoS), reduced latency, and better policy control. Carriers will create application services that make higher quality transport available. They will also figure out how to monetize location and presence information, but with subscriber opt in as a mandatory feature.

16.4.2   Technology

In addition to more robust and efficient radio technology, LTE brings VoIP, IP Multimedia Subsystem (IMS), and IPv6. VoIP will take several years before it becomes the dominant mobile voice transport protocol, but when it does Session Initiation Protocol (SIP) end to end services can open up new opportunities for application developers and for mobile wireless convergence. IPv6 also brings the possibility of better peering services and push services, but application developers won't find those markets attractive until the number of LTE devices is substantially higher. That is likely to take a few years.

IMS was introduced with LTE. It is a complex technology with highly distributed functions, separating applications from network elements with the goal of enabling application introduction without touching the voice and packet cores. The technology and business model calls for application developers to build on top of, or at least interwork with, IMS Application Programming Interfaces (APIs) to take advantage of QoS, charging, and other functions only available through those APIs. It goes in an opposite direction, or at least orthogonal to HTML5 and over the top services. The key may be how Apple, Google, Microsoft, and Facebook embrace IMS and whether they incorporate those functions natively in their operating systems (OSs) or otherwise advocate the adoption of IMS services. Without support from the OS providers, IMS may have a small development community with little added value. Certainly VoIP over LTE could be accomplished without the complexity of IMS.

After IMS, the biggest technology challenge facing carriers in the next several years are the LTE – 3G interworking issues. Circuit switched fallback will be around several years for some carriers. It increases the number of Inter-Radio Access Technology (IRAT) transfers significantly, potentially affecting service reliability. As applications continue to use more Internet Protocol (IP) sessions, and as they begin to use IPv6 services in a few years, IRAT transfers become even more resource intensive and some IPv6 services may not transfer to 3G cleanly. Ideally carriers will be able to speed the transition to VoLTE and limit the use of circuit switched fallback.

16.5   Backbone Networks

Backbone networks are moving to 100G transport using Optical Transport Network (OTN). As 2020 approaches, 400G will be needed. The economics of these upgrades are daunting. In recent history revenue growth hasn't kept pace with transport costs. Carriers have done an excellent job of driving cost reductions back into their supply chains, but much of that ore has been mined. New revenues have to come from either subscribers or content providers and applications if data rates are to continue their historical climb. Data tiers and usage based pricing for wireless and wireline data services are likely to expand further, but consumers and enterprises have set budgets for communication as a percentage of their total spend, and that's not likely to change. Carriers can also improve efficiency and offer unique services by taking advantage of intelligent routing services, described in Chapter 7.

Security is likely to continue to be a unique and important service provided by carriers. Security technology positioned in the network enables solutions that intercept problems before they reach the enterprise. Carriers can provide secure cloud services that are better defended than those offered by hosting providers. Increasing attacks from the Internet also make Virtual Private Network (VPN) and switched Ethernet services more attractive. Ethernet in particular is far less exposed to potential security intrusions and hacks.

The shift to IPv6 over the next decade is an additional cost and operational complexity that will be lived out over many years. Much of the work has been done in the core of the network and at provider edges. But many access networks will remain largely IPv4 for some time. In the beginning tunneling and technologies like 6rd will enable transitions. Once IPv6 gains significant share, 10–20% of traffic, those solutions cease to scale. Access networks will need to be upgraded to dual stack operation or potentially capped to new services.

16.6   Science and Technology Matter

The precise nature of the next technology transformation for carriers is unclear but is emerging along the lines described previously. While uncertainty will always be part of network evolution, we can be sure sustained success for carriers depends most directly not on marketing or finance acumen, but on the ability to design, engineer, and manage efficient and reliable networks. Mistakes in other fields can be recovered in months. Errors in science, technology, and network design can remain unresolved until the next network transformation, often 10 years or more. Reliability analysis, system testing and integration, and network optimization are topics guaranteed to bring yawns outside of engineering circles, but unless they are practiced with rigor by skilled engineering teams, network performance, quality, and time to market will suffer severely. The ability to build and sustain science and engineering teams with the range of skills and experience needed to design and integrate voice, data, and video over wireless, wireline, and optical networks directly translates into the quality of the services and the reputation of the carrier; Ultimately, the success of the carrier, the industry, and the society they serve depend on it.

References

1. Joel, A.E. Jr., (1982) *A History of Engineering and Science in the Bell System, Switching Technology (1925–1975)*, Bell Telephone Laboratories.
2. Binmore, K. and Klemperer, P. (2002) The biggest auction ever: the sale of the British 3G telecom licenses. *The Economic Journal*, **112**, C74–C96.
3. Cable Europe (2011) Broadband On Demand, Cable's 2020 Vision, Cable Europe presentation at the US Cable Show.
4. The German Federal Ministry of Economics and Technology (2009) The Federal Government's Broadband Strategy, February 2009.
5. Sugayu, M. (2012) Regulation and Competition in the JP Broadband Market, Presentation at the Pacific Telecommunications Council, January 2012.
6. Dolan, R.E. and Worden, R.L. (1994) *Japan: A Country Study*, Washington, DC, GPO for the Library of Congress.

Appendix A

IPv6 Technologies

Table A.1 is a list of technologies from which operators can choose. The table is meant to be as complete as possible at this point in time, although some of the techniques, such as Silkroad and Network Address Translation/Protocol Translation (NAT-PT) are not likely to see wide adoption. The various techniques generally rely on transition mechanisms defined in Request for Comments (RFCs) 3056 and 4213. No doubt other technologies are entering the market or will in the near future. The table omits comments on security, routing protocols (e.g., BGP4+ (Border Gateway Protocol)), and Internet Control Message Protocol (ICMP) issues because they are beyond the scope of this paper. Designers need to explore them in detail because virtually every technology has issues in these areas. The IETF IPv6 Operations Group is a good source for monitoring industry progress.

Table A.1 IPv6 transition technologies

Technology	Application	Comments
	Tunneling	
ISATAP Intra-Site Automatic Tunnel Addressing Protocol	Enables IPv6 hosts to communicate with each other over an IPv4 network.	Requires all IPv6 nodes (routers or hosts) to implement the protocol. It treats the IPv4 network as a single link. See RFC 5214 and isatap.org.
	Works with public and private IPv4 addresses.	It can be used for intranet communication between sites. It can also be used to access the IPv6 Internet by configuring an ISATAP border router.
	ISATAP has been widely implemented in Linux, Windows, FreeBSD, and by router suppliers.	Microsoft uses ISATAP for their DirectAccess technology offering.

(continued overleaf)

Global Networks: Engineering, Operations and Design, First Edition. G. Keith Cambron.
© 2013 John Wiley & Sons, Ltd. Published 2013 by John Wiley & Sons, Ltd.

Table A.1 (*continued*)

Technology	Application	Comments
Teredo	A client–server tunnel technology that enables IPv6 hosts to access IPv6 servers behind a Teredo server over an IPv4 network.	Has the advantage that the client does not need a public IPv4 address, requires no special NAT, but operates over layers of NATs. Because it uses UDP and requires deployment of Teredo servers or Teredo relays, its application is likely to be somewhat limited. See RFC 4380.
	Teredo is useful in applications where the IPv4 network sits behind layers of NATs.	Each client must be configured for Teredo service. The IPv6 portal has a good description of the types of NATs that can be used with a Teredo implementation.
	Microsoft is the principal proponent of Teredo but clients are also available for Linux and BSD.	May be used more for testing over the public Internet. Probably not a good implementation strategy since it bypasses firewalls.
Silkroad	A client–server tunnel technology that enables IPv6 hosts to access IPv6 servers over an IPv4 network. I found few references of implementations of Silkroad.	Has similarities to Teredo, but requires three elements instead of Teredo's two. A Silkroad client, router, and navigator are required. Has the advantage over Teredo in that it works over a wider range of IPv4 NATs. Like Teredo it uses UDP. See the Silkroad IETF draft.
Tunnel broker	Provides a shared tunnel over an IPv4 network allowing IPv6 hosts to communicate with IPv6 servers.	This client–server design allows clients to setup and configure shared tunnels to the server. The advantage over Teredo is that Teredo is generally an unshared tunnel. See RFC 3053.
	The client must have a public IPv4 address and be authenticated.	This solution offers automatic configuration and dynamic channel management and updating DNS automatically.
	Tunnel brokers are often provided by ISPs or hosting providers. This technology is supported on Windows, Linux, and other OSs, depending on which ISP you choose.	Multiple technologies, such as Proto 41 Forwarding are often included in this category.

Table A.1   (*continued*)

Technology	Application	Comments
6to4, 6to4 Router and 6to4 Relay Router	Enables IPv6 hosts to communicate over an IPv4 network. No changes are needed in the IPv6 hosts. It eliminates complex tunnel management. 6to4 is well supported in the router community	Generally deployed by an enterprise at the customer gateway. The 6to4 gateway encapsulates the IPv6 packet and sends it over an IPv4 network to a 6to4 relay router. The relay router reconstructs the IPv6 message and routes it on the IPv6 Internet. See RFCs 3056 and 3068.
6rd (IPv6 rapid deployment). Similar technologies are Address plus Port (A + P) and 4rd (IPv4 residual deployment).	Enables dual stack hosts to communicate with IPv6 servers.	Generally deployed by a carrier at the provider edge or core, to enable transition of Ipv4 consumer services, or offered as an enterprise service. Enables dual stack hosts with an IPv4 only Internet access to reach the IPv6 Internet. See RFC 5569.
DSTM – Dual Stack IPv6 Dominant Transition Mechanism	Enables IPv4 hosts to communicate with each other over an IPv6 network.	In this client–server technology the client encapsulates IPv4 packets in IPv6 and forwards them to a DSTM server over an IPv6 network. The terminating server reconstructs and forwards the IPv4 message and maintains a stateful mapping for responses. A DSTM address server is needed to manage the IPv4 address pool.
	DSTM is supported on Linux, Windows, and FreeBSD.	DSTM is described in an IETF draft and was evaluated by HP and others.
MPLS – Multi-Protocol Label Switching	If you have MPLS service, adoption of IPv6 is reasonably straightforward. Dual stack routing at the provider and customer edge enable the service.	MPLS networks have several advantages in adoption of IPv6. Dual stack routing at the edge allows staged introduction of IPv6 across a common wide area network. Wide Area VPLS service over an MPLS network is a quick way to connect smaller IPv6 islands into a single subnet using VLANs, simplifying administration and enabling sharing of DNS64, tunnel brokers, or other transition technologies.

(*continued overleaf*)

Table A.1   (*continued*)

Technology	Application	Comments
	Translation	
NAT64	Enables IPv6 hosts to access IPv4 sites.	NAT64 can be deployed by an enterprise at the customer gateway. It uses a statically assigned pool of source IPv4 addresses when accessing the IPv4 Internet from an IPv6 LAN. A DNS64 is used to embed the destination IPv4 address in an IPv6 address range. The common prefix is assigned to the NAT64. The NAT64 caches the two addresses and creates a stateful connection to insure returning IPv4 messages are appropriately mapped. See RFCs 6144, 6145, 6146, 6147, and the IETF draft.
	NAT64 is generally preferred over NAT-PT which is deprecated. Windows and Linux can work in a NAT64 environment.	NAT64 implementations depend on the DNS64 resolver, which should be verified when choosing solutions.
DNS64	Enables IPv6 hosts to address IPv4 end points.	Embeds an IPv4 address in an IPv6 address and returns an AAAA record using a well-defined algorithm. Requires an IPv4 URI and will not work with IPv4 literals.
Carrier Grade NAT 64 (CGN64)	A carrier provided version of NAT64. It allow IPv6 hosts to communicate with IPv4 servers.	The term CGN is overloaded, so in this paper I'll use the term CGN64 to identify a design acting as an IPv6 – IPv4 translation service. See RFCs 6144, 6145, 6146, and 6147.
NAT444, LSN – Large Scale NAT	Uses double NAT44 to conserve IPv4 addresses.	Two levels of private IPv4 addressing are used, one on customer premises and one on the CE-PE access. An IPv4 public address is assigned at the carrier gateway when accessing the IPv4 Internet.

Table A.1   (*continued*)

Technology	Application	Comments
NAT444, LSN – Large Scale NAT	This is a transition strategy when public IPv4 addresses are no longer available, but the primary service desired is IPv4 Internet access.	This is compatible with basic services, but peer to peer services in particular may break.
Dual Stack HTTP Proxy	Enables IPv4 hosts to view IPv6 web content.	Practical solution in the short term for enabling access to IPv6 web content.
Network Address Translation/Protocol Translation (NAT-PT), Network Address Port Translation/Protocol Translation (NAPT-PT) and Transport Relay Translation	Enables IPv6 hosts to communicate with IPv4 servers.	Some combinations of these technologies are generally deployed by an enterprise at the customer gateway. They are deployed with a DNS Application Level Gateway (DNS-ALG) and require protocol specific ALGs (ftp, http, etc.). Enough issues were encountered to cause the RFC 2766 to be deprecated. See RFC 2694 and 4966.
Stateless IP/ICMP Transition (SIIT)	An IPv4 to IPv6 mapping algorithm that enables IPv6 hosts to communicate with IPv4 hosts. Can be used in NAT64/DNS64 in stateless deployments if static IPv4 addresses can be assigned to the hosts, which is seldom possible.	Address assignment and routing are not defined.

Dual stack operation

| Dual stack Hosts | Hosts with operating systems capable of assigning either IPv4 or IPv6 addresses to their interfaces. Virtually all popular operating systems provide some level of support for dual stack operation. | Dual stack hosts may use tunneling or translation to communicate with a remote host using a protocol different than the one supported on the direct interface. It's important to understand how the host or applications choose which destination to use when DNS returns both A and AAAA records. Both the DNS resolver and the application may need to be configured to achieve the desired routing. |

(*continued overleaf*)

Table A.1 (*continued*)

Technology	Application	Comments
Dual stack lite (DS-Lite)	Enables IPv4 only hosts to communicate with IPv4 servers over an IPv6 access network.	An IPv4 packet is encapsulated in IPv6 either at the host or a gateway and addressed to an address translator, or CGN, in a carrier network. The CGN terminates the IPv6 tunnel providing IPv4 Internet access to the host. The main advantage is that it does not require a separate NAT64 at the premises.
Dual Stack Interconnection	Internet Service Providers (ISPs) offer dual stack Internet service that enables a customer gateway router to send both IPv4 and iPv6 messages on the same access link.	

UDP: User Datagram Protocol; BSD: Berkeley Software Distribution; DNS: Domain Name System; OS: Operating System; VPLS: Virtual Private LAN Service; VLAN: Virtual Local Area Network; LAN: Local Area Network; AAAA: Quad A DNS Record; URI: Uniform Resource Identifier; CE: Customer Edge; PE: Provider Edge; http: Hyper Text Transfer Protocol.

Appendix B

The Next Generation Network and Why We'll Never See It[1]

I have been waiting decades for the emergence of the Next Generation Network (NGN). The promise of a network being born from inspiration and insight, and sweeping aside all before it is compelling, but without precedent and unlikely in the extreme. Conferences and technology visionaries thrive on the idea. The catchphrase promotes a notion that networks travel serially through time, each one spreading out and dominating until it is supplanted by the next great idea. Bold claims are made about the benefits of the transmission and switching technology ushering in the NGN. The revolutionary nature of the network enables new "paradigms" in the development and application of software technology to manage them. Capital costs and operational expense are reduced by orders of magnitude by the uniformity of the technology and ease of management, using the most conservative estimates. The economics of the NGN are in fact, so compelling, that operators replace legacy networks that have been paid for, just for the operational advantages of the NGN.

Unified communication is a coconspirator in this long running play. The reason the NGN is next, is because it has the ability to transport all current and future services on a common infrastructure, rendering ineffective and inefficient the incumbent networks. If the NGN couldn't claim service and transport unification, it would be Another Really Good Network (ARGN). ARGN has a ring to it, but can't command the imagination like NGN. Using unified communication, the NGN is able to rationalize all preceding technology and services. Once the technology is unified, like Lego blocks, we only have to plug the components into each other. Services find the transport and switching technology "agnostic," and the new network fills the needs of all, and at a reduced cost.

B.1 Claims of the Next Generation Network

- **Unified communications** – We can collapse all of our network and services on a single infrastructure. Operational costs are reduced to the point we can afford to replace existing

---

[1] First published in IEEE *Communications* magazine, October 2006.

Global Networks: Engineering, Operations and Design, First Edition. G. Keith Cambron.
© 2013 John Wiley & Sons, Ltd. Published 2013 by John Wiley & Sons, Ltd.

legacy networks. The new technology is so efficient, the capital costs to deploy it are insignificant compared to the cost of growing legacy networks.

- **Convergence** – All services ride the single network and operations support is simplified, yielding capital and expense savings. The current and future demands of the varied services are met.

While each promising NGN has fallen short of these claims, we have seen new network technologies succeed and work in concert with existing technologies. Like the new kid at school, the new technology must find a way to fit in with the existing crowd. This trend, in fact moves us further away from unification, not toward it.

The three real forces that drive the emergence of broad new network technologies are quite different than the ones touted for the NGN.

B.2   Forces of Network Transformation

- **The changing nature of traffic** – Beyond content type (voice, data, and video), other traffic characteristics such as symmetry, multicasting, latency, quality, security, demand, and ever lower sustainable error rates all play a role. When any or all of these change dramatically, new network technologies rise to meet the need if the existing technologies fall short.
- **Network entropy** – I use the term entropy loosely to connote complexity that arises in an almost random way. New networks are born with a simplicity that gives them a natural advantage to serve the changing nature of traffic. To achieve unification and meet new traffic demands, designers, often with different objectives, begin to make changes to expand the base technology, and thereby increase the complexity of the new network. As each change is made, the network begins to be weighed down by its own success. Every design choice removes a degree of freedom, solving an immediate problem, but eliminating potential solutions to other problems that lie in the future. Then, once again the nature of traffic changes and the entropy of the network makes it brittle, incapable of flexing to meet the new need. The aging NGN is recast as a legacy network, only suited for the old traffic types. We then search for the next NGN.
- **Component transformation** – Significant, and often unforeseen advances in component technology shift the earth beneath existing networks. The integrated circuit eliminated space division switching by making time division switching possible. Ultra Large Scale Integration and Digital Signal Processing technology gave us sufficient processing power to enable Orthogonal Frequency Division Multiplexing (OFDM) and Digital Subscriber Line (DSL). Advances in optical technology continue to shake our network world, and much more is in store.

To test this thesis, let's examine some NGNs and the claims that promoted them.

The Integrated Services Digital Network (ISDN) brought voice and data together by bonding voice channels to yield Nx64 data channels. Other channels provided the associated signaling and it all rode on the emerging all digital voice network. There was little to compete with the ubiquity and capability of that network at the time. ISDN failed to achieve true next generation status because of the forces of network transformation. Instead of making it a simple service, scores of features, and complexity were added from the outset, and government regulation gave us the T and S interfaces, that added more unneeded complexity with no benefit. ISDN was in effect, born at a ripe old age because of the complexity added at the outset. X.25 and later Frame Relay Service proved more practical for business applications. In consumer and small business applications, ISDN suffered because it was designed to carry symmetrical traffic, just when the Internet and asymmetrical traffic grew quickly into dominance. V.90 voice band modems, emerging in the same era, were designed to carry asymmetrical traffic and boosted throughput with compression; they claimed comparability to ISDN, but were cheaper and simpler. What if ISDN had emerged more quickly and delivered a 192 kbps down/64 kbps up connection to the Internet in a simple and reliable way? Its fate might have been quite different.

Asynchronous Transfer Mode (ATM) was designed as a unified NGN from the beginning. To meet the goal of unification, ATM restricted the cell payload to 48 octets, a payload size marginally acceptable for voice transport and inefficient for data. That yielded multiple adaptation rules for segmentation and reassembly of data, voice and video, and increased network complexity. It's hard to believe now, but the hope for ATM was so great that carriers began to replace Time Division Multiplexed (TDM) network elements with Voice and Telephony Over ATM (VTOA) switching and transport. A few VTOA switches are still in service and are orphans that will be a burden until replaced. Even more surprising was the introduction of ATM on the desktop. Network interface devices were produced, at considerable expense, in a belief that the Local Area Network (LAN) would convert to the new unified ATM technology. ATM succeeded for data in the edge and core, primarily because of the concatenated physical layer format adopted in the ubiquitous TDM network. ATM generally failed for voice. I doubt there is one native voice (G.711) cell transported for every million data cells traversing our network. Almost all ATM networks spend their time segmenting and reassembling Ethernet frames, and collecting the associated 10% cell tax in the process.

The Internet Protocol (IP) is recently discovered as the true NGN; it is an overnight success 20 years after its broad acceptance in the scientific and engineering communities. But does the IP truly define a network? The expression "the IP Network" is broadly used throughout the industry and press, but is it meaningful? The IP's ubiquity and universal acceptance are unquestioned and it will not be displaced as the lingua franca in my lifetime. However, IP's greatest contribution is its ability to switch information across diverse networks, independent of the underlying switching technology; the greatest legacy of IP is the universal acceptance of the address scheme and message structure, much like the Public Switched Telephone Network (PSTN) is ubiquitous because of E.164 numbering and G.711 encoding. IP routing and route distribution protocols may not fare as well over time. They are advertised as standard and they are. Pick from the many available IETF drafts, and if you don't like it, wait several months and another will emerge. Together, these protocol variations and their interactions weigh down IP. IP routing's entropy is growing, and the variations arising from the need for bilingual IPV4-IPV6 networks pile on more weight.

Perhaps we should discard the term "IP Network" altogether. To warrant the title of a network, the technology should dominate routing and switching, if not end to end, then certainly in the core of the network, where the heavy lifting is done. Full IP routing dominates at LAN/WAN (Wide Area Network) interfaces and the interfaces between autonomous systems. IP was never dominant in LAN switching, and IP routing is losing out in core networks to label switching. IP forwarding protocols are heavy and in retreat in the face of lighter weight label switching in the core. Much of this retreat is attributable to the changing nature of traffic. IP forwarding could not readily respond to the demand for Virtual Private Networks (VPNs) on a common transport. Label switching rose to meet the need in a relatively simple way, using connection oriented virtual circuits borrowed from other protocols. No doubt, as large enterprises demand VPNs and extranets across carrier networks, label switching will make inroads at enterprise-carrier boundaries. Carriers are likely to forge agreements on label switching at network boundaries, pushing IP further to end customer edges.

As the entropy of IP grows, the nature of traffic is also changing in other ways. The asymmetrical delivery of Internet Protocol Television (IPTV) may dwarf other traffic in large segments of our networks. We are facing 2-hour sessions above 3 Mbps, not just short video clips at 128 kbps. IP addressing and messaging is well suited to MPEG4 transport and the flexibility of IP connectivity enables an array of video services not previously possible. But ironically, the IP protocol may not be well suited to route and switch IPTV. Optical solutions, Ethernet, and Multiprotocol Label Switching (MPLS) solutions are proving more economic and more reliable. IPTV tolerates the 50 ms reconfiguration times guaranteed by Synchronous Optical Network (SONET) and the fast reroute times of layer 2 MPLS traffic engineered circuits. But it may not tolerate the substantial reconvergence times of IP forwarding and multicasting. Packet jitter experienced in large IP networks poses additional design issues for IPTV. This change in the nature of traffic provides an opening for new technologies, like Reconfigurable Optical Add Drop Multiplexers (ROADMs), to bypass IP switching, and possibly SONET STS-1 (Synchronous Transport Signal) distribution. To grasp the opportunity, ROADMs need to improve the granularity of switchable wavelengths and add control plane features suited to the purpose. Each technology must find its place.

At the physical layer some see GigE and 10GigE native transport on fiber as the optical NGN, displacing SONET on a broad scale in the near future. While GigE is emerging in the distribution network and in some metro applications, the SONET protocol remains the technology of choice for mixed services and medium and long haul transport, using G.709 wrappers. In fact, we are currently deploying OC768 systems in our core, substantially increasing our SONET capacity, not reducing it. Granted it is increasingly being used for router connectivity, but replacing SONET TDM services with Circuit Emulation Service (CES) on more than a selective basis looks unlikely. Over time, Voice over Internet Protocol (VoIP) services appear likely to displace TDM as we move to IP, thereby decreasing the amount of native TDM traffic, and making the widespread adoption of CES to interconnect TDM switching systems moot.

The general notion that a NGN foreshadows the early demise of existing networks is faulty in the extreme, if we see history as a guide to the future. For example, there is a continued perception that wireless phone service is bringing about the early demise of the wired PSTN. No doubt there is a loss of some wired access lines to wireless phones, but the total minutes carried by the PSTN has risen, in part because wireless has driven up total voice usage significantly (remember that annoying cell conversation at the next table). Quite often the person on the other end of a wireless call is on a wired phone. Even when they are not, there's a good chance the traffic is carried over the PSTN. On a similar note, Wi-Fi has not displaced DSL, it has stimulated it. Public Wi-Fi access points are best served by DSL. In a similar way WiMax may stimulate Ethernet access.

At the next conference watch for the NGN slides that show the cloud at the center, or the metallic grid, uniform and expansive, covering the globe. Our global network is not a cloud, nor is it a uniform grid. If we must use a metaphor, a mixed fabric may serve us best. Like threads of different colors and grade, they are woven together to suit a purpose, but no one pattern dominates, and no single weaver sets the weave.

Acknowledgments

My thanks to Chuck Kalmanek, Mike Pepe, and Steve Weinstein for reviewing this article and providing constructive comments.

Index

Global Networks: Engineering, Operations and Design, First Edition. G. Keith Cambron.
© 2013 John Wiley & Sons, Ltd. Published 2013 by John Wiley & Sons, Ltd.